Springer Series in Optical Sciences

Volume 230

Springer Series in Optical Sciences is led by Editor-in-Chief William T. Rhodes, Florida Atlantic University, USA, and provides an expanding selection of research monographs in all major areas of optics:

- lasers and quantum optics
- ultrafast phenomena
- optical spectroscopy techniques
- optoelectronics
- information optics
- applied laser technology
- industrial applications and
- other topics of contemporary interest.

With this broad coverage of topics the series is useful to research scientists and engineers who need up-to-date reference books.

More information about this series at http://www.springer.com/series/624

Afshin Moradi

Canonical Problems
in the Theory of Plasmonics

From 3D to 2D Systems

 Springer

Afshin Moradi
Kermanshah University of Technology
Kermanshah, Iran

ISSN 0342-4111   ISSN 1556-1534   (electronic)
Springer Series in Optical Sciences
ISBN 978-3-030-43838-8   ISBN 978-3-030-43836-4   (eBook)
https://doi.org/10.1007/978-3-030-43836-4

This Springer imprint is published by the registered company Springer Nature Switzerland AG.
The registered company address is: Gewerbestrasse 11, 6330 Cham, Switzerland

To my daughter, Helia
for her future

Preface

The theory of plasmonics is built on the interaction of electromagnetic radiation and conduction electrons at metallic interfaces or in metallic nanostructures. Plasmonics (or, more accurately in the nano-world, nanoplasmonics) is a component of nano-optics, which is optics at the nanoscale dimensions. This theory merges electronics and photonics at the nanometer scale. Although many books have been published in this field in recent years, there is no available book about the theoretical account of many of the basic properties of plasmonics. The aim of the present text is to provide such accounts. To do this and describe the fundamental concepts of plasmonics, canonical boundary-value problems (BVPs) must be solved, using electromagnetic wave theory based on Maxwell's equations and also electrostatic approximation based on Poisson's equation. Therefore, a collection of BVPs is presented here, which may describe the fundamental aspects of the fascinating field of plasmonics.

In order that the system of field equations be closed, one needs the so-called material equations, which relate the induced densities of charge and current to the fields. Such relations essentially also determine the electromagnetic properties of the media under consideration. Strictly, one should use a quantum many-body technique to determine the charge response due to the application of fields, but this theory is very complicated. A common approach, however, is to form simple equations relating only the macroscopic variables, e.g., for metals, the hydrodynamic approximation of Bloch. This approximation has proved to be a useful guide to many metal properties. In this text as a simple model of conduction electrons of a metallic structure, we consider a degenerate electron gas (EG), whose dynamics may be described by means of the hydrodynamic theory, while keeping in mind the theory can be applied to highly doped semiconductors and potentially other conducting systems. In this way, in Chap. 1 the basic concepts and formalism of electromagnetic and electrostatic theories and also the local and spatial nonlocal plasma models of an EG that will be needed in subsequent chapters are briefly reviewed.

The content is divided into two parts and split by dimensionality. The BVPs in three-dimensional EGs (3DEGs) are presented in Part I (Chaps. 2–5) and Part II for BVPs of two-dimensional EGs (2DEGs) from Chap. 6–10. Chapter 2 discusses the electrostatic BVPs in the bounded EGs with different geometries, using the local dielectric function model. Chapter 3 presents an analysis of local electromagnetic BVPs in some bounded EGs, such as planar, cylindrical, and spherical 3DEGs. In Chap. 4 and also Chap. 5, by considering the spatial nonlocal effects, some electrostatic and electromagnetic BVPs involving the bounded EG are studied. Then in Chap. 6, the electrostatic BVPs in planar 2DEGs are discussed, while Chap. 7 considers electromagnetic BVPs in layered planar EG structures. Chapters 8 and 9 present the electrostatic and electromagnetic BVPs involving 2DEGs in cylindrical geometry, and Chap. 10 deals with some plasmonic BVPs involving 2DEGs in spherical geometry. Finally, it should be noted that many of the results described in the text are original to me. Therefore, I have striven to make sure of the accuracy of the equations, derivations, and descriptions of numerical results, and any errors that remain are solely my responsibility.

Throughout the text, the International System of Units (SI) is used and attention is only confined to the linear phenomena. Also, all media under consideration are nonmagnetic and have a magnetic permeability μ_0. Furthermore, for brevity, the $\exp(-i\omega t)$ time factor is suppressed in almost all sections of the text and vectors are denoted with bold fonts (not to be confused with matrix notation), such as **a**, **A**, etc.

I sincerely hope that the final outcome of this book, with the theme of electromagnetic and electrostatic BVPs in 3D and 2D plasmonic systems, will help students, researchers, and other users who are assumed to have general knowledge in electromagnetism and wish to enter the field of plasmonics or to gain insight into the subject. I would appreciate being informed of errors or receiving other comments about the book. Please write to me at the Kermanshah University of Technology address or send e-mail to: a.moradi@kut.ac.ir or afshin551@gmail.com.

Finally, it should be noted that the completion of this book would not have been possible without the patience and support of my wife, Maryam. Her understanding and sacrifice are greatly appreciated.

Kermanshah, Iran   Afshin Moradi
February 2020

Contents

List of Acronyms

1D	One dimensional
2D	Two dimensional
3D	Three dimensional
EG	Electron gas
MNS	Metallic nanostructure
LHD	Local hydrodynamic
SHD	Standard hydrodynamic
QHD	Quantum hydrodynamic
SP	Surface plasmon
SPP	Surface plasmon polariton
ABC	Additional boundary condition
BVP	Boundary-value problem
BC	Boundary condition
SMP	Surface magneto plasmon
SMPP	Surface magneto plasmon polariton
PDE	Partial differential equation
Im	Imaginary part
Re	Real part
TF	Thomas–Fermi
2DEG	Two-dimensional electron gas
3DEG	Three-dimensional electron gas
CNT	Carbon nanotube
ATR	Attenuated total reflection

Part I
Three-Dimensional Electron Gases

Chapter 1
Basic Concepts and Formalism

Abstract The book aims to present a systemic and self-contained guide to the canonical electromagnetic and electrostatic boundary-value problems in metallic nanostructures. In this way, the conduction electrons of a metallic medium are modeled as a degenerate electron gas, whose dynamics may be described by means of the hydrodynamic theory. Therefore, at first we need to know something about the hydrodynamic model of an electron gas. Then, we need to know something about the basic concepts and formalism of electromagnetic and electrostatic theories of an electron gas that will be used later in the book. For brevity, in many sections of this chapter the $\exp(-i\omega t)$ time factor is suppressed. Furthermore, all media under consideration are nonmagnetic and attention is only confined to the linear phenomena.

1.1 Introduction

The main chapters of the present book are related to the canonical electromagnetic and electrostatic boundary-value problems (BVPs) in 3D and 2D metallic nanostructures (MNSs).[1] To study the BVPs in MNSs, *Maxwell's equations* are fundamental and will be introduced in the present chapter. Also, we consider the conduction electrons of a MNS as a *degenerate* or metallic electron gas (EG), globally neutralized by the lattice ions and use the *plasma model*. A degenerate EG is a medium that involves the coexistence in free space of high-density and low-temperature quasi-free electrons and fixed positive ions of the lattice, which can exhibit collective effects arising from the Coulomb force between the charges.

The elementary and local plasma model used to describe the dynamics of an EG is the *Drude model* [1]. However, within the Drude model, all electrons

---

The original version of this chapter was revised. The correction to this chapter is available at https://doi.org/10.1007/978-3-030-43836-4_11

[1]We use the term metallic structures throughout this book, while keeping in mind the analysis can be applied to highly doped semiconductors and potentially other conducting systems.

are displaced collectively and therefore this model does not allow for the local fluctuations of the electron density. But, at room temperature and standard metallic densities, fluctuations in the electron density of MNSs are present [2–4], so that a MNS can sustain plasma waves, charge density waves with longitudinal electric fields.[2] These waves are homogeneous solutions of Maxwell's equations and may therefore be included in the general solution and consequently be considered in the optics of the MNSs. An extension of the Drude model can include the local fluctuations of the density of the free valence electrons and hence electron plasma waves in a metal. The so-called *hydrodynamic model* provides this extension. As a result, various versions of the hydrodynamic model have been proposed to describe the behavior of EG of MNSs. More recently, Kupresak et al. [6, 7] compared four macroscopic hydrodynamic versions, with four different boundary conditions (BCs), to model the electromagnetic response of MNSs; the standard hydrodynamic (SHD) model [8–20] with the Sauter [21] *additional boundary condition* (ABC), the curl-free hydrodynamic model [13, 22–24] with the Pekar ABC [25], the shear forces hydrodynamic model[3] [26] with the specular reflection ABC [27, 28], and the quantum hydrodynamic (QHD) model [29–33] with the corresponding ABC [34].

According to the results in [6, 7], SHD and QHD models are the most accurate hydrodynamic descriptions for studying the plasmonic properties of MNSs. Hence, in several chapters of the present book we will deal with SHD and/or QHD models of MNSs.

1.2 Hydrodynamic Model of Three-Dimensional Electron Gases

We consider a uniform positive neutralizing background for a 3DEG. Also, we assume that the EG in the closed region V is surrounded by a dielectric material with the relative dielectric constant ε_r that may depend on the position vector $\mathbf{r} = (x, y, z)$. Let $-e$ and m_e be the charge and mass of an electron,[4] and let n_0 be the quasi-free electron density (per unit volume).

As remarked above, we assume that the system is, on the average, neutral, where this assumption is only true for volumes whose linear dimensions are substantially larger than the *Thomas–Fermi screening length* λ_F that is equivalent with the *Debye screening length* for a classical or non-degenerate EG [35]. For volumes whose linear dimensions are substantially smaller than λ_F, the electrons motion causes the

---

[2]In solid state physics the quanta of longitudinal charge density waves is called a *plasmon*. The first theoretical description of plasmons was presented by Pines and Bohm in 1952 [5]. In plasma physics these charge density oscillations are referred to as plasma waves.

[3]This model has not been applied yet to practical MNSs.

[4]$e \approx 1.60 \times 10^{-19}$ C, and $m_e \approx 9.11 \times 10^{-31}$ kg.

electron density and the ion density to be unequal. The expression for the Thomas–Fermi (TF) distance is [36–38]

$$\lambda_F = \left(\frac{2\varepsilon_0 E_F}{3n_0 e^2} \right)^{1/2} , \tag{1.1}$$

where ε_0[5] is the electric *permittivity* of vacuum and $E_F = m_e v_F^2/2$ and $v_F = \hbar(3\pi^2 n_0)^{1/3}/m_e$ are the *Fermi energy* and the *Fermi velocity* of the 3DEG, respectively [37, 38]. Also, we introduce $k_s \equiv \lambda_F^{-1}$ that is known as the inverse screening length in the TF model of an EG. We shall be employing elements of volume of a MNS with linear dimensions appreciably greater than the TF screening length. Another relevant parameter of an EG near the Fermi energy is the de Broglie (or Fermi) wavelength of the electrons, as

$$\lambda_B \equiv k_F^{-1} = \frac{\hbar}{m_e v_F} , \tag{1.2}$$

where k_F is the Fermi wavenumber and $\hbar$ is equal to the Planck constant[6] divided by 2π. We note that QHD model may be employed for an EG when the de Broglie wavelength associated with the electrons is comparable to dimension of the EG system.

In the hydrodynamic scheme, the equation of *drift motion*,[7] Maxwell's equations (or *Poisson's equation* in the electrostatic approximation) and the *continuity equation* determine the dynamics of a metallic system. Therefore, at first the two macroscopic SHD equations for an EG, i.e., equation of drift motion and the continuity equation are introduced.

1.2.1 Equation of Drift Motion

Under the influence of a small perturbation such as an electromagnetic wave, the electrons in a macroscopic element of volume acquire a drift motion in addition to the electrons random motion. The drift velocity is obtained by averaging the motions of all the electrons in the element of volume; this averages out the random motion. By defining $n(\mathbf{r}, t)$ as the electron density and $\mathbf{v}(\mathbf{r}, t)$ as the macroscopic hydrodynamic electron velocity, the equation of drift motion for electrons has the form [2]

$$\left(\frac{\partial}{\partial t} + \mathbf{v} \cdot \nabla \right) \mathbf{v} = -\frac{e}{m_e} (\mathbf{E} + \mathbf{v} \times \mathbf{B}) - \gamma \mathbf{v} - \frac{v_F^2}{2n_0^{2/3}} \nabla n^{2/3} , \tag{1.3}$$

---

[5]$\varepsilon_0 \approx 8.854 \times 10^{-12}$ F/m.

[6]$h \approx 6.63 \times 10^{-34}$ J.s $\approx 4.14 \times 10^{-15}$ eV.s.

[7]The equation of drift motion is also known as the hydrodynamic equation or Newton's equation of motion.

where $\nabla = (\partial/\partial x, \partial/\partial y, \partial/\partial z)$ is the gradient vector in Cartesian coordinates and the electromagnetic fields $\mathbf{E}(\mathbf{r}, t)$ and $\mathbf{B}(\mathbf{r}, t)$ are not the microscopic fields that exit at each point in the system, but are the smoothed-out macroscopic fields together with any external fields that may be present.

The second term in right-hand side of (1.3) represents the damping term of electrons, where γ is a phenomenological *collision frequency* and the appropriate choice of this value (that is assumed to be constant) is important. We note that free path of the quasi-free electrons in an EG is given by $l = v_F \tau$, where τ is the electron collision time and we have $\gamma = 1/\tau$. Typical values of l for metals at room temperature are of the order of $10 \, \text{nm}$. In bulk metals the mean free path is determined by short-range interactions with the background, i.e., the interaction of the electrons with ions, impurities, and lattice defects. For a MNS comparable with l, the collisions of the conduction electrons with the surface of the MNS have to be included as an additional collision process [4, 8, 39]. The mean free path due to the latter effect only has been calculated by Euler [40] for a spherical particle of radius R. He has found that the mean free path was equal to the radius R. Therefore, in a MNS we may modify the bulk damping parameter γ as[8]

$$\gamma \to \gamma + \frac{v_F}{R} . \tag{1.4}$$

The third term in right-hand side of (1.3), that is related to the so-called *Thomas–Fermi pressure* in a degenerate EG, is of great interest because the fluctuation of the electron density or *spatial dispersion* in an EG arises from it [2–4]. Also, there may be a steady imposed magnetic field of flux density $\mathbf{B}_0$, converting the EG into what is known as a *magnetoplasma*. If so, we have

$$\textit{Total magnetic field} = \mathbf{B} + \mathbf{B}_0 .$$

In (1.3), the term $\mathbf{v} \cdot \nabla \mathbf{v}$ constitutes a nonlinear effect; so does the term $\mathbf{v} \times \mathbf{B}$ and also third term in right-hand side of (1.3). Therefore, (1.3) may be linearized by expanding n and $\mathbf{v}$, as

$$n(\mathbf{r}, t) = n_0(\mathbf{r}) + n_1(\mathbf{r}, t) + n_2(\mathbf{r}, t) + \ldots , \tag{1.5}$$

$$\mathbf{v}(\mathbf{r}, t) = \mathbf{v}_1(\mathbf{r}, t) + \mathbf{v}_2(\mathbf{r}, t) + \ldots , \tag{1.6}$$

where $n_0(\mathbf{r})$ is the equilibrium electronic concentration which is assumed to be constant n_0 throughout the homogeneous and infinite EG and we have $\mathbf{v}_0 = 0$ in equilibrium state. Furthermore, for weak perturbation we have $n_0 \gg n_1 \gg n_2$,

---

[8]In general, we have $\gamma \to \gamma + A\dfrac{v_F}{R}$, where A is a constant, which is related to the probability of the free electrons scattering off the surface of the MNS. However, experimental observations and advanced theoretical calculations show that in most cases $A \approx 1$.

so that the first-order perturbation terms are sufficient to describe the system. By equating separately the terms of various orders we obtain in first order [18]

$$\frac{\partial \mathbf{v}}{\partial t} = -\frac{e}{m_e}(\mathbf{E} + \mathbf{v} \times \mathbf{B}_0) - \gamma \mathbf{v} - \frac{\alpha^2}{n_0}\nabla n \,, \tag{1.7}$$

where we are considering $\mathbf{v}_1 = \mathbf{v}$ and $n_1 = n$ to the first order. The first-order quantities are assumed then to vary as $\exp[i(\mathbf{k} \cdot \mathbf{r} - \omega t)]$, where $\mathbf{k}$ is the wavevector and ω is the angular frequency of the small monochromatic perturbation. Also, α is a characteristic velocity that represents the speed of propagation of the density disturbances in the 3DEG due to restoring effect of the TF pressure. Let us note that in the macroscopic hydrodynamic theory, we obtain α^2 as $v_F^2/3$ (instead of $3v_F^2/5$ in the dielectric approach[9] [41]) which is the value for a low-frequency EG, i.e., $\omega \ll \gamma$ (system close to static equilibrium). But many authors have argued that the theory can be used to describe high-frequency EG oscillations also, and in doing so one has simply to replace $v_F/\sqrt{3}$ by $\sqrt{3/5}v_F$[10] as the value of α. As a matter of fact, both limits, i.e., $\omega \ll \gamma$ and $\omega \gg \gamma$ have to break down when $\omega \approx \gamma$ and there exists a further generalization, based on the Boltzmann equation, which results in a complex-valued, frequency-dependent $\alpha(\omega)$. The theory is discussed in detail by Halevi in [42] and he found

$$\alpha^2(\omega) = \frac{\frac{3}{5}\omega + \frac{1}{3}i\gamma}{\omega + i\gamma}v_F^2 \,, \tag{1.8}$$

where for $\omega \ll \gamma$ or $\omega \gg \gamma$, this formula tends to the two known real-valued limiting cases discussed here.

1.2.2   Equation of Continuity

The principle of conservation of charge indicates that the temporal and spatial variations of electron density n induced by a perturbation such as an electromagnetic wave satisfy equation of continuity. This equation states that the time-rate of decrease of the number of particles in a macroscopic element of volume is equal to the rate at which particles are leaving the element of volume across its surface and therefore, we have

$$\frac{\partial n}{\partial t} + \nabla \cdot (n\mathbf{v}) = 0 \,. \tag{1.9}$$

Using (1.5) and (1.6), so that this equation becomes

---

[9]Thus, the present SHD model yields a coefficient that is by a factor 9/5 too small.

[10]We note that for typical metals the values of α are of the order of the *Bohr velocity*.

$$\frac{\partial n_1}{\partial t} + \nabla \cdot [(n_0 + n_1) \mathbf{v}_1] = 0 . \tag{1.10}$$

The term $n_1 \mathbf{v}_1$ is nonlinear, but its effect is usually small. Dropping this term, we have the equation of continuity for the electrons in the linearized form

$$\frac{\partial n}{\partial t} + n_0 \nabla \cdot \mathbf{v} = 0 , \tag{1.11}$$

where again we are considering $\mathbf{v}_1 = \mathbf{v}$ and $n_1 = n$ to the first order. There is also a continuity equation for charge density ρ and for current density $\mathbf{J}$, and this may be derived from (1.11). We have $\rho = -en$ and the drift velocity of the electrons gives a current density as $\mathbf{J} = -en_0 \mathbf{v}$. Now, by multiplying (1.11) by $-e$, we obtain

$$\frac{\partial \rho}{\partial t} + \nabla \cdot \mathbf{J} = 0 . \tag{1.12}$$

This is the continuity relation of electromagnetic theory.

1.2.3 Gradient Correction: Bohm Potential

Although, the linearized equation of drift motion in the presence of the TF pressure, i.e., (1.7) together with (1.11) and Maxwell's equations may be used for dynamics investigation of a metallic EG, the appropriate hydrodynamic model should, in principle, incorporate the *gradient correction*. Gradient correction manifests through a *Bohm potential* term in right-hand side of (1.3). In the process of linearization, one can find

$$\frac{\partial \mathbf{v}}{\partial t} = -\frac{e}{m_e} (\mathbf{E} + \mathbf{v} \times \mathbf{B}_0) - \gamma \mathbf{v} - \frac{\alpha^2}{n_0} \nabla n + \xi \frac{\beta^2}{n_0} \nabla \nabla^2 n , \tag{1.13}$$

where $\beta = \hbar / 2m_e$ and the coefficient ξ in the Bohm potential term sensitively depends on the considered values of the wavenumber and frequency [43]. Now, (1.13) together with (1.11) show the linearized macroscopic QHD equations of an EG. We may express the last two terms in (1.13) together as $-\frac{\alpha^2}{n_0} \left(1 - \xi l_c^2 \nabla^2\right) \nabla n$, where $l_c = \beta / \alpha$. We note that the critical parameter l_c provides a characteristic length scale for the density variations in the EG for which the effects of the Bohm potential become important. Using the definition $\alpha = v_F / \sqrt{3}$, we may also express the critical parameter l_c in terms of the inverse Fermi wavenumber of the EG via $l_c = \sqrt{3} \lambda_B / 2$. Also, we define a critical wavenumber as $k_c \equiv l_c^{-1} = 2k_F / \sqrt{3}$. Note that the gradient correction can be neglected, when the inverse TF screening length of the EG, i.e., k_s is very smaller than the critical wavenumber, i.e., $k_s \ll k_c$ or $l_c k_s \ll 1$.

1.3 Maxwell's Equations

In order to discuss the electromagnetic properties of an EG, the equations developed in the previous section must be combined with Maxwell's equations. These involve not only the time-varying electromagnetic field of the wave but also the current density $\mathbf{J}$ and the charge density ρ created by the wave in EG.

Instead of expressing Maxwell's equations in terms of current and charge densities, it is also possible to express them in terms of the induced polarization. At the first, we describe the method that employs electron current density and total charge density.

1.3.1 Maxwell's Equations in Terms of Current and Charge Densities

In the absence of the bound charged particles in a *linear*[11] and *homogeneous*[12] EG, the electromagnetic fields are caused by the external charges and currents and by the induced counterparts, arise from the response of quasi-free electrons of the plasma itself, so they may be called self-consistent. In this description, electromagnetic field may be described using the charges in vacuum model. This means that in this model electromagnetic fields always occur in free space (i.e., there is no effective medium), but Maxwell's equations are modified by the presence of total charges and currents.

Explicitly excluding the ideas of external charge and current densities, which can represent, for example, wave excitations, Maxwell's equations can be written as

$$\varepsilon_0 \nabla \cdot \mathbf{E} = \rho , \tag{1.14a}$$

$$\mu_0 \nabla \cdot \mathbf{H} = 0 , \tag{1.14b}$$

$$\nabla \times \mathbf{E} = -\mu_0 \frac{\partial \mathbf{H}}{\partial t} , \tag{1.14c}$$

$$\nabla \times \mathbf{H} = \mathbf{J} + \varepsilon_0 \frac{\partial \mathbf{E}}{\partial t} , \tag{1.14d}$$

which are considered in the spatial $\mathbf{r}$ and time t domains. As mentioned before, $\rho(\mathbf{r}, t)$ and $\mathbf{J}(\mathbf{r}, t)$ are the quasi-free electron charge and current densities, respec-

---

[11]Materials whose *constitutive parameters* (for instant dielectric function and conductivity, which are, in general, functions of the applied field strength, the position within the medium, the direction of the applied field and the frequency of operation) are not functions of the applied field are usually known as linear; otherwise they are nonlinear.

[12]When the constitutive parameters of media are not functions of position, the materials are called homogeneous.

tively. Furthermore, $\mathbf{E}(\mathbf{r}, t)$ is the electric-field-intensity (or electric field) vector, $\mathbf{H}(\mathbf{r}, t)$ is the magnetic-field-intensity (or magnetic field) vector, $c = (\varepsilon_0 \mu_0)^{-1/2}$ is the light speed in vacuum, and μ_0 [13] is magnetic *permeability* of vacuum.

Since this description applies only to charges in vacuum, the vectors $\mathbf{D}$, $\mathbf{E}$, $\mathbf{B}$, and $\mathbf{H}$ are connected, respectively, through ε_0 and μ_0 by

$$\mathbf{D} = \varepsilon_0 \mathbf{E} , \tag{1.15}$$

$$\mathbf{B} = \mu_0 \mathbf{H} , \tag{1.16}$$

where $\mathbf{D}(\mathbf{r}, t)$ and $\mathbf{B}(\mathbf{r}, t)$ are the electric-flux-density (or electric displacement) and magnetic-flux-density (or magnetic induction) vectors, respectively. Now, it is necessary to express the electron current density $\mathbf{J}$ in terms of the electric field $\mathbf{E}$. One may make this replacement using the electrical *conductivity* $\sigma(\mathbf{r}, t)$, as

$$\mathbf{J} = \sigma \mathbf{E} , \tag{1.17}$$

where $\sigma(\mathbf{r}, t)$ in (1.17) is not a function of $\mathbf{E}$ or $\mathbf{B}$ that is valid when the response of EG will be linear. In general, we may assume that EG is *anisotropic*, so that the vector $\mathbf{J}$ and $\mathbf{E}$ are not generally colinear and σ is a tensor. [14]

Let us note that the relation given by (1.17) (that shows the quasi-free electron current density $\mathbf{J}$ is proportional to the *local* electric-field-intensity vector $\mathbf{E}$) is not the most general one for a linear homogeneous EG. Such material is also known as linear media that do not exhibit temporal or spatial dispersion. In other words, (1.17) states that the current in a point $\mathbf{r}$ and at a certain time t do only depend on the values of electric field at this position and time.

If only temporal dispersion is considered, the equation relating $\mathbf{J}$ and $\mathbf{E}$ for a linear and anisotropic can be written as

$$J_i(\mathbf{r}, t) = \sum_j \int_{-\infty}^{t} dt' \sigma_{ij}(\mathbf{r}, t - t') E_j(\mathbf{r}', t') , \tag{1.18}$$

where i and j are Cartesian indices. The above equation shows $J_i(\mathbf{r}, t)$ depends on $E_j(\mathbf{r}', t')$ at time t' earlier that t. Since (1.18) is a so-called *constitutive relation*, a *Fourier transform* with respect to time of this relation yields the familiar equation in the Fourier domain, as

$$J_i(\mathbf{r}, \omega) = \sum_j \sigma_{ij}(\mathbf{r}, \omega) E_j(\mathbf{r}, \omega) , \tag{1.19}$$

---

[13] $\mu_0 = 4\pi \times 10^{-7}$ H/m.
[14] When the medium is *isotropic*, the vector $\mathbf{J}$ and $\mathbf{E}$ are colinear.

where $\sigma_{ij}(\mathbf{r}, \omega)$ is the Fourier transform[15] of $\sigma_{ij}(\mathbf{r}, t-t')$. In the absence of temporal dispersion, EG would respond instantaneously to the applied field $\mathbf{E}$, so $\sigma_{ij}(\mathbf{r}, t-t')$ would be different from zero only for $t = t'$ and $\sigma_{ij}(\mathbf{r}, \omega) = \sigma_{ij}(\mathbf{r})$.

In the general case, when an EG that has the spatial and temporal dispersion, the current is not a function of only the local electric field, but $J_i(\mathbf{r}, t)$ also depends on the electric field $E_j(\mathbf{r}', t' < t)$, at any other place $\mathbf{r}'$ and time $t' < t$. Therefore (1.18) and (1.19) must be replaced by

$$J_i(\mathbf{r}, t) = \sum_j \int d\mathbf{r}' \int_{-\infty}^{t} dt' \sigma_{ij}(\mathbf{r}, \mathbf{r}', t - t') E_j(\mathbf{r}', t') , \qquad (1.20)$$

$$J_i(\mathbf{r}, \omega) = \sum_j \int d\mathbf{r}' \sigma_{ij}(\mathbf{r}, \mathbf{r}', \omega) E_j(\mathbf{r}', \omega) , \qquad (1.21)$$

respectively, where $d\mathbf{r}' \equiv dx' \, dy' \, dz'$. Now, the tensor $\underline{\sigma}(\mathbf{r}, \mathbf{r}', t - t')$ or $\underline{\sigma}(\mathbf{r}, \mathbf{r}'\omega)$ may be called *spatially nonlocal* conductivity tensor.[16] The range of this nonlocality is the range of distances $|\mathbf{r}-\mathbf{r}'|$ for which $\underline{\sigma}(\mathbf{r}, \mathbf{r}', t-t')$ is appreciably different from zero. In the interaction of electromagnetic wave with an EG, in general the range of nonlocality is usually much smaller than the wavelength of electromagnetic wave in the system and it may be a good approximation to use the spatially local response via (1.18) and (1.19).

If the EG is homogeneous and infinite so that the bulk metal can be taken as a translationally invariant system. Then σ_{ij} depends only on the difference $\mathbf{r} - \mathbf{r}'$ and we have $\sigma_{ij}(\mathbf{r}, \mathbf{r}', t - t') = \sigma_{ij}(\mathbf{r} - \mathbf{r}', t - t')$. In this case, the Fourier transform of (1.21) gives

$$\mathbf{J}(\mathbf{k}, \omega) = \underline{\sigma}(\mathbf{k}, \omega) \cdot \mathbf{E}(\mathbf{k}, \omega) , \qquad (1.22)$$

where $\mathbf{k} = (k_x, k_y, k_z)$. Here, spatial dispersion appears as a $\mathbf{k}$-dependence of $\underline{\sigma}(\mathbf{k}, \omega)$, and can be simplified to the limit of a spatially local response via $\underline{\sigma}(\mathbf{k} = 0, \omega)$.

---

[15]The Fourier transform can be defined in different ways. We define the Fourier transforms in 3D with respect to position and time and their inverses in the following way:

$$f(\mathbf{k}) = \int f(\mathbf{r}) e^{-i\mathbf{k}\cdot\mathbf{r}} \, d\mathbf{r} , \qquad f(\mathbf{r}) = \frac{1}{(2\pi)^3} \int f(\mathbf{k}) e^{i\mathbf{k}\cdot\mathbf{r}} \, d\mathbf{k} ,$$

$$f(\omega) = \int f(t) e^{i\omega t} \, dt , \qquad f(t) = \frac{1}{2\pi} \int f(\omega) e^{-i\omega t} \, d\omega .$$

[16]We indicate a tensor by bold italics but, where Greek characters are used, as in the above case, we add underlining for additional clarity.

We close this section by this note that Maxwell's equations are linear with respect to the various fields, and that neither the space coordinate $\mathbf{r}$ nor time t appears explicitly in these equations. Therefore, the Fourier transforms in space and time can be used, to transform Maxwell's equations into a set of linear algebraic equations. If $\mathbf{E}$ has the integral representation

$$\mathbf{E}(\mathbf{r}, t) = (2\pi)^{-4} \int d\mathbf{k} \int d\omega e^{i(\mathbf{k}\cdot\mathbf{r}-\omega t)} \mathbf{E}(\mathbf{k}, \omega) , \qquad (1.23)$$

with analogous forms for the other quantities, the Fourier transforms of Maxwell's equations in terms of total current and charge densities give

$$\varepsilon_0 \mathbf{k} \cdot \mathbf{E}(\mathbf{k}, \omega) = -i\rho(\mathbf{k}, \omega) , \qquad (1.24a)$$

$$\mu_0 \mathbf{k} \cdot \mathbf{H}(\mathbf{k}, \omega) = 0 , \qquad (1.24b)$$

$$\mathbf{k} \times \mathbf{E}(\mathbf{k}, \omega) = \mu_0 \omega \mathbf{H}(\mathbf{k}, \omega) , \qquad (1.24c)$$

$$\mathbf{k} \times \mathbf{H}(\mathbf{k}, \omega) = -\varepsilon_0 \omega \mathbf{E}(\mathbf{k}, \omega) - i\mathbf{J}(\mathbf{k}, \omega) , \qquad (1.24d)$$

where the space and time differential operators ∇ and $\partial/\partial t$ are replaced by $i\mathbf{k}$ and $-i\omega$, respectively. Also, the Fourier transforms of the linearized QHD equations, i.e., (1.11) and (1.13), give

$$-\omega n(\mathbf{k}, \omega) + n_0 \mathbf{k} \cdot \mathbf{v}(\mathbf{k}, \omega) = 0 , \qquad (1.25)$$

$$i\omega \mathbf{v}(\mathbf{k}, \omega) = \frac{e}{m_e} [\mathbf{E}(\mathbf{k}, \omega) + \mathbf{v}(\mathbf{k}, \omega) \times \mathbf{B}_0] + \gamma \mathbf{v}(\mathbf{k}, \omega) + i\frac{\alpha^2 + \xi\beta^2 k^2}{n_0} \mathbf{k} n(\mathbf{k}, \omega) , \qquad (1.26)$$

where $k = |\mathbf{k}|$.

1.3.2 Maxwell's Equations in Terms of Induced Polarization

Displacement of bound charges in an EG leads to a bound polarization $\mathbf{P}_b$, where $\mathbf{P}_b$ describes the bound electric dipole moment per unit volume inside the EG, caused by the alignment of microscopic dipoles with the electric field. In terms of $\mathbf{P}_b$, the electric current density associated with time-variation of the electronic displacement is given by $\mathbf{J}_b = \partial\mathbf{P}_b/\partial t$, and the associated distribution of electric charge density is $\rho_b = -\nabla \cdot \mathbf{P}_b$. It is easy to find that $\mathbf{J}_b$ and ρ_b in these equations satisfy the continuity relation, i.e., $\partial\rho_b/\partial t + \nabla \cdot \mathbf{J}_b = 0$.

Also, in an EG, drift displacement of free electrons leads to a free electric dipole moment per unit volume $\mathbf{P}$. In terms of $\mathbf{P}$ the electric current density and charge

density associated with time-variation of the drift displacement of the free electrons
are given by

$$J = \frac{\partial P}{\partial t} \, , \tag{1.27}$$

$$\rho = -\nabla \cdot P \, . \tag{1.28}$$

Let u be the drift displacement of the free electrons, so that $v = \partial u / \partial t$.
Displacement of an electron through u creates an electric dipole of moment $-eu$,
and there are n_0 such dipoles per unit volume. Hence the free electric moment per
unit volume of the EG is $P = -en_0 u$. Therefore, for current density and charge
density in the EG, we obtain

$$J = -en_0 \frac{\partial u}{\partial t} \, , \tag{1.29}$$

$$\rho = en_0 \nabla \cdot u \, . \tag{1.30}$$

Equation (1.29) should be the same as $J = -en_0 v$, and we see that this is so.
Equation (1.30) should be the same as $\rho = -en$, and we see that this is so if

$$n = -n_0 \nabla \cdot u \, . \tag{1.31}$$

This is the continuity relation for the electrons written in terms of drift displacement
rather than drift velocity; the time derivative of (1.31) gives (1.11).

If we use drift displacement u rather than drift velocity v, we may write the
equation of drift motion in a new form. Furthermore, substitution from (1.27)
and (1.28) into (1.14a) and (1.14d), respectively, gives

$$\nabla \cdot (\varepsilon_0 E + P) = 0 \, , \tag{1.32a}$$

$$\nabla \times H = \frac{\partial}{\partial t} (\varepsilon_0 E + P) \, . \tag{1.32b}$$

Hence, Maxwell's equations for the system under consideration may be written as

$$\nabla \cdot D = 0 \, , \tag{1.33a}$$

$$\nabla \cdot B = 0 \, , \tag{1.33b}$$

$$\nabla \times E = -\frac{\partial B}{\partial t} \, , \tag{1.33c}$$

$$\nabla \times H = \frac{\partial D}{\partial t} \, , \tag{1.33d}$$

with **B** defined by (1.16) and **D** as

$$\mathbf{D} = \varepsilon_0 \mathbf{E} + \mathbf{P} . \tag{1.34}$$

For linear, anisotropic, and homogeneous EG, we can relate the electric displacement and the induced polarization to the electric field through the constitutive relations

$$D_i(\mathbf{r}, t) = \varepsilon_0 \sum_j \int d\mathbf{r}' \int_{-\infty}^t dt' \varepsilon_{ij}(\mathbf{r} - \mathbf{r}', t - t') E_j(\mathbf{r}', t') , \tag{1.35a}$$

$$P_i(\mathbf{r}, t) = \varepsilon_0 \sum_j \int d\mathbf{r}' \int_{-\infty}^t dt' \chi_{ij}(\mathbf{r} - \mathbf{r}', t - t') E_j(\mathbf{r}', t') , \tag{1.35b}$$

where in general the relative dielectric tensor $\underline{\varepsilon}$ and *susceptibility* tensor $\underline{\chi}$ of the EG are considered as nonlocal functions. Equations (1.35a) and (1.35b) can be Fourier transformed as

$$\mathbf{D}(\mathbf{k}, \omega) = \varepsilon_0 \underline{\varepsilon}(\mathbf{k}, \omega) \cdot \mathbf{E}(\mathbf{k}, \omega) , \tag{1.36a}$$

$$\mathbf{P}(\mathbf{k}, \omega) = \varepsilon_0 \underline{\chi}(\mathbf{k}, \omega) \cdot \mathbf{E}(\mathbf{k}, \omega) . \tag{1.36b}$$

Also, from (1.34) and (1.27) after Fourier transforms in space and time, we find a general relation between the electrical conductivity tensor and the relative permittivity (or relative *dielectric function*) tensor, as

$$\underline{\varepsilon}(\mathbf{k}, \omega) = I + \underline{\chi}(\mathbf{k}, \omega), \tag{1.37}$$

with

$$\underline{\chi}(\mathbf{k}, \omega) = i \frac{\underline{\sigma}(\mathbf{k}, \omega)}{\omega \varepsilon_0} , \tag{1.38}$$

where I is just the identity tensor or, in index notation, the matrix with ones along the main diagonal and zeros elsewhere, δ_{ij}. If, for an EG, it were desired to take account not only of the free electrons but also of the bound charges, use would be made of the electric displacement defined as[17]

$$\mathbf{D} = \varepsilon_0 \mathbf{E} + \mathbf{P} + \mathbf{P}_b . \tag{1.39}$$

---

[17]If we consider $\mathbf{D} = \varepsilon_0 \mathbf{E} + \mathbf{P}_b$, then (1.33a) and (1.33d) must be read as $\nabla \cdot \mathbf{D} = \rho$ and $\nabla \times \mathbf{H} = \mathbf{J} + \frac{\partial \mathbf{D}}{\partial t}$, respectively.

In a homogeneous EG, we have $\mathbf{P}_b = \varepsilon_0 \underline{\chi_b} \cdot \mathbf{E}$, where $\underline{\chi_b}$ in general is frequency-dependent susceptibility tensor of the bound charges. In this case (1.37) must be read as

$$\underline{\varepsilon}(\mathbf{k}, \omega) = \underline{\varepsilon_b} + \underline{\chi_e}(\mathbf{k}, \omega) , \qquad (1.40)$$

where $\underline{\varepsilon_b} = I + \underline{\chi_b}$. These equations show that by incorporating all responding currents and charges into the polarization, we can treat an EG as a *dielectric*, such that the electric displacement in (1.39) satisfies (1.33a). Also, from the Fourier transforms of Maxwell's equations in terms of induced polarization, we get

$$\mathbf{k} \cdot \mathbf{D}(\mathbf{k}, \omega) = 0 , \qquad (1.41a)$$

$$\mathbf{k} \cdot \mathbf{B}(\mathbf{k}, \omega) = 0 , \qquad (1.41b)$$

$$\mathbf{k} \times \mathbf{E}(\mathbf{k}, \omega) = \omega \mathbf{B}(\mathbf{k}, \omega) , \qquad (1.41c)$$

$$\mathbf{k} \times \mathbf{H}(\mathbf{k}, \omega) = -\omega \mathbf{D}(\mathbf{k}, \omega) , \qquad (1.41d)$$

where $\mathbf{D} = \varepsilon_0 \underline{\varepsilon} \cdot \mathbf{E}$. We close this section by this note that relative dielectric function $\underline{\varepsilon}(\mathbf{k}, \omega)$, which is in general a complex function of the variable $\mathbf{k}$ and ω, obeys several symmetry conditions [44]. For instance, the requirement that $\mathbf{D}(\mathbf{r}, t)$ be real if $\mathbf{E}(\mathbf{r}', t')$ is real implies $\underline{\varepsilon}(\mathbf{k}, \omega) = \underline{\varepsilon}^*(-\mathbf{k}^*, -\omega^*)$, where * denotes complex conjugation and we have allowed $\mathbf{k}$ and ω to be complex variables.

1.4 Optical Properties of Collisionless, Isotropic, and Homogeneous Three-Dimensional Electron Gases

In this section, we study the pure electromagnetic and pure electrostatic properties of a collisionless, isotropic[18], and homogeneous 3DEG.[19] This separation into pure electromagnetic and pure electrostatic properties is in general only possible for a homogeneous EG. Hence, it is helpful to separate the vector quantities into transverse and longitudinal parts [45, 46], for example $\mathbf{E} = \mathbf{E}_T + \mathbf{E}_L$, where we have $\nabla \times \mathbf{E}_L = 0$ (or $\mathbf{k} \times \mathbf{E}_L = 0$) and $\nabla \cdot \mathbf{E}_T = 0$ (or $\mathbf{k}.\mathbf{E}_T = 0$).

By definition, pure electromagnetic wave has pure traverse electric fields. This means that, $\mathbf{E}_T$ is perpendicular to $\mathbf{k}$ so that $\mathbf{k} \cdot \mathbf{E}_T = 0$, and (1.24a) and (1.41a) give

---

[18] Here, this is an electron plasma that is not subjected to a steady imposed magnetic field.

[19] We note that the present study is based on the non-relativistic QHD model, where the phase velocities of the waves (as well as the particle velocities) are non-relativistic. In relativistic EGs, the relativistic effects may greatly modify the behavior of plasma waves.

$\rho = 0$. Thus, there is no space charge induced in such a wave. This is the classical behavior of a wave in vacuum.

Also, pure electrostatic waves have no electric field component perpendicular to the direction of propagation. Hence, since $\mathbf{k}$ is in the direction of propagation, $\mathbf{E}_L$ and $\mathbf{k}$ are parallel. Therefore, from (1.24c) and (1.41c) we see that $\mathbf{B}$ must be zero.[20] This means that a purely electrostatic wave has no magnetic effect associated with it. Such a wave does not always exist. For example, in vacuum by (1.15) $\mathbf{D}_L$ is parallel to $\mathbf{E}_L$, which is parallel to $\mathbf{k}$. Therefore, we see from (1.24a) and (1.41a) that $\mathbf{E}_L$ and $\mathbf{D}_L$ must be zero in vacuum. But in a homogeneous EG, pure electrostatic waves can exist.

1.4.1   Conductivity

We recall that the oscillating field quantities were assumed to vary as $\exp[i\,(\mathbf{k}\cdot\mathbf{r} - \omega t)]$. This means that rather than inverting of an arbitrary quantity, such as $\mathbf{E}(\mathbf{k}, \omega)$ through Fourier transform to find $\mathbf{E}(\mathbf{r}, t)$, a shortcut representation (when dealing with single values for $\mathbf{k}$ and ω) can be written simply, as $\mathbf{E}(\mathbf{r}, t) \sim \mathbf{E}(\mathbf{k}, \omega)e^{i(\mathbf{k}\cdot\mathbf{r}-\omega t)}$. Now, by eliminating the perturbed electron density from (1.25) and (1.26), for a collisionless, isotropic ($B_0 = 0$) and homogeneous EG, one can obtain the following equations:

$$\mathbf{v}_T(\mathbf{k}, \omega) = -i\,\frac{e}{m_e\omega}\mathbf{E}_T(\mathbf{k}, \omega)\,, \tag{1.42a}$$

$$\mathbf{v}_L(\mathbf{k}, \omega) = -\frac{ie\omega/m_e}{\omega^2 - \left(\alpha^2 + \xi\beta^2k^2\right)k^2}\mathbf{E}_L(\mathbf{k}, \omega)\,. \tag{1.42b}$$

The electron current density is $\mathbf{J} = -en_0\mathbf{v}$ and may therefore be written

$$\mathbf{J}_T(\mathbf{k}, \omega) = i\,\frac{n_0e^2}{m_e\omega}\mathbf{E}_T(\mathbf{k}, \omega)\,, \tag{1.43a}$$

$$\mathbf{J}_L(\mathbf{k}, \omega) = \frac{in_0e^2\omega/m_e}{\omega^2 - \left(\alpha^2 + \xi\beta^2k^2\right)k^2}\mathbf{E}_L(\mathbf{k}, \omega)\,. \tag{1.43b}$$

Then, by Ohm's law it is easy to find that the pure electromagnetic and pure electrostatic conductivity of the system are

$$\sigma_T(\omega) = i\,\frac{n_0e^2}{m_e\omega}\,, \tag{1.44a}$$

---

[20]However, in an anisotropic EG we can still have a steady imposed magnetic field $\mathbf{B}_0$.

$$\sigma_L(k, \omega) = \frac{in_0 e^2 \omega / m_e}{\omega^2 - (\alpha^2 + \xi\beta^2 k^2)\, k^2} \,, \tag{1.44b}$$

that are purely imaginary. It is clear that conductivity of a pure electromagnetic wave is not sensitive to the terms with α and β. Also, we note that for the isotropic EG under consideration, $\underline{\sigma}$ is a symmetric tensor containing only diagonal elements σ_T and σ_L. If the propagation of bulk pure electrostatic waves along the z-axis is assumed, we obtain

$$\underline{\sigma} = \begin{pmatrix} \sigma_T & 0 & 0 \\ 0 & \sigma_T & 0 \\ 0 & 0 & \sigma_L \end{pmatrix}. \tag{1.45}$$

1.4.2   Susceptibility and Dielectric Function

At high frequencies it is usually more convenient to use dielectric terminology rather than conduction terminology. We then use the susceptibility χ expressed in terms of conductivity σ, i.e., (1.38). Comparing (1.38) with (1.44a) and (1.44b), we find the pure electromagnetic and pure electrostatic electric susceptibility of the EG, as

$$\chi_T(\omega) = -\frac{n_0 e^2}{\varepsilon_0 m_e \omega^2}\,, \tag{1.46a}$$

$$\chi_L(k, \omega) = -\frac{n_0 e^2 / \varepsilon_0 m_e}{\omega^2 - (\alpha^2 + \xi\beta^2 k^2)\, k^2}\,. \tag{1.46b}$$

It is often convenient to specify the electron density with the aid of electron plasma frequency. The electron plasma frequency is defined by $\omega_p = (n_0 e^2 / \varepsilon_0 m_e)^{1/2}$.[21] The electron density is therefore proportional to the square of the electron plasma frequency. In terms of the electron plasma frequency, (1.46a) and (1.46b) may be written either as

$$\chi_T(\omega) = -\frac{\omega_p^2}{\omega^2}\,, \tag{1.47a}$$

$$\chi_L(k, \omega) = -\frac{\omega_p^2}{\omega^2 - (\alpha^2 + \xi\beta^2 k^2)\, k^2}\,. \tag{1.47b}$$

Hence, using (1.40) the relative dielectric function for an isotropic EG is

---

[21] In the literature, the values of plasma frequency are usually given by $\nu_p = \omega_p / 2\pi$.

$$\varepsilon_T(\omega) = \varepsilon_b - \frac{\omega_p^2}{\omega^2} , \qquad (1.48a)$$

$$\varepsilon_L(k, \omega) = \varepsilon_b - \frac{\omega_p^2}{\omega^2 - \left(\alpha^2 + \xi\beta^2 k^2\right) k^2} . \qquad (1.48b)$$

One can see that the transverse dielectric function is the same as the local dielectric function in the Drude model. Clearly the nonlocal terms do not affect the components of ε_T that describe the transverse motion, where there is no density change.

1.4.3   Dispersion Relation

Equations (1.41a)–(1.41d) constitute a set of linear and homogeneous equations with respect to the fields for any four quantities (k_x, k_y, k_z, ω). Also, this set of linear and homogeneous equations is of first degree with respect to the four fields **E**, **H**, **B**, and **D**, and we have the same number of equations as variables. In general, for any given four quantities (k_x, k_y, k_z, ω), there will exist non-trivial solutions for the fields only if the determinant of the coefficients is zero. This introduces a functional relationship between ω and k_x, k_y, and k_z. Such a relationship is a condition of compatibility between ω and k_x, k_y, and k_z for having non-zero fields in the system, i.e., for having waves. This relationship is called the *dispersion relation*. The dispersion relation is very important because it contains the linear properties of the medium with respect to waves.

For a collisionless and isotropic EG to be discussed here, cross-multiplying (1.41c) on the left by the vector **k** and using (1.41d) and $\mathbf{D} = \varepsilon_0 \underline{\varepsilon} \cdot \mathbf{E}$, leads to the wave equation

$$\mathbf{k}(\mathbf{k} \cdot \mathbf{E}) - k^2\mathbf{E} = -\frac{\omega^2}{c^2}\underline{\varepsilon} \cdot \mathbf{E} , \qquad (1.49)$$

in the Fourier domain. Now, by introducing the so-called *refractive index* vector $\mathbf{n} = c\mathbf{k}/\omega$, for pure electromagnetic (transverse) waves, we get

$$n^2 = \varepsilon_T(\omega) , \qquad (1.50)$$

where $\underline{\varepsilon} \cdot \mathbf{E} = \varepsilon_T \mathbf{E}_T$ and $n = |\mathbf{n}|$. From (1.48a), the dispersion relation of the pure electromagnetic waves is given by the solution of (1.50), as

$$\omega^2 = \frac{1}{\varepsilon_b}\left(\omega_p^2 + c^2k^2\right) , \qquad (1.51)$$

which gives a condition on the allowed values of ω and k. At $\omega = \omega_p/\sqrt{\varepsilon_b}$ the pure electromagnetic wave exhibits a *cutoff*.[22] In other words, no propagation occurs for frequencies less than $\omega = \omega_p/\sqrt{\varepsilon_b}$. Also, no *resonance*[23] is observed for this wave.

For pure electrostatic (longitudinal) waves, (1.49) implies that

$$\varepsilon_L(k, \omega) = 0 \,, \tag{1.52}$$

and yields the well-known bulk plasmon dispersion

$$\omega^2 = \frac{\omega_p^2}{\varepsilon_b} + \left(\alpha^2 + \beta^2 k^2\right) k^2 \,, \tag{1.53}$$

for $\xi = 1$ [47].

1.4.4 Phase and Group Velocities

The *phase velocity* of a wave is the rate at which the phase of the wave propagates in space. By definition, the propagation velocity of phase is $v_p = c/n$, and in general, we have

$$v_p = \frac{\omega}{k} \,. \tag{1.54}$$

Also, an EG is a dispersive medium, and consequently the *group velocity* v_g for a pulse traveling through it differs from the phase velocity associated with individual wave crests. Whereas the phase velocity is given by (1.54), the group velocity can be obtained as

$$v_g = \frac{\partial \omega}{\partial k} \,. \tag{1.55}$$

Hence, using (1.51) and (1.53), when $\varepsilon_b = 1$, we find

$$v_{pT}(\omega) = c\left(1 - \frac{\omega_p^2}{\omega^2}\right)^{-1/2} \,, \tag{1.56a}$$

$$v_{pL}(k, \omega) = k^{-1}\left[\omega_p^2 + \left(\alpha^2 + \beta^2 k^2\right) k^2\right]^{1/2} \,, \tag{1.56b}$$

---

[22] When the index of refraction n goes to zero, we say that there is a cutoff. This occurs when the phase velocity $v_p = \omega/k$ goes to infinity.

[23] When the index of refraction n goes to infinity, we say that there is a resonance. This occurs when the phase velocity goes to zero.

and

$$v_{gT}(\omega) = c \left(1 - \frac{\omega_p^2}{\omega^2}\right)^{1/2}, \qquad (1.57a)$$

$$v_{gL}(k, \omega) = \frac{\left(\alpha^2 + 2\beta^2 k^2\right) k}{\left[\omega_p^2 + \left(\alpha^2 + \beta^2 k^2\right) k^2\right]^{1/2}}. \qquad (1.57b)$$

Comparison of (1.56a) and (1.57a) shows that the group velocity of the pure electromagnetic wave in an EG is less than c, and the phase velocity is greater than c. Also, we see that $v_{pT} v_{gT} = c^2$. However, for pure electrostatic waves, we find

$$v_{gL} v_{pL} = \alpha^2 + 2\beta^2 k^2, \qquad (1.58)$$

that shows in the presence of Bohm potential term, $v_{gL} v_{pL}$ is not equal to a constant and is dependent on the wavenumber k [47].

1.4.5 Poynting's Theorem: Pure Electromagnetic Wave

From Maxwell's equations, in terms of electron current and electron charge densities, i.e., (1.14a)–(1.14d), we obtain the standard form of *Poynting's theorem*, as

$$-\frac{\partial U_{EM}}{\partial t} = \nabla \cdot \mathbf{S}_T + \mathbf{E}_T \cdot \mathbf{J}_T, \qquad (1.59)$$

where $\mathbf{S}_T = \mathbf{E}_T \times \mathbf{H}$ is known as the Poynting vector (in watts per square meter), which is a power density vector associated with the electromagnetic field. Also, $U_{EM} = U_{ET} + U_M$ is the stored energy density of electromagnetic wave (in joules per cubic meter), where $U_{ET} = \frac{1}{2}\varepsilon_0 \varepsilon_b \mathbf{E}_T \cdot \mathbf{E}_T$ is the stored *electric energy* density and $U_M = \frac{1}{2}\mu_0 \mathbf{H} \cdot \mathbf{H}$ is the stored *magnetic energy* density. But there is also the stored *kinetic energy* density associated with the drift motion of the electrons due to the propagation of pure electromagnetic waves in an EG. To find this kinetic energy density, we note that (1.59) is usually interpreted by saying that the decrease of electromagnetic energy density on the left is caused by the energy flow leaving the volume element plus the Ohmic loss $\mathbf{E}_T \cdot \mathbf{J}_T$ in the volume element. But for a collisionless EG, the term $\mathbf{E}_T \cdot \mathbf{J}_T$ is not the energy loss density and actually shows the kinetic energy density. To show this, from (1.13) we have

$$\frac{\partial \mathbf{v}_T}{\partial t} = -\frac{e}{m_e}\mathbf{E}_T. \qquad (1.60)$$

From (1.60) in combination with the electron current density, i.e., $\mathbf{J}_T = -en_0\mathbf{v}_T$, we rewrite the second term in the right-hand side of (1.59) as

$$\mathbf{E}_T \cdot \mathbf{J}_T = \frac{m_e n_0}{2} \frac{\partial}{\partial t} (\mathbf{v}_T \cdot \mathbf{v}_T) \ . \tag{1.61}$$

In consequence, the kinetic energy density of the system is defined by

$$U_{KT} = \frac{1}{2} m_e n_0 \mathbf{v}_T \cdot \mathbf{v}_T \ , \tag{1.62}$$

in the real number representation. Therefore, we have

$$U_T = U_{EM} + U_{KT} \ . \tag{1.63}$$

It follows that Poynting's theorem for the pure electromagnetic waves involves, as it should, the sum of the electric energy, the magnetic energy, and the kinetic energy.

It is likewise true that, for an EG, the stored kinetic energy enters into the complex version of Poynting's theorem. We are then considering an oscillatory electromagnetic field of angular frequency ω, and we are making use of complex electromagnetic vectors. The time-averaged Poynting vector in the complex number representation is defined by

$$\mathbf{S}_T = \frac{1}{2} \mathrm{Re} \left[\mathbf{E}_T \times \mathbf{H}^* \right] \ , \tag{1.64}$$

where Re denotes real part and the time-averaged electric, magnetic, and kinetic energy densities are given by

$$U_{ET} = \frac{1}{4} \varepsilon_0 \varepsilon_b \mathbf{E}_T \cdot \mathbf{E}_T^* \ , \tag{1.65a}$$

$$U_M = \frac{1}{4} \mu_0 \mathbf{H} \cdot \mathbf{H}^* \ , \tag{1.65b}$$

$$U_{KT} = \frac{1}{4} m_e n_0 \mathbf{v}_T \cdot \mathbf{v}_T^* \ , \tag{1.65c}$$

in the complex number representation. Now from (1.60), (1.63), and (1.65a)–(1.65c), one can find

$$U_T = \frac{1}{4} \varepsilon_0 \left(\varepsilon_b + \frac{\omega_p^2}{\omega^2} \right) \mathbf{E}_T \cdot \mathbf{E}_T^* + \frac{1}{4} \mu_0 \mathbf{H} \cdot \mathbf{H}^* \ , \tag{1.66}$$

in the complex number representation. At this stage, from (1.48a), it is easy to find [48]

$$U_T = \frac{1}{4}\varepsilon_0 \left(\varepsilon_T + \omega \frac{d\varepsilon_T}{d\omega} \right) \mathbf{E}_T \cdot \mathbf{E}_T^* + \frac{1}{4}\mu_0 \mathbf{H} \cdot \mathbf{H}^* . \tag{1.67}$$

1.4.6   Poynting's Theorem: Pure Electrostatic Wave

To find Poynting's theorem in the electrostatics approximation, we note that the magnetic energy density must be neglected and we have

$$U_{EL} = \frac{1}{2}\varepsilon_0 \varepsilon_b \mathbf{E}_L \cdot \mathbf{E}_L , \tag{1.68}$$

where $\mathbf{E}_L$ is irrotational ($\nabla \times \mathbf{E}_L = 0$) and can be written as

$$\mathbf{E}_L = -\nabla \Phi. \tag{1.69}$$

Here, Φ is the *scalar electric potential*. Also, in the electrostatics approximation, we can find an alternative expression for **S** that does not involve **H**. To do this, from standard form of Poynting's theorem, as

$$-\frac{\partial U_{EL}}{\partial t} = \nabla \cdot (\mathbf{E}_L \times \mathbf{H}) + \mathbf{E}_L \cdot \mathbf{J}_L , \tag{1.70}$$

we obtain

$$\mathbf{E}_L \times \mathbf{H} = -\nabla \Phi \times \mathbf{H}. \tag{1.71}$$

Next, we use the vector identity, as

$$\nabla \times (\Phi \mathbf{H}) = \nabla \Phi \times \mathbf{H} + \Phi \nabla \times \mathbf{H} . \tag{1.72}$$

Thus, by using (1.72) in (1.71), we find

$$\mathbf{E}_L \times \mathbf{H} = -\nabla \times (\Phi \mathbf{H}) + \Phi \nabla \times \mathbf{H} . \tag{1.73}$$

At this stage, we eliminate the first term on the right-hand side of (1.73) because divergence of the curl of $\Phi \mathbf{H}$ is zero, i.e., $\nabla \cdot \nabla \times (\Phi \mathbf{H}) = 0$. Now, modified Ampere's law, i.e., (1.14d), can be used to eliminate $\nabla \times \mathbf{H}$ from the second term of (1.73). Therefore, we find

$$\mathbf{E}_L \times \mathbf{H} = \Phi \left(\mathbf{J}_L - \varepsilon_0 \varepsilon_b \frac{\partial}{\partial t} \nabla \Phi \right) . \tag{1.74}$$

Finally, similar to the method discussed in previous section, from (1.11), for propagation of a pure electrostatic wave in a collisionless, isotropic, and homogeneous EG, we obtain

$$\mathbf{E}_L = -\frac{m_e}{e} \left[\frac{\partial}{\partial t} \mathbf{v}_L + \frac{\alpha^2 + \xi \beta^2 k^2}{n_0} \nabla n \right] , \tag{1.75}$$

and we rewrite the second term in the right-hand side of (1.70) as

$$\mathbf{E}_L \cdot \mathbf{J}_L = \frac{m_e n_0}{2} \frac{\partial}{\partial t} (\mathbf{v}_L \cdot \mathbf{v}_L) + m_e \left(\alpha^2 + \xi \beta^2 k^2 \right) \nabla n \cdot \mathbf{v}_L . \tag{1.76}$$

Next, we use the vector identity, as

$$\nabla \cdot (n \mathbf{v}_L) = n (\nabla \cdot \mathbf{v}_L) + \mathbf{v}_L \cdot \nabla n . \tag{1.77}$$

Then, by using (1.9) and (1.77) in (1.76), we find

$$\mathbf{E}_L \cdot \mathbf{J}_L = \frac{m_e n_0}{2} \frac{\partial}{\partial t} (\mathbf{v}_L \cdot \mathbf{v}_L) + \frac{m_e}{2n_0} \left(\alpha^2 + \xi \beta^2 k^2 \right) \frac{\partial}{\partial t} n^2 + m_e \left(\alpha^2 + \xi \beta^2 k^2 \right) \nabla \cdot (n \mathbf{v}_L) . \tag{1.78}$$

Thus, we have the *energy density* U_L and the *power flow* density vector $\mathbf{S}_L$ in the EG in the real number representation, as

$$U_L = \frac{1}{2} \varepsilon_0 \varepsilon_b \mathbf{E}_L \cdot \mathbf{E}_L + \frac{m_e n_0}{2} (\mathbf{v}_L \cdot \mathbf{v}_L) + \frac{m_e}{2n_0} \left(\alpha^2 + \xi \beta^2 k^2 \right) n^2 , \tag{1.79}$$

$$\mathbf{S}_L = \Phi \left(\mathbf{J}_L - \varepsilon_0 \varepsilon_b \frac{\partial}{\partial t} \nabla \Phi \right) + m_e \left(\alpha^2 + \xi \beta^2 k^2 \right) n \mathbf{v}_L , \tag{1.80}$$

where the second term in the right-hand side of (1.79) shows the kinetic energy density due to the drift velocity of the current carrying electrons, and the third term represents the *potential energy* density or *hydrodynamic compressional energy* density of an EG with perturbed density n in direct analogy to the electrostatic wave field. Also, in the right-hand side of (1.80), the first term is the electrostatic power flow density vector and the second term is the nonlocal power flow density vector. Furthermore, the time-averaged power flow density vector is defined by

$$\mathbf{S}_L = \frac{1}{2} \mathrm{Re} \left[\Phi \left(\mathbf{J}_L^* - \varepsilon_0 \varepsilon_b \frac{\partial}{\partial t} \nabla \Phi^* \right) + m_e \left(\alpha^2 + \xi \beta^2 k^2 \right) n \mathbf{v}_L^* \right] , \tag{1.81}$$

in the complex number representation and the time-averaged electric, magnetic, and kinetic energy densities are given by

$$U_{EL} = \frac{1}{4}\varepsilon_0\varepsilon_b \mathbf{E}_L \cdot \mathbf{E}_L^* \,, \tag{1.82a}$$

$$U_{KL} = \frac{1}{4}m_e n_0 \mathbf{v}_L \cdot \mathbf{v}_L^* \,, \tag{1.82b}$$

$$U_P = \frac{1}{4}\frac{m_e}{n_0}\left(\alpha^2 + \xi\beta^2 k^2\right)nn^* \,, \tag{1.82c}$$

in the complex number representation.

1.5   Bounded Homogeneous Electron Gases

In Sect. 1.4, we described the basic equations in an infinite space of homogeneous EG. However, there are another important class of problems, known as the BVPs, where we deal with a bounded homogeneous EG. Although the treatment of the field equations inside and outside of EG are exactly the same as the procedures we followed in previous sections, it is necessary to have some prescribed BCs at the surface of the EG to allow matching of the solutions.

1.5.1   Boundary Conditions

Let us consider a boundary surface separating two media with different electrical conductivity and relative permittivity, as shown in Fig. 1.1. Also, let $\mathbf{n}$ denotes the unit normal to the surface directed from medium 1 into medium 2. Furthermore, in general, let $\underline{\sigma_j}$ and $\underline{\varepsilon_j}$ (with $j = 1$ and 2) be the electrical conductivity tensor and the relative permittivity tensor of medium j, respectively. To obtain the usually electromagnetic BCs (in the absence of the spatial nonlocal effects) for $\mathbf{D}$ and $\mathbf{H}$, we construct a thin cylinder over a unit area of the surface, as shown in panel (a) of Fig. 1.1. The end faces of the cylinder are parallel to the surface. We now apply Gauss's divergence theorem, as

$$\int \nabla \cdot \mathbf{F}\, dv = \oint \mathbf{F} \cdot \mathbf{n}\, ds \,, \tag{1.83}$$

to both sides of $\nabla \cdot \mathbf{D} = \rho$ and $\nabla \cdot \mathbf{H} = 0$. The surface integral reduces, in the limit as the height of the cylinder approaches zero, to an integral over the end surfaces only. This leads to

$$\mathbf{n} \cdot (\mathbf{H}_2 - \mathbf{H}_1) = 0 \,, \tag{1.84a}$$

$$\mathbf{n} \cdot (\mathbf{D}_2 - \mathbf{D}_1) = \sigma \,, \tag{1.84b}$$

Fig. 1.1 (a) A small cylinder
about the interface between
two media: S is the surface of
this cylinder. (b) A narrow
rectangle about the interface
between two media; C is the
boundary of this rectangle

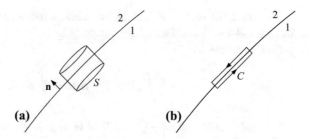

where σ (not to be confused with conductivity) is the free surface charge density (in
coulomb per square meter), and the subscripts refer to values at the surfaces in the
two media. The BCs in (1.84a) and (1.84b) are often written as

$$H_{1n} = H_{2n} \,, \tag{1.85a}$$

$$D_{2n} - D_{1n} = \sigma \,, \tag{1.85b}$$

where $H_{1n} = \mathbf{H}_1 \cdot \mathbf{n}$, $H_{2n} = \mathbf{H}_2 \cdot \mathbf{n}$, $D_{1n} = \mathbf{D}_1 \cdot \mathbf{n}$, and $D_{2n} = \mathbf{D}_2 \cdot \mathbf{n}$.

The fact that, in general, σ is not zero introduces some complexity in (1.84b);
however, noting that charge must be conserved, that is, that

$$\nabla \cdot \mathbf{J} = -\frac{\partial \rho}{\partial t} \,, \tag{1.86}$$

makes possible certain simplifications. If we integrate the above equation as we did
for $\nabla \cdot \mathbf{D} = \rho$, and shrink the thin cylinder in the same way, we obtain

$$\mathbf{n} \cdot (\mathbf{J}_2 - \mathbf{J}_1) = -\frac{\partial \sigma}{\partial t} \,. \tag{1.87}$$

Now, we consider a Fourier component of a small perturbation, therefore the right
side of (1.87) can be written as $i\omega\sigma$. Also, using the constitutive relations $\mathbf{D} = \varepsilon_0 \underline{\varepsilon} \cdot \mathbf{E}$
and $\mathbf{J} = \underline{\sigma} \cdot \mathbf{E}$, then (1.84b) and (1.87) can be written as

$$\mathbf{n} \cdot \left(\underline{\varepsilon_2} \cdot \mathbf{E}_2 - \underline{\varepsilon_1} \cdot \mathbf{E}_1 \right) = \frac{\sigma}{\varepsilon_0} \,, \tag{1.88a}$$

$$\mathbf{n} \cdot \left(\underline{\sigma_2} \cdot \mathbf{E}_2 - \underline{\sigma_1} \cdot \mathbf{E}_1 \right) = i\omega\sigma \,. \tag{1.88b}$$

Now, σ can be eliminated from (1.88a) and (1.88b). The result of this elimination is

$$\mathbf{n} \cdot \left[\left(\underline{\varepsilon_2} - \frac{\underline{\sigma_2}}{i\omega\varepsilon_0} \right) \cdot \mathbf{E}_2 - \left(\underline{\varepsilon_1} - \frac{\underline{\sigma_1}}{i\omega\varepsilon_0} \right) \cdot \mathbf{E}_1 \right] = 0 \,. \tag{1.89}$$

This equation is useful as it stands in providing a BC.

For the field vectors **E** and **H**, we draw a rectangular contour with two long sides parallel to the surface of discontinuity, as shown in Fig. 1.1. Now, by applying Stokes' theorem, as

$$\int (\nabla \times \mathbf{F}) \cdot \mathbf{n} \, ds = \oint \mathbf{F} \cdot \mathbf{dl} \,, \tag{1.90}$$

to both sides of $\nabla \times \mathbf{E} = -\frac{\partial \mathbf{B}}{\partial t}$ and $\nabla \times \mathbf{H} = \mathbf{J} + \frac{\partial \mathbf{D}}{\partial t}$, the contour integral reduced, in the limit as the width of the rectangle approaches zero, to an integral over the two long sides only. This leads to

$$\mathbf{n} \times (\mathbf{E}_2 - \mathbf{E}_1) = 0 \,, \tag{1.91a}$$

$$\mathbf{n} \times (\mathbf{H}_2 - \mathbf{H}_1) = \mathbf{J}_{s\perp} \,, \tag{1.91b}$$

where $\mathbf{J}_{s\perp}$ is the component of the (free) surface current density (in ampere per meter) perpendicular to the direction of the **H**-component which is being matched. Again, the BCs for the electric and magnetic field vectors, i.e., (1.91a) and (1.91b) are often written as

$$\mathbf{E}_{2t} = \mathbf{E}_{1t} \,, \tag{1.92a}$$

$$\mathbf{H}_{2t} - \mathbf{H}_{1t} = \mathbf{J}_{s\perp} \,, \tag{1.92b}$$

where the subscript t means the tangential component of a field vector.[24]

1.5.2 Electrostatic Approximation

Again, let us consider a boundary surface separating two media with different relative permittivity, as shown in Fig. 1.2. To obtain the usually electrostatic BCs (in the absence of the spatial nonlocal effects), at first we note that the electrostatic potential is continuous across any boundary. To show this point, by using Fig. 1.2 we have

$$\Phi_2 - \Phi_1 = -\int_a^b \mathbf{E} \cdot \mathbf{dl} \,, \tag{1.93}$$

as the path length shrinks to zero, so too does the integral, as

$$\Phi_1 = \Phi_2 \,. \tag{1.94}$$

---

[24]The tangential components of these field vectors to the boundary surface are still vectors in the tangential plane of the surface.

Fig. 1.2 An interface
between two media with the
free surface charge density σ

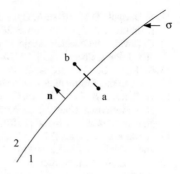

However, the gradient of Φ inherits the discontinuity in $\mathbf{E}$, since $\mathbf{E} = -\nabla\Phi$. Therefore (1.88a) implies that

$$\underline{\varepsilon_2} \cdot \nabla\Phi_2 - \underline{\varepsilon_1} \cdot \nabla\Phi_1 = -\frac{\sigma}{\varepsilon_0}\mathbf{n}, \tag{1.95}$$

and finally from (1.89), we obtain

$$\mathbf{n} \cdot \left[\left(\underline{\varepsilon_2} - \frac{\sigma_2}{i\omega\varepsilon_0}\right) \cdot \nabla\Phi_2 - \left(\underline{\varepsilon_1} - \frac{\sigma_1}{i\omega\varepsilon_0}\right) \cdot \nabla\Phi_1\right] = 0. \tag{1.96}$$

The above equation is also useful as it stands in providing a BC.

References

1. P. Drude, Zur ionentheorie der metalle. Phys. Z. **1**, 161–165 (1900)
2. I. Villo-Perez, Z.L. Mišković, N.R. Arista, Plasmon spectra of nano-structures: a hydrodynamic model, in *Trends in Nanophysics*, ed. by A. Barsan, V. Aldea (Springer, Berlin, 2010)
3. C. Ciraci, J.B. Pendry, D.R. Smith, Hydrodynamic model for plasmonics: a macroscopic approach to a microscopic problem. ChemPhysChem **14**, 1109–1116 (2013)
4. S. Raza, S.I. Bozhevolnyi, M. Wubs, N.A. Mortensen, Nonlocal optical response in metallic nanostructures. J. Phys. Condens. Matter **27**, 183204 (2015)
5. D. Pines, D. Bohm, A collective description of electron interactions: II. Collective vs individual particle aspects of the interactions. Phys. Rev. **85**, 338–353 (1952)
6. M. Kupresak, X. Zheng, G.A.E. Vandenbosch, V.V. Moshchalkov, Comparison of hydrodynamic models for the electromagnetic nonlocal response of nanoparticles. Adv. Theory Simul. **1**, 1800076 (2018)
7. M. Kupresak, X. Zheng, G.A.E. Vandenbosch, V.V. Moshchalkov, Appropriate nonlocal hydrodynamic models for the characterization of deep-nanometer scale plasmonic scatterers. Adv. Theory Simul. **3**, 1900172 (2020)
8. R. Ruppin, Optical properties of a plasma sphere. Phys. Rev. Lett. **31**, 1434–1437 (1973)
9. R. Ruppin, Optical properties of small metal spheres. Phys. Rev. B **11**, 2871–2876 (1975)
10. R. Ruppin, Extinction properties of thin metallic nanowires. Opt. Commun. **190**, 205–209 (2001)

11. V. Datsyuk, O.M. Tovkach, Optical properties of a metal nanosphere with spatially dispersive permittivity. J. Opt. Soc. Am. B **28**, 1224–1230 (2011)
12. C. David, F.J. Garcia de Abajo, Spatial nonlocality in the optical response of metal nanoparticles. J. Phys. Chem. C **115**, 19470–19475 (2011)
13. S. Raza, G. Toscano, A.-P. Jauho, M. Wubs, N.A. Mortensen, Unusual resonances in nanoplasmonic structures due to nonlocal response. Phys. Rev. B **84**, 121412 (2011)
14. G. Toscano, S. Raza, W. Yan, C. Jeppesen, S. Xiao, M. Wubs, A.-P. Jauho, S.I. Bozhevolnyi, N.A. Mortensen, Nonlocal response in plasmonic waveguiding with extreme light confinement. Nanophotonics **2**, 161–166 (2013)
15. A. Moreau, C. Ciraci, D.R. Smith, Impact of nonlocal response on metallodielectric multilayers and optical patch antennas. Phys. Rev. B **87**, 045401 (2013)
16. S. Raza, W. Yan, N. Stenger, M. Wubs, N.A. Mortensen, Blueshift of the surface plasmon resonance in silver nanoparticles: substrate effects. Opt. Express **21**, 27344–27355 (2013)
17. T. Christensen, W. Yan, S. Raza, A.-P. Jauho, N.A. Mortensen, M. Wubs, Nonlocal response of metallic nanospheres probed by light, electrons, and atoms. ACS Nano **8**, 1745–1758 (2014)
18. A. Moradi, E. Ebrahimi, Plasmon spectra of cylindrical nanostructures including nonlocal effects. Plasmonics **9**, 209–218 (2014)
19. M. Wubs, Classification of scalar and dyadic nonlocal optical response models. Opt. Express **23**, 31296–31312 (2015)
20. C. Tserkezis, J.R. Maack, Z. Liu, M. Wubs, N.A. Mortensen, Robustness of the far-field response of nonlocal plasmonic ensembles. Sci. Rep. **6**, 28441 (2016)
21. F. Sauter, Der einfl uss von plasmawellen auf das refl exionsverm o gen von metallen (I). Z. Phys. **203**, 488–494 (1967)
22. R. Ruppin, Optical properties of spatially dispersive dielectric spheres. J. Opt. Soc. Am. **71**, 755–758 (1981)
23. R. Ruppin, Mie theory with spatial dispersion. Opt. Commun. **30**, 380–382 (1979)
24. R. Ruppin, Optical properties of a spatially dispersive cylinder. J. Opt. Soc. Am. B **6**, 1559–1563 (1989)
25. S.I. Pekar, The theory of electromagnetic waves in a crystal in which excitons are produced. J. Phys. Chem. Solids **6**, 785–796 (1958)
26. F. Forstmann, R.R. Gerhardts, *Metal Optics Near Plasma Frequency* (Springer, Berlin, 1986)
27. K.L. Kliewer, R. Fuchs, Anomalous skin effect for specular electron scattering and optical experiments at non-normal angles of incidence. Phys. Rev. **172**, 607–624 (1968)
28. D.L. Johnson, P.R. Rimbey, Aspects of spatial dispersion in the optical properties of a vacuum-dielectric interface. Phys. Rev. B **14**, 2398–2410 (1976)
29. A. Moradi, Quantum nonlocal effects on optical properties of spherical nanoparticles. Phys. Plasmas **22**, 022119 (2015)
30. A. Moradi, Quantum effects on propagation of bulk and surface waves in a thin quantum plasma film. Phys. Lett. A **379**, 1139–1143 (2015)
31. A. Moradi, Plasmon modes of spherical nanoparticles: the effects of quantum nonlocality. Surf. Sci. **637**, 53–57 (2015)
32. A. Moradi, Maxwell-Garnett effective medium theory: quantum nonlocal effects. Phys. Plasmas **22**, 042105 (2015)
33. A. Moradi, Quantum nonlocal polarizability of metallic nanowires. Plasmonics **10**, 1225–1230 (2015)
34. Y.-Y. Zhang, S.-B. An, Y.-H. Song, N. Kang, Z.L. Mišković, Y.-N. Wang, Plasmon excitation in metal slab by fast point charge: the role of additional boundary conditions in quantum hydrodynamic model. Phys. Plasmas **21**, 102114 (2014)
35. H.G. Booker, *Cold Plasma Waves* (Martinus Nijhoff Publishers, Dordrecht, 1984)
36. G. Manfredi, How to model quantum plasmas. Fields Inst. Commun. **46**, 263–287 (2005)
37. M. Bonitz, N. Horing, P. Ludwig, *Introduction to Complex Plasmas* (Springer, Berlin, 2010)
38. F. Haas, *Quantum Plasmas: An Hydrodynamic Approach* (Springer, New York, 2011)
39. U. Kreibig, C. von Fragstein, The limitation of electron mean free path in small silver particles. Z. Phys. **224**, 307–323 (1969)

40. J. Euler, Ultraoptische eigenschaften von metallen und mittlere freie weglange der leitungselektronen. Z. Phys. **137**, 318–332 (1954)
41. D. Pines, *Elementary Excitations in Solids* (Benjamin, New York, 1963)
42. P. Halevi, Hydrodynamic model for the degenerate free-electron gas: generalization to arbitrary frequencies. Phys. Rev. B **51**, 7497–7499 (1995)
43. Z.A. Moldabekov, M. Bonitz, T.S. Ramazanov, Theoretical foundations of quantum hydrodynamics for plasmas. Phys. Plasmas **25**, 031903 (2018)
44. P. Halevi, Spatial dispersion in solids and plasmas, in *Electromagnetic Waves: Recent Developments in Research*, ed. by P. Halevi, vol. 1 (North-Holland, Amsterdam, 1992)
45. J. D. Jackson, *Classical Electrodynamics* (Wiley, New York, 1962)
46. W.D. Jones, H.J. Doucet, J.M. Buzzi, *An Introduction to the Linear Theories and Methods of Electrostatic Waves in Plasma* (Plenum Press, New York, 1985)
47. A. Moradi, Propagation of electrostatic energy through a quantum plasma. Contrib. Plasma Phys. **59**, 173–180 (2019)
48. L.D. Landau, E.M. Lifshitz, *Electrodynamics of Continuous Media* (Pergamon, Oxford, 1971)

Chapter 2
Problems in Electrostatic Approximation

Abstract In this chapter, some electrostatic boundary-value problems involving bounded electron gases with different geometries are studied. By considering the local Drude model, where the dielectric function is calculated by neglecting the spatial nonlocal effects, Laplace's equation in planar, cylindrical, and spherical geometries is solved, using the separation of variables technique. For brevity, in many sections of this chapter the $\exp(-i\omega t)$ time factor is suppressed. Furthermore, all media under consideration are nonmagnetic and attention is only confined to the linear phenomena.

2.1 Plasmonic Properties of Semi-Infinite Electron Gases

The simplest configuration for guided *surface plasmon* (SP) propagation is a plane interface between an *insulator*[1] (or a vacuum) and a semi-infinite EG. The study of SP propagation in this configuration is important. Firstly, it helps to understand the characteristics of SP propagation in more complicated layered structures. Moreover, the semi-infinite EG geometry serves as a good model for the experimental investigations of SP propagation on thick metallic substrates.

2.1.1 Surface Plasmon Frequency

Consider a semi-infinite EG occupying the half-space $z > 0$ in Cartesian coordinates (x, y, z). The plane $z = 0$ is the EG-insulator interface. Here we consider the case of

---

The original version of this chapter was revised. The correction to this chapter is available at https://doi.org/10.1007/978-3-030-43836-4_11

[1] Here, an insulator means a medium whose relative dielectric constant is assumed to be frequency independent.

A. Moradi, *Canonical Problems in the Theory of Plasmonics*, Springer Series in Optical Sciences 230, https://doi.org/10.1007/978-3-030-43836-4_2

a sharp boundary[2] [1–3] between the EG and insulator. The electrostatic potential generated by the charge fluctuations satisfies *Laplace's equation* inside and outside the separation surface, as

$$\nabla^2 \Phi(\mathbf{r}) = 0 , \qquad z \neq 0 , \tag{2.1}$$

where Φ and $\mathbf{E} = -\nabla\Phi$ must be finite everywhere and $\nabla^2 = (\partial^2/\partial x^2) + (\partial^2/\partial y^2) + (\partial^2/\partial z^2)$ is the Laplace operator in Cartesian coordinates. From (1.94) and (1.95), the BCs for the present BVP may be written as

$$\Phi_1\big|_{z=0} = \Phi_2\big|_{z=0} , \tag{2.2a}$$

$$\varepsilon_1 \frac{\partial \Phi_1}{\partial z}\bigg|_{z=0} = \varepsilon_2 \frac{\partial \Phi_2}{\partial z}\bigg|_{z=0} , \tag{2.2b}$$

where subscripts 1 and 2 refer to outside and inside the EG, respectively. As mentioned in Sect. 1.5, Eq. (2.2a) ensures the continuity of the tangential components of the corresponding electric field $\mathbf{E}$ across this plane. Furthermore, (2.2b) indicates that the normal component of the electric displacement be continuous across the plane $z = 0$.

The solution of Laplace's equation[3] is wavelike in the x- and y-directions, and decays exponentially as $|z| \to \infty$. We now study a Fourier component of electrostatic waves at the interface of the system that propagates in the both x- and y-directions. Hence, the electric potential is represented in the form

$$\Phi(\mathbf{r}) = \tilde{\Phi}(z) \exp\left(i\mathbf{k}_\parallel \cdot \mathbf{r}_\parallel\right) , \tag{2.3}$$

where $\mathbf{k}_\parallel$ is the longitudinal wavevector and $\mathbf{r}_\parallel = (x, y)$. After substitution $\Phi(\mathbf{r})$ in Laplace's equation, one finds

$$\left(\frac{d^2}{dz^2} - k_\parallel^2\right) \tilde{\Phi}(z) = 0 , \qquad z \neq 0 . \tag{2.4}$$

The solution of (2.4) has the form

$$\tilde{\Phi}(z) = \begin{cases} A_1 e^{+k_\parallel z} , & z \leq 0 , \\ A_2 e^{-k_\parallel z} , & z \geq 0 , \end{cases} \tag{2.5}$$

---

[2] In the sharp boundary model the thickness of the transition layer of the EG density is much shorter than the wavelength.

[3] The method of separation of variables is applied to Laplace's equation inside and outside the separation surface. This means that the solution for the electrostatic potential Φ can be written as the product of single-variable functions [4].

where the relations between the coefficients A_1 and A_2 can be determined from the matching BCs at the separation surface. On applying the electrostatic BCs at $z = 0$, we find

$$\varepsilon_1 + \varepsilon_2 = 0 \,. \tag{2.6}$$

Obviously since $\varepsilon_1 > 0$, the existence of an electrostatic surface wave is a matter of whether ε_2 can be negative or not. For part or all of the optical frequency range it is evidently true for metals that $\varepsilon_2 < 0$. It is also true in the infrared for semiconductors, so both metals and semiconductors will support surface modes. If the EG is characterized by the local Drude dielectric function, as[4]

$$\varepsilon(\omega) = 1 - \frac{\omega_p^2}{\omega^2} \,, \tag{2.7}$$

then the frequency of the SP resonance is given by

$$\omega = \frac{\omega_p}{\sqrt{1 + \varepsilon_1}} \,, \tag{2.8}$$

which in the special case $\varepsilon_1 = 1$ can be written in a more familiar form, as

$$\omega = \frac{\omega_p}{\sqrt{2}} \,. \tag{2.9}$$

Equation (2.8) shows the simple theory of electrostatic oscillation of a semi-infinite EG in which the spatial dispersion of the electrons are neglected and gives the electrostatic oscillation at a single frequency only that is, $\omega_s = \omega_p/\sqrt{1 + \varepsilon_1}$, say, the SP[5] frequency, but permits all wavelengths and wavenumber $k_{\parallel}$ can take any value whatsoever. There is thus no dispersion relation in the usual sense.

   It is natural to expect that (2.8) would not represent the true dispersion in a small wavenumber region since it has been obtained in the electrostatic approximation, i.e., $k_{\parallel} \gg \omega_p/c$, where the speed of light can be taken to be infinitely large. In order to obtain the correct dispersion relation for the (modified) electrostatic waves and other possible fast waves, we carry out full electromagnetic analysis in the next chapter. Furthermore, (2.8) would not represent the true dispersion in a large wavenumber region since it is based on the assumption that the spatial variations in the electron density are small enough such that $k_{\parallel} \ll k_s$, where k_s is the inverse TF screening length of the EG, as introduced in Chap. 1. This problem will be addressed in Chap. 4.

---

[4]Here, we consider the case, where $\varepsilon_b = 1$.

[5]The first theoretical description of SPs was presented by Ritchie in 1957 [5].

2.1.2 Power Flow

A practically important characteristic of surface waves is the power flow density
associated with them [6–8]. This is given in terms of the Poynting vector that
has the dimensions of energy per unit area per unit time. Under the electrostatic
approximation, the formula of the power flow density vector is given by (1.80). For
the power flow density vector associated with the SPs of a semi-infinite EG, using
$\mathbf{J} = -\left(i\varepsilon_0\omega_p^2/\omega\right)\nabla\Phi_2$, and also (1.80), we have, in the two media,

$$\mathbf{S} = -\varepsilon_0 \begin{cases} \varepsilon_1\Phi_1\dfrac{\partial}{\partial t}\nabla\Phi_1\,, & z \leq 0\,, \\[3mm] \Phi_2\left(i\dfrac{\omega_p^2}{\omega} + \dfrac{\partial}{\partial t}\right)\nabla\Phi_2\,, & z \geq 0\,, \end{cases} \qquad (2.10)$$

where these vectors in the two media have components in the x- and z-directions,
but their z-components vanish on averaging over a cycle of oscillation of the fields.
The cycle-averaged x-components are

$$S_x = -\frac{\varepsilon_0}{2}\mathrm{Re} \begin{cases} \varepsilon_1\Phi_1\dfrac{\partial}{\partial t}\dfrac{\partial}{\partial x}\Phi_1^*\,, & z \leq 0\,, \\[3mm] \Phi_2\left(-i\dfrac{\omega_p^2}{\omega} + \dfrac{\partial}{\partial t}\right)\dfrac{\partial}{\partial x}\Phi_2^*\,, & z \geq 0\,. \end{cases} \qquad (2.11)$$

After substitution (2.5) into (2.11) and using (2.8), we obtain

$$S_x = \frac{1}{2}\varepsilon_0\varepsilon_1\omega k_{\parallel}A_1^2 \begin{cases} -e^{+2k_{\parallel}z}\,, & z \leq 0\,, \\[2mm] +e^{-2k_{\parallel}z}\,, & z \geq 0\,, \end{cases} \qquad (2.12)$$

where A_1 is a real constant. We note that the distribution (2.12) is discontinuous
at the interface $z = 0$. One can find that power flow densities are largest at the
boundary $z = 0$, and their amplitudes decay exponentially with increasing distance
into each medium from the interface. It is seen that the power flow density for $z > 0$
is positive, while in the region $z < 0$, it is negative.

   The total power flow associated with the SPs is determined by an integration over
z. If the integrated Poynting vectors are denoted by angle brackets, and if we assume
that SPs are propagating in the x-direction, the power flow through an area in the yz
plane of infinite length in the z-direction and unit width in the y direction is

$$\langle S_x \rangle = 0\,, \qquad (2.13)$$

where $\langle \cdots \rangle \equiv \int_{-\infty}^{+\infty} \cdots \mathrm{d}z$. This means that the total power flow density (per unit
width) associated with the SPs of a semi-infinite EG is zero.

2.1.3 Energy Distribution

Now we consider the energy density distribution in the transverse direction. As described in Sect. 1.4.6, the energy density of the electrostatic field in a lossless medium can be expressed by (1.79). In the absence of the spatial dispersion, for the cycle-averaged of energy density distribution associated with the SPs of a semi-infinite EG, we have, in the two media

$$
U = \frac{\varepsilon_0}{4}
\begin{cases}
\varepsilon_1 |\nabla \Phi_1|^2 \,, & z \leq 0 \,, \\
(2 - \varepsilon_2) |\nabla \Phi_2|^2 \,, & z \geq 0 \,.
\end{cases}
\tag{2.14}
$$

After substitution (2.7) into (2.14) and using (2.8), we obtain

$$
U = \frac{1}{2} \varepsilon_0 k_\parallel^2 A_1^2
\begin{cases}
\varepsilon_1 e^{+2k_\parallel z} \,, & z \leq 0 \,, \\
(2 + \varepsilon_1)\, e^{-2k_\parallel z} \,, & z \geq 0 \,.
\end{cases}
\tag{2.15}
$$

From (2.15), it is clear that both contributions to the energy density are positive, as expected on physical grounds. The total energy density associated with the SPs is again determined by integration over z, the energy per unit surface area being

$$
\langle U \rangle = \frac{1}{2} \varepsilon_0 k_\parallel \left(1 + \varepsilon_1\right) A_1^2 \,.
\tag{2.16}
$$

Also, from the distribution of energy density, we can find the position of the energy center along the z-axis, as

$$
\langle z \rangle \equiv \frac{1}{\langle U \rangle} \int_{-\infty}^{+\infty} z\, U \, dz = \frac{1}{2k_\parallel \left(1 + \varepsilon_1\right)} \,,
\tag{2.17}
$$

that shows the center of energy is located in the EG. Also, it can be seen from (2.17) that the center of energy depends on ε_1.

2.1.4 Quantization of Surface Plasmon Fields

In the previous section, the electric field $\mathbf{E} = -\nabla \Phi$ associated with SPs has been treated as classical variable. It can also be treated as quantum-mechanical observable represented by operators. The canonical procedure for *quantization* of SPs can be used here [9–11], but we leave it for the reader as a BVP [see Sects. 6.3 and 8.2] and follow an alternative approach to show an application of the energy density formula derived in the previous section. To derive the electric field from the scalar electric potential Φ, the potential should be represented by different operators in media 1 and 2. The scalar electric potential operator associated with an excitation of wavevector $\mathbf{k}_\parallel$ can be written in the form

$$\hat{\Phi}_{\mathbf{k}_\parallel} = f_{\mathbf{k}_\parallel} \left[\hat{a}_{\mathbf{k}_\parallel} e^{i\left(\mathbf{k}_\parallel \cdot \mathbf{r}_\parallel - \omega_s t\right)} + \hat{a}_{\mathbf{k}_\parallel}^\dagger e^{-i\left(\mathbf{k}_\parallel^* \cdot \mathbf{r}_\parallel - \omega_s t\right)} \right] \begin{cases} e^{+k_\parallel z} , z \leq 0 , \\ \\ e^{-k_\parallel z} , z \geq 0 . \end{cases} \tag{2.18}$$

Here, the circumflex identifies quantum-mechanical operators, $\omega_s = \omega_p / \sqrt{1 + \varepsilon_1}$, $f_{\mathbf{k}_\parallel}$ is a factor to be determined, $\hat{a}_{\mathbf{k}_\parallel}$ and $\hat{a}_{\mathbf{k}_\parallel}^\dagger$ are annihilation and creation operators for SP quanta, where by their definitions, they must have unit real commutators and the following commutation rules are understood

$$\left[\hat{a}_{\mathbf{k}_\parallel}, \hat{a}_{\mathbf{k}_\parallel'}^\dagger \right] = \delta_{\mathbf{k}_\parallel \mathbf{k}_\parallel'} , \tag{2.19}$$

$$\left[\hat{a}_{\mathbf{k}_\parallel}, \hat{a}_{\mathbf{k}_\parallel'} \right] = 0 = \left[\hat{a}_{\mathbf{k}_\parallel}^\dagger, \hat{a}_{\mathbf{k}_\parallel'}^\dagger \right] . \tag{2.20}$$

Also, $\mathbf{k}_\parallel^*$ is the complex conjugate of $\mathbf{k}_\parallel$, although here we have $\mathbf{k}_\parallel^* = \mathbf{k}_\parallel$. With the complex conjugations shown in (2.18), $\hat{\Phi}_{\mathbf{k}_\parallel}$ is a *Hermitian* operator.

Let $|n_{\mathbf{k}_\parallel}\rangle$ be a state in which $n_{\mathbf{k}_\parallel}$ quanta of SPs are excited, and let A be the area of the boundary between the two media. The correspondence between the SP energy given by the usual quantum-harmonic-oscillator expression and the classical expression obtained from (2.16) is

$$A \langle U \rangle = \left(n_{\mathbf{k}_\parallel} + \frac{1}{2} \right) \hbar \omega_s . \tag{2.21}$$

In making the transition from classical to quantum mechanics, the classical cycle-averaged square of the electric field must be replaced by the quantum-mechanical *expectation value* of the square of the electric field operator according to

$$\frac{1}{2}|\mathbf{E}|^2 \rightarrow \langle n_{\mathbf{k}_\parallel} | \hat{\mathbf{E}}^2 | n_{\mathbf{k}_\parallel} \rangle = 4k_\parallel^2 f_{\mathbf{k}_\parallel}^2 \left(n_{\mathbf{k}_\parallel} + \frac{1}{2} \right) e^{2k_\parallel z} . \tag{2.22}$$

In the final step it has been assumed that the evaluation of the expectation value uses the electric field operator obtained from $\hat{\mathbf{E}} = -\nabla \hat{\Phi}$ and (2.18) together with the usual properties of the shift operators. With the above replacement (2.16) becomes

$$\langle U \rangle = 2\varepsilon_0 k_\parallel (1 + \varepsilon_1) f_{\mathbf{k}_\parallel}^2 \left(n_{\mathbf{k}_\parallel} + \frac{1}{2} \right) . \tag{2.23}$$

Comparison of (2.21) and (2.23) now determines the unknown factor in (2.18) as

$$f_{\mathbf{k}_\parallel} = \left(\frac{\hbar \omega_s}{2\varepsilon_0 (1 + \varepsilon_1) A k_\parallel} \right)^{1/2} . \tag{2.24}$$

Each quantum of SP excites electrostatic fields in both media 1 and 2, and the contributions to the field energy from both media have accordingly been included in the present calculation of the appropriate factor in the quantized scalar electric potential (2.18). Now, the electric potential operators in the two media are obtained by

$$\hat{\Phi}(\mathbf{r}, t) = \sum_{\mathbf{k}_\parallel} f_{\mathbf{k}_\parallel} \left[\hat{a}_{\mathbf{k}_\parallel} e^{i(\mathbf{k}_\parallel \cdot \mathbf{r}_\parallel - \omega_s t)} + \hat{a}_{\mathbf{k}_\parallel}^\dagger e^{-i\left(\mathbf{k}_\parallel^* \cdot \mathbf{r}_\parallel - \omega_s t\right)} \right] \begin{cases} e^{+k_\parallel z} , & z \leq 0 , \\ e^{-k_\parallel z} , & z \geq 0 , \end{cases}$$

(2.25)

which shows the potential field in terms of $\hat{a}_{\mathbf{k}_\parallel}$ and $\hat{a}_{\mathbf{k}_\parallel}^\dagger$. The *Hamiltonian* for the system can then be written as

$$\mathscr{H}_{sp} = \sum_{\mathbf{k}_\parallel} \hbar \omega_s \left(\hat{a}_{\mathbf{k}_\parallel}^\dagger \hat{a}_{\mathbf{k}_\parallel} + \frac{1}{2} \right) .$$

(2.26)

2.2 Surface Magneto Plasmon Frequency of Semi-Infinite Magnetized Electron Gases

In this section, we study the propagation of a *surface magneto plasmon* (SMP) along the interface between an insulator and a semi-infinite magnetized EG in the presence of a static magnetic field $\mathbf{B}_0$ in the special directions, where $\mathbf{B}_0$ is parallel to the coordinate axis, i.e., $\mathbf{B}_0 = B_0 \mathbf{e}_x$, $B_0 \mathbf{e}_y$, and $B_0 \mathbf{e}_z$ ($\mathbf{e}_j$ being the unit vector along the j-axis, with $j = x, y$, and z), as shown in Fig. 2.1. Again, we consider a semi-infinite EG occupying the half-space $z > 0$. Also, it can be assumed without loss of generality that a SP is propagating in the x-direction, so the electric potential is represented in the form $\Phi(x, z) = \tilde{\Phi}(z) \exp(i k_x x)$.

2.2.1 Faraday Configuration

In the *Faraday configuration*, the applied magnetic field is parallel to the surface and the direction of propagation of the wave [12–16], therefore we have $\mathbf{B}_0 = B_0 \mathbf{e}_x$. In this case, from (1.7) with $\gamma = 0$ and $\mathbf{J} = -e n_0 \mathbf{v}$, we find $\mathbf{J} = \underline{\sigma} \cdot \mathbf{E}$, where the conductivity tensor of the system is

$$\underline{\sigma}(\omega) = i \varepsilon_0 \frac{\omega_p^2}{\omega} \begin{pmatrix} 1 & 0 & 0 \\ 0 & \dfrac{\omega^2}{\omega^2 - \omega_c^2} & -\dfrac{i \omega \omega_c}{\omega^2 - \omega_c^2} \\ 0 & \dfrac{i \omega \omega_c}{\omega^2 - \omega_c^2} & \dfrac{\omega^2}{\omega^2 - \omega_c^2} \end{pmatrix} ,$$

(2.27)

Fig. 2.1 Illustration of an EG occupying the half-space $z > 0$ on the top of an insulator with relative dielectric constant ε_1 in the presence of a static magnetic field $\mathbf{B}_0$ in the special directions. The different panels refer to (**a**) Faraday geometry, (**b**) Voigt geometry, and (**c**) perpendicular geometry

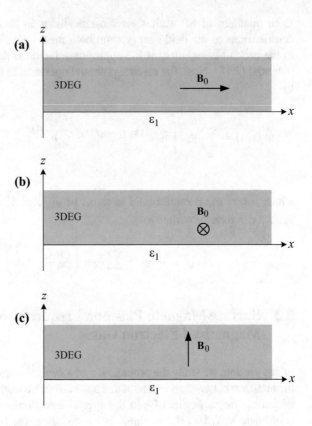

and $\omega_c = eB_0/m_e$ is the *cyclotron frequency* of an electron. Now, by using (1.37) and (1.38), one can obtain the relative dielectric tensor of the system, as

$$\underline{\varepsilon}(\omega) = \begin{pmatrix} \varepsilon_{xx} & 0 & 0 \\ 0 & \varepsilon_{yy} & \varepsilon_{yz} \\ 0 & \varepsilon_{zy} & \varepsilon_{zz} \end{pmatrix} , \qquad (2.28)$$

where

$$\varepsilon_{xx} = 1 - \frac{\omega_p^2}{\omega^2} ,$$

$$\varepsilon_{yy} = \varepsilon_{zz} = 1 - \frac{\omega_p^2}{\omega^2 - \omega_c^2} ,$$

$$\varepsilon_{yz} = -\varepsilon_{zy} = \frac{i\omega_c\omega_p^2}{\omega\left(\omega^2 - \omega_c^2\right)} .$$

After substitution $\Phi(x, z) = \tilde{\Phi}(z) \exp(ik_x x)$ into $\nabla \cdot \mathbf{D} = 0$, where $\mathbf{D} = \underline{\varepsilon} \cdot \mathbf{E}$, we find

$$\left(\frac{d^2}{dz^2} - \kappa^2 \right) \tilde{\Phi}(z) = 0 , \quad z \geq 0 , \tag{2.29}$$

where

$$\kappa = k_x \sqrt{\frac{\varepsilon_{xx}}{\varepsilon_{zz}}} = k_x \sqrt{\left(1 - \frac{\omega_p^2}{\omega^2} \right) \bigg/ \left(1 - \frac{\omega_p^2}{\omega^2 - \omega_c^2} \right)} . \tag{2.30}$$

Therefore, the solutions of (2.4) for $z \leq 0$ with $k_\parallel = k_x$ and (2.29) for $z \geq 0$ have the form

$$\tilde{\Phi}(z) = \begin{cases} A_1 e^{+k_x z} , & z \leq 0 , \\ A_2 e^{-\kappa z} , & z \geq 0 . \end{cases} \tag{2.31}$$

But, (2.2b) must be read as

$$\varepsilon_1 \frac{\partial \Phi_1}{\partial z} \bigg|_{z=0} = \varepsilon_{zz} \frac{\partial \Phi_2}{\partial z} \bigg|_{z=0} , \tag{2.32}$$

where again subscripts 1 and 2 refer to outside and inside the EG, respectively. On applying the electrostatic BCs at $z = 0$, i.e., (2.2a) and (2.32), we find

$$\varepsilon_{zz} \kappa + \varepsilon_1 k_x = 0 , \tag{2.33}$$

and also for the particular case, when $\varepsilon_1 = 1$, we get

$$\omega^2 = \frac{\omega_p^2 + \omega_c^2}{2} . \tag{2.34}$$

2.2.2 Voigt Configuration

In the *Voigt configuration*, a surface wave propagates across an external magnetic field that is parallel to the interface [17–20], therefore we have $\mathbf{B}_0 = B_0 \mathbf{e}_y$. In this case, the conductivity tensor of the system is

$$\underline{\sigma}(\omega) = i\varepsilon_0 \frac{\omega_p^2}{\omega} \begin{pmatrix} \dfrac{\omega^2}{\omega^2 - \omega_c^2} & 0 & \dfrac{i\omega\omega_c}{\omega^2 - \omega_c^2} \\ 0 & 1 & 0 \\ -\dfrac{i\omega\omega_c}{\omega^2 - \omega_c^2} & 0 & \dfrac{\omega^2}{\omega^2 - \omega_c^2} \end{pmatrix} . \tag{2.35}$$

Also, the relative dielectric tensor of the system is

$$\underline{\varepsilon}(\omega) = \begin{pmatrix} \varepsilon_{xx} & 0 & \varepsilon_{xz} \\ 0 & \varepsilon_{yy} & 0 \\ \varepsilon_{zx} & 0 & \varepsilon_{zz} \end{pmatrix} , \tag{2.36}$$

where

$$\varepsilon_{yy} = 1 - \frac{\omega_p^2}{\omega^2} ,$$

$$\varepsilon_{xx} = \varepsilon_{zz} = 1 - \frac{\omega_p^2}{\omega^2 - \omega_c^2} ,$$

$$\varepsilon_{xz} = -\varepsilon_{zx} = -\frac{i\omega_c\omega_p^2}{\omega\left(\omega^2 - \omega_c^2\right)} .$$

For the present case, we find (2.4) for $z \leq 0$ and (2.29) for $z \geq 0$ with the solutions of the form of (2.31). But, (2.2b) must be read as

$$\varepsilon_1 \frac{\partial \Phi_1}{\partial z}\bigg|_{z=0} = \varepsilon_{zx} \frac{\partial \Phi_2}{\partial x}\bigg|_{z=0} + \varepsilon_{zz} \frac{\partial \Phi_2}{\partial z}\bigg|_{z=0} . \tag{2.37}$$

On applying the electrostatic BCs at $z = 0$, i.e., (2.2a) and (2.37), we obtain

$$(-i\varepsilon_{zx} + \varepsilon_1) + \varepsilon_{zz} = 0 . \tag{2.38}$$

If we assume a backward-going MSP, so the electric potential must be represented in the form $\Phi(x, z, t) = \tilde{\Phi}(z) \exp[i(-k_x x - \omega t)]$. In this case, we find $(i\varepsilon_{zx} + \varepsilon_1) + \varepsilon_{zz} = 0$. Therefore, in the general form of (2.38), we have

$$(\mp i\varepsilon_{zx} + \varepsilon_1) + \varepsilon_{zz} = 0 . \tag{2.39}$$

We note that the upper and lower signs refer to wave propagating in the positive and negative x-directions, respectively. If $\varepsilon_1 = 1$, Eq. (2.39) yields

$$\omega = \frac{1}{2}\left[\left(\omega_c^2 + 2\omega_p^2\right)^{1/2} \mp \omega_c\right] . \tag{2.40}$$

It can be seen that for a forward-going surface wave, the increasing of ω_c red-shifts the SP frequency, while for a backward-going wave the frequency of surface wave is blue-shifted [17, 21].

2.2.3 Perpendicular Configuration

In the *perpendicular configuration*, the applied magnetic field is perpendicular to the surface and the direction of propagation of the wave [22–24], therefore we have $\mathbf{B}_0 = B_0\mathbf{e}_z$. In this case, the conductivity tensor of the system is

$$\underline{\sigma}(\omega) = i\varepsilon_0 \frac{\omega_p^2}{\omega} \begin{pmatrix} \dfrac{\omega^2}{\omega^2 - \omega_c^2} & -\dfrac{i\omega\omega_c}{\omega^2 - \omega_c^2} & 0 \\ \dfrac{i\omega\omega_c}{\omega^2 - \omega_c^2} & \dfrac{\omega^2}{\omega^2 - \omega_c^2} & 0 \\ 0 & 0 & 1 \end{pmatrix}. \tag{2.41}$$

Also, the relative dielectric function tensor of the system is

$$\underline{\varepsilon}(\omega) = \begin{pmatrix} \varepsilon_{xx} & \varepsilon_{xy} & 0 \\ \varepsilon_{yx} & \varepsilon_{yy} & 0 \\ 0 & 0 & \varepsilon_{zz} \end{pmatrix}, \tag{2.42}$$

where

$$\varepsilon_{zz} = 1 - \frac{\omega_p^2}{\omega^2},$$

$$\varepsilon_{xx} = \varepsilon_{yy} = 1 - \frac{\omega_p^2}{\omega^2 - \omega_c^2},$$

$$\varepsilon_{xy} = -\varepsilon_{yx} = \frac{i\omega_c\omega_p^2}{\omega\left(\omega^2 - \omega_c^2\right)}.$$

Again, we find (2.29) for $z \geq 0$, but here we have

$$\kappa = k_x\sqrt{\frac{\varepsilon_{xx}}{\varepsilon_{zz}}} = k_x\sqrt{\left(1 - \frac{\omega_p^2}{\omega^2 - \omega_c^2}\right) \Big/ \left(1 - \frac{\omega_p^2}{\omega^2}\right)}. \tag{2.43}$$

Also, the solutions of (2.4) for $z \leq 0$ and (2.29) for $z \geq 0$ have the form of (2.31). On applying the electrostatic BCs at $z = 0$, i.e., (2.2a) and (2.32), one obtains an equation exactly identical to (2.33). It is interesting that the frequency of a SMP is exactly same for the two different configurations, i.e., Faraday and perpendicular configurations provided the field strength B_0 is the same. Furthermore (2.30) and (2.43) show that the decay constants κ are inverse to each other. Also, we note that SMPs in the Faraday and perpendicular configurations are complementary to each other in the sense that when the SMP in the Faraday configuration has strong surface wave characteristics, the SMP in the perpendicular configuration acts like a bulk wave and vice versa.

2.3 Plasmonic Properties of Planar Electron Gas Slabs

Planar EG slabs and their corresponding channels represent significant structures in the plasmonics and used in many practical systems. The objective in this section is to introduce and analyze some plasmonic properties of them in the electrostatic approximation.

2.3.1 Dispersion Relation and Electrostatic Potential Distribution

As shown in Fig. 2.2, we consider a planar EG slab of thickness d along the z-direction and infinitely wide in the x- and y-directions, bounded between two thick layers of insulator material. Again, we consider the propagation of an electrostatic surface wave in the x-direction. For this geometry, we have

$$\tilde{\Phi}(z) = \begin{cases} A_1 e^{+k_x z}, & z \le 0, \\ A_2 e^{+k_x z} + A_3 e^{-k_x z}, & 0 \le z \le d, \\ A_4 e^{-k_x z}, & z \ge d, \end{cases} \quad (2.44)$$

where the relations between the coefficients $A_1 - A_4$ can be determined from the matching BCs at the separation surfaces. The BCs at $z = 0$ and $z = d$ are

$$\Phi_1\big|_{z=0} = \Phi_2\big|_{z=0}, \quad (2.45a)$$

$$\Phi_2\big|_{z=d} = \Phi_3\big|_{z=d}, \quad (2.45b)$$

Fig. 2.2 Schematic representation of a planar EG slab capable of supporting SP modes at each of its interfaces

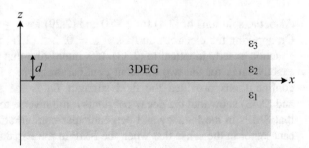

$$\varepsilon_1 \left. \frac{\partial \Phi_1}{\partial z} \right|_{z=0} = \varepsilon_2 \left. \frac{\partial \Phi_2}{\partial z} \right|_{z=0} , \qquad (2.45\text{c})$$

$$\varepsilon_2 \left. \frac{\partial \Phi_2}{\partial z} \right|_{z=d} = \varepsilon_3 \left. \frac{\partial \Phi_3}{\partial z} \right|_{z=d} , \qquad (2.45\text{d})$$

where subscripts 1 and 3 denote the regions outside the EG slab, while subscript 2 denotes the region inside the slab. The application of the electrostatic BCs at $z = 0$, and $z = d$, gives rise to the following dispersion relation:

$$\begin{vmatrix} 1 & -1 & -1 & 0 \\ 0 & e^{2k_x d} & 1 & -1 \\ \varepsilon_1 & -\varepsilon_2 & \varepsilon_2 & 0 \\ 0 & \varepsilon_2 e^{2k_x d} & -\varepsilon_2 & \varepsilon_3 \end{vmatrix} = 0 . \qquad (2.46)$$

Also, we note that the three unknown amplitudes $A_2 - A_4$ can be related to the amplitude A_1 through the BCs. In this way, by applying the mentioned BCs we will be able to find the relationship between the constants $A_1 - A_4$, as

$$A_1 = e^{-2k_x d} \frac{\varepsilon_2 - \varepsilon_3}{\varepsilon_1 + \varepsilon_2} A_4 , \qquad (2.47\text{a})$$

$$A_2 = e^{-2k_x d} \frac{\varepsilon_2 - \varepsilon_3}{2\varepsilon_2} A_4 , \qquad (2.47\text{b})$$

$$A_3 = \frac{\varepsilon_2 + \varepsilon_3}{2\varepsilon_2} A_4 . \qquad (2.47\text{c})$$

In the simple case, when $\varepsilon_1 = 1 = \varepsilon_3$ and using (2.7) for ε_2, we find two branches for ω defining the resonant frequencies of the SP excitations in an EG slab, as

$$\omega_\pm = \frac{\omega_p}{\sqrt{2}} \sqrt{1 \pm e^{-k_x d}} . \qquad (2.48)$$

The SP frequencies of an EG slab[6] free-standing in vacuum can now be understood as the coupling of the SPs on the two EG-vacuum interfaces as indicated in the *plasmon hybridization theory*.[7] This coupling or hybridization results in a splitting

---

[6]If we reverse the EG slab geometry and study a vacuum gap in an EG, we find the same dispersion relation as for an EG slab, i.e., (2.48). However, we note that in the presence of retardation and/or spatial nonlocal effects they are not.

[7]The plasmon hybridization theory (that is a nonretard method) can predict the location of plasmon resonances in complex systems based on knowledge of the resonant behavior of elementary building blocks. It thus provides a powerful tool for optical engineers in the design of functional plasmonic nanostructures. This method was first presented by Prodan et al. [25].

Fig. 2.3 Dispersion curves of SP modes of a planar EG slab free-standing in vacuum, as given by (2.48). The dashed horizontal line indicates the asymptotic SP frequency on a single EG-vacuum interface, i.e., $\omega_s = \omega_p/\sqrt{2}$. Because of the use of appropriate units shown along the axes, these curves do not depend on any physical parameters

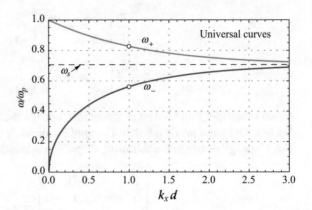

of the SP into a high-frequency mode ω_+ and a low-frequency mode ω_-, as indicated by the subscripts.

A number of different terminologies such as *symmetric* and *anti-symmetric* modes with respect to field/potential distribution, symmetric and anti-symmetric modes with respect to charge distribution and even and odd modes are used in the literature to distinguish between the two structural modes [26–29]. In the present problem, SP modes classified as an odd mode, where the electric potential is anti-symmetric with respect to the plane $z = d/2$; and an even mode, in which electric potential is symmetric relative to the $z = d/2$ plane.

According to (2.48), the two SP modes of a planar EG slab in vacuum are plotted in Fig. 2.3, where an appropriate choice of units (ω is measured in ω_p units and k_x in $1/d$ units) yields *universal* curves independent of any choice of specific parameters such as d. If we let the thickness go to infinity the modes decouple and there are two independent modes, one on each face of the EG slab. Since *retardation effect* has not been considered here, the range of validity of the present results may be characterized by $k_x \gg \omega_p/c$, as mentioned in Sect. 2.1.

Also, by using (2.44) and (2.47a)–(2.47c), we calculate the normalized profiles of electrostatic potentials in Fig. 2.4, when $k_x d = 1$ corresponding to the two labeled points in Fig. 2.3.

2.3.2 Induced Surface Charge

Now that we know the electrostatic potential of the system, it is a straightforward matter to compute the surface charge σ_{ind} induced on the two surfaces of the system. Integrating Gauss's law for the electric field, $\varepsilon_0 \nabla \cdot \mathbf{E} = \rho_{ind}$[8] we find

---

[8]The external free charge densities on the separation surfaces of slab are zero.

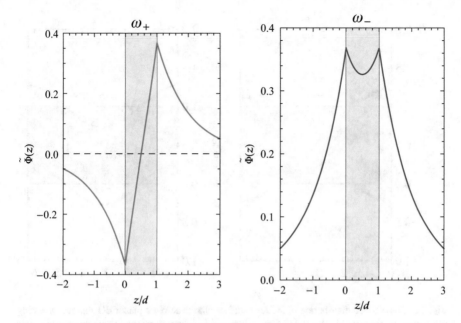

Fig. 2.4 Normalized profile $\tilde{\Phi}(z)$ of SP modes of a planar EG slab free-standing in vacuum, when $k_x d = 1$ corresponding to the two labeled points in Fig. 2.3. Panels (ω_+) and (ω_-) show the results for anti-symmetric (odd) and symmetric (even) modes, respectively

$$\sigma_{ind}(x)\big|_{z=0} = \varepsilon_0 \left[\frac{\partial \tilde{\Phi}_1}{\partial z}\bigg|_{z=0} - \frac{\partial \tilde{\Phi}_2}{\partial z}\bigg|_{z=0} \right] \cos k_x x \ , \qquad (2.49a)$$

$$\sigma_{ind}(x)\big|_{z=d} = \varepsilon_0 \left[\frac{\partial \tilde{\Phi}_2}{\partial z}\bigg|_{z=d} - \frac{\partial \tilde{\Phi}_3}{\partial z}\bigg|_{z=d} \right] \cos k_x x \ , \qquad (2.49b)$$

where $\Phi(x, z)$ is considered as $\tilde{\Phi}(z) \cos k_x x$[9] that shows a *standing wave* in the x-direction. Therefore, by substituting (2.44) into (2.49a)–(2.49b) and using (2.47a)–(2.47c), we arrive to

$$\sigma_{ind}(x)\big|_{z=0} = \varepsilon_0 k_x \frac{\varepsilon_2 - \varepsilon_3}{2\varepsilon_2} \left(\frac{\varepsilon_2 + \varepsilon_3}{\varepsilon_2 - \varepsilon_3} - e^{-2k_x d} \frac{\varepsilon_1 - \varepsilon_2}{\varepsilon_1 + \varepsilon_2} \right) A_4 \cos k_x x \ , \qquad (2.50a)$$

$$\sigma_{ind}(x)\big|_{z=d} = \varepsilon_0 k_x e^{-k_x d} \frac{\varepsilon_2 - \varepsilon_3}{\varepsilon_2} A_4 \cos k_x x \ . \qquad (2.50b)$$

---

[9]We note by choosing $\Phi(x, z)$ as $-\dfrac{1}{k_x} \tilde{\Phi}(z) \sin k_x x$, the results of induced charge density correspond to the nonretarded results in Sect. 3.2.2. However, here the qualitative description of the problem is desired.

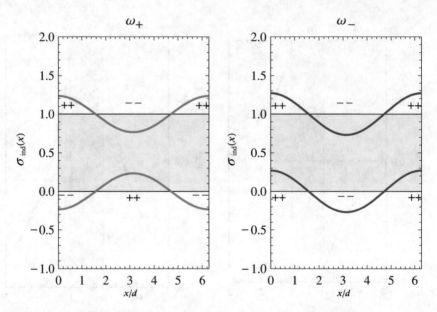

Fig. 2.5 Normalized distribution of induced surface charge across a planar EG slab free-standing in vacuum, as given by (2.50a) and (2.50b), when $k_x d = 1$ corresponding to the two labeled points in Fig. 2.3. Panels (ω_+) and (ω_-) show the results for anti-symmetric (odd) and symmetric (even) modes, respectively

Figure 2.5 gives the normalized induced charge density σ_{ind} as a function x/d, when $k_x d = 1$, corresponding to the two labeled points in Fig. 2.3. The distribution of induced charge gives rise to two standing surface charge modes and shows that the high-frequency mode ω_+ is anti-symmetric and low-frequency mode ω_- is symmetric.

2.3.3 Group Velocity

We now consider the group velocity of SP modes of an EG slab. Using (1.55) and (2.48), we obtain

$$v_g = \frac{\mathrm{d}\omega}{\mathrm{d}k_x} = \mp \frac{\omega_p}{2\sqrt{2}} \frac{d e^{-k_x d}}{\sqrt{1 \pm e^{-k_x d}}} . \tag{2.51}$$

According to the above equation, group velocity curves of two SP modes of a planar EG slab in vacuum are plotted in Fig. 2.6, where an appropriate choice of units (v_g is measured in $\omega_p d$ units and k_x in $1/d$ units) yields universal curves independent of any choice of specific parameters such as d. One can see that group velocity of a symmetric mode is always positive and decreases monotonically with increasing the value of $k_x d$ and finally becomes zero, when $k_x d \to \infty$

Fig. 2.6 Group velocity curves of SP modes of a planar EG slab free-standing in vacuum, as given by (2.51). Because of the use of appropriate units shown along the axes, these curves do not depend on any physical parameters

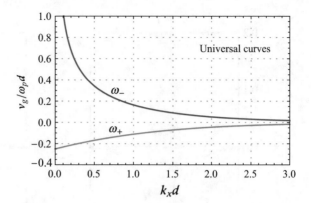

2.3.4 Power Flow

Now, it might be interesting to see how the power flow density varies. To show this, for the cycle-averaged x-components of power flow density associated with the SP modes of an EG slab in the three media we have

$$
S_x = -\frac{\varepsilon_0}{2}\mathrm{Re}\left\{
\begin{array}{ll}
\varepsilon_1 \Phi_1 \dfrac{\partial}{\partial t}\dfrac{\partial}{\partial x}\Phi_1^*, & z \leq 0 , \\[2ex]
\Phi_2\left(-i\dfrac{\omega_p^2}{\omega}+\dfrac{\partial}{\partial t}\right)\dfrac{\partial}{\partial x}\Phi_2^*, & 0 \leq z \leq d , \\[2ex]
\varepsilon_3 \Phi_3 \dfrac{\partial}{\partial t}\dfrac{\partial}{\partial x}\Phi_3^*, & z \geq d .
\end{array}
\right.
\tag{2.52}
$$

After substitution (2.44) into (2.52), we obtain

$$
S_x = -\frac{\varepsilon_0 \omega k_x}{2}\left\{
\begin{array}{ll}
\varepsilon_1 A_1^2 e^{2k_x z}, & z \leq 0 , \\[1.5ex]
\varepsilon_2\left(A_2 e^{k_x z}+A_3 e^{-k_x z}\right)^2, & 0 \leq z \leq d , \\[1.5ex]
\varepsilon_3 A_4^2 e^{-2k_x z}, & z \geq d ,
\end{array}
\right.
\tag{2.53}
$$

where $A_1 - A_4$ are real and ε_2 is characterized by the local dielectric function, i.e., (2.7). Note that the distribution (2.53) is discontinuous at the interfaces $z = 0$ and $z = d$. By using (2.53) and (2.47a)–(2.47c), in Fig. 2.7, we calculate normalized profiles of S_x of SP modes in a planar EG slab free-standing in vacuum, when $k_x d = 1$ corresponding to the two labeled points in Fig. 2.3. It can be seen that power flow densities are largest at the boundaries, and their amplitudes decay exponentially with increasing distance into each medium from the interfaces. Also, it is evident that the power flow in the EG region occurs in the $+x$-direction, while in the dielectric regions, the power flow occurs in the $-x$-direction, i.e., opposite to the direction of phase propagation. Also, the total power flow density (per unit width) associated with the SPs is

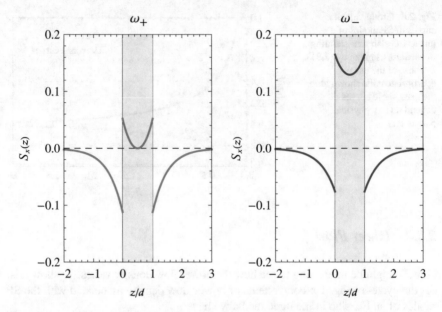

Fig. 2.7 Normalized profile $S_x(z)$ of SP modes of a planar EG slab free-standing in vacuum, when $k_x d = 1$ corresponding to the two labeled points in Fig. 2.3. Panels (ω_+) and (ω_-) show the results for anti-symmetric (odd) and symmetric (even) modes, respectively

$$\langle S_x \rangle = -\frac{\varepsilon_0 \omega}{4} \left\{ \varepsilon_1 A_1^2 + \varepsilon_3 A_4^2 e^{-2k_x d} \right.$$
$$\left. + \varepsilon_2 \left[A_2^2 \left(e^{2k_x d} - 1 \right) - A_3^2 \left(e^{-2k_x d} - 1 \right) + 4 k_x d A_2 A_3 \right] \right\} . \quad (2.54)$$

This total power flow density (per unit width) is positive (negative) for symmetric (anti-symmetric) mode, when k_x is positive. Note that for anti-symmetric mode the power flow densities in the insulator regions (here vacuum) are greater than that in EG slab, given a net power flow density to the $-x$-direction. This means that group velocity should be negative, as can be seen in Fig. 2.6. For symmetric mode, the net power flow density occurs in the direction of phase propagation; however, the power flow density in each insulator region occurs in the opposite direction.

2.3.5   Energy Distribution

In the absence of the spatial dispersion, for the cycle-averaged of energy density distribution associated with the SP modes of an EG slab, we have, in the three media,

$$U = \frac{\varepsilon_0}{4} \begin{cases} \varepsilon_1 |\nabla \Phi_1|^2 , & z \leq 0 , \\ (2 - \varepsilon_2) |\nabla \Phi_2|^2 , & 0 \leq z \leq d , \\ \varepsilon_3 |\nabla \Phi_3|^2 , & z \geq d . \end{cases} \quad (2.55)$$

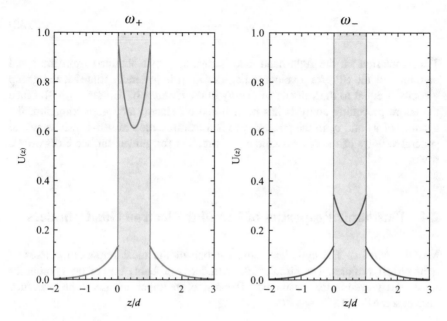

Fig. 2.8 Normalized profile $U(z)$ of SP modes of a planar EG slab free-standing in vacuum, when $k_x d = 1$ corresponding to the two labeled points in Fig. 2.3. Panels (ω_+) and (ω_-) show the results for anti-symmetric (odd) and symmetric (even) modes, respectively

After substitution (2.44) into (2.55), we obtain

$$U = \frac{\varepsilon_0 k_x^2}{2} \begin{cases} \varepsilon_1 A_1^2 e^{2k_x z} , & z \le 0 , \\ (2 - \varepsilon_2) \left[A_2^2 e^{2k_x z} + A_3^2 e^{-2k_x z} \right] , & 0 \le z \le d , \\ \varepsilon_3 A_4^2 e^{-2k_x z} , & z \ge d . \end{cases} \tag{2.56}$$

From Fig. 2.8, it is clear that all contributions to the energy density are positive, as expected on physical grounds. The total energy density associated with the SPs is again determined by integration over z, the energy per unit surface area being

$$\langle U \rangle = \frac{\varepsilon_0 k_x}{4} \left\{ \varepsilon_1 A_1^2 + \varepsilon_3 A_4^2 e^{-2k_x d} \right.$$
$$\left. + (2 - \varepsilon_2) \left[A_2^2 \left(e^{2k_x d} - 1 \right) - A_3^2 \left(e^{-2k_x d} - 1 \right) \right] \right\} . \tag{2.57}$$

2.3.6   Energy Velocity

The energy velocity of the SPs is given as the ratio of the total power flow density (per unit width) and total energy density (per unit area), such as

$$v_e = \frac{\langle S_x \rangle}{\langle U \rangle} \, . \tag{2.58}$$

The expression on the right-hand side is precisely that obtained from the usual definition of the SP group velocity, i.e., (2.51). It is generally true that the group velocity is equal to the velocity of energy in the absence of damping [6–8]. Since the entire preceding analysis has been based on electrostatic approximation, the equality of v_g and v_e in the present case is a manifestation of self-consistency and general validity of the electrostatic approximation for guided surface electrostatic waves.

2.4 Plasmonic Properties of Circular Electron Gas Cylinders

Metallic wires or EG cylinders represent one of the most important classes of geometrical surfaces. The circular EG cylinder is probably one of the geometries most widely used in the plasmonics. Because of its important, it will be examined here in some detail.

2.4.1 Surface Plasmon Modes

Here, by using cylindrical coordinates (ρ, ϕ, z) the electrostatic surface modes of a cylindrical EG of radius a, embedded in an insulator, are obtained from Laplace's equation, in terms of *cylindrical Bessel functions* subject to the BCs at $\rho = a$. The electrostatic potential of the system is represented in the form

$$\Phi(\rho, \phi, z) = \tilde{\Phi}(\rho) \exp(iqz) \exp(im\phi) \, , \tag{2.59}$$

where m being an integer (known as the azimuthal quantum number), q is the longitudinal wavenumber along the axial direction of cylindrical EG denoted by z. In general, we note that the solution of Laplace's equation in the cylindrical symmetry can be expanded in the following Fourier–Bessel form:

$$\Phi(\rho, \phi, z) = \sum_{m=-\infty}^{\infty} \int_{-\infty}^{\infty} \frac{dq}{(2\pi)^2} \tilde{\Phi}(\rho) \exp(iqz) \exp(im\phi) \, , \tag{2.60}$$

but, here we consider a Fourier–Bessel component, as shown in (2.59). After substitution (2.59) into Laplace's equation, one finds

$$\frac{d^2 \tilde{\Phi}(\rho)}{d\rho^2} + \frac{1}{\rho} \frac{d\tilde{\Phi}(\rho)}{d\rho} - \left(q^2 + \frac{m^2}{\rho^2} \right) \tilde{\Phi}(\rho) = 0 \, , \qquad \rho \neq a \, . \tag{2.61}$$

The solution of (2.61) has the form

$$\tilde{\Phi}(\rho) = \begin{cases} A_1 I_m(q\rho) \,, & \rho \leq a \,, \\ A_2 K_m(q\rho) \,, & \rho \geq a \,, \end{cases} \tag{2.62}$$

where $I_m(x)$ and $K_m(x)$ are modified Bessel functions with $m = 0, \pm 1, \pm 2, \ldots$, and the relations between the coefficients A_1 and A_2 can be determined from the matching BCs at the separation surface. From (1.94) and (1.95), the BCs are

$$\Phi_1|_{\rho=a} = \Phi_2|_{\rho=a} \,, \tag{2.63a}$$

$$\varepsilon_1 \frac{\partial \Phi_1}{\partial \rho}\bigg|_{\rho=a} = \varepsilon_2 \frac{\partial \Phi_2}{\partial \rho}\bigg|_{\rho=a} \,. \tag{2.63b}$$

where subscript 1 denotes the region inside the EG wire and subscript 2 denotes the region outside the system. Equation (2.63b) at $\rho = a$ implies

$$A_1 = \frac{\varepsilon_2}{\varepsilon_1} \frac{K'_m(qa)}{I'_m(qa)} A_2 \,, \tag{2.64}$$

while employing (2.63a) leads to

$$\frac{\varepsilon_1}{\varepsilon_2} = \frac{I_m(qa)}{I'_m(qa)} \frac{K'_m(qa)}{K_m(qa)} \,, \tag{2.65}$$

where $I'_m(x) = dI_m(x)/dx$, and $K'_m(x) = dK_m(x)/dx$. Equation (2.65) provides the dispersion relation of the SP modes of a cylindrical EG embedded in an insulator. Using (2.7) and $\varepsilon_2 = 1$, the dispersion relation of SP modes becomes [30]

$$\omega^2 = \omega_p^2 qa I'_m(qa) K_m(qa) \,. \tag{2.66}$$

Here the Wronskian property $I'_m(x) K_m(x) - I_m(x) K'_m(x) = 1/x$ has been used. To see clearly the character of the dispersion relation of SP modes of a cylindrical EG free-standing in vacuum, we illustrate in panel (a) of Fig. 2.9 the dependence of the dimensionless frequency ω/ω_p on the dimensionless variable qa for different values of m. Also, the group velocity curves of them are plotted in panel (a) of Fig. 2.10.

Furthermore, the normalized profile of electrostatic potentials of the SP modes are plotted in Fig. 2.11, when $qa = 2$. In this example, we have taken the constant A_2 as independent and equal to one. Also, for the surface charge σ_{ind} induced on the surface of the system, we have

$$\sigma_{ind}(\phi, z)|_{\rho=a} = \varepsilon_0 \left[\frac{\partial \tilde{\Phi}_1}{\partial \rho}\bigg|_{\rho=a} - \frac{\partial \tilde{\Phi}_2}{\partial \rho}\bigg|_{\rho=a} \right] \cos qz \, \cos m\phi \,. \tag{2.67}$$

Fig. 2.9 Dispersion curves of SP modes of (**a**) an isolated cylindrical EG free-standing in vacuum, as given by (2.66) and (**b**) a cylindrical vacuum in an EG, as given by (2.69) for $m = 0$, 1, 2, and 3. The dashed horizontal line indicates the asymptotic SP frequency on a single EG-vacuum interface, i.e., $\omega_s = \omega_p/\sqrt{2}$. Because of the use of appropriate units shown along the axes, these curves do not depend on any physical parameters

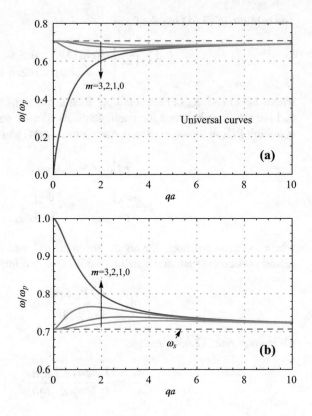

Here, $\Phi(\rho, \phi, z)$ is considered as $\tilde{\Phi}(\rho) \cos qz \ \cos m\phi$, where $\cos qz$ represents a standing wave parallel to the axial direction of the system and $\cos m\phi$ shows a *periodic wave* in the azimuthal direction. Therefore, by substituting (2.62) into (2.67) and using (2.64), we obtain

$$\sigma_{ind}(\phi, z)\big|_{\rho=a} = \varepsilon_0 q A_2 \left(\frac{\varepsilon_2}{\varepsilon_1} - 1\right) K'_m(qa) \cos qz \ \cos m\phi . \qquad (2.68)$$

To show the behavior of the induced surface charge density σ_{ind}, we note that the behavior of function $\cos qz$ is well-known, as can be seen in Fig. 2.5. Thus, it only remains for us to show the behavior of the function $\cos m\phi$, which determine the ϕ dependence of the induced charge density. Polar plots of $\cos m\phi$ for $m = 0-5$ are shown in Fig. 2.12. These plots are more pleasing to the eye if we allow ϕ to range from 0 to 360°. Note that $\cos m\phi$ for all value of m, except $m = 0$ that is constant, take on both positive and negative values. For example, the function $\cos \phi$ is positive from 0 to 90°, negative from 90 to 270°, and positive from 270 to 360°. Also, the function $\cos 2\phi$ is positive from 0 to 45°, negative from 45 to 135°, positive from 135 to 225°, negative from 225 to 315°, and positive from 315 to 360°. As m increases, the number of lobes increases. Therefore, using (2.68) and

Fig. 2.10 Group velocity
curves of SP modes of (**a**) an
isolated cylindrical EG
free-standing in vacuum and
(**b**) a cylindrical vacuum in an
EG, for $m = 0$, 1, 2, and 3.
Because of the use of
appropriate units shown along
the axes, these curves do not
depend on any physical
parameters

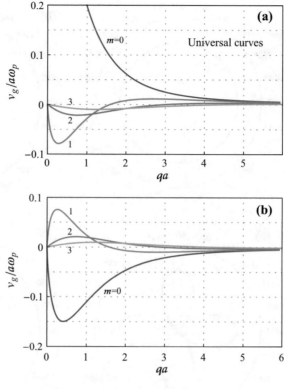

Fig. 2.11 Normalized profile
$\tilde{\Phi}(\rho)$ of SP modes of a
cylindrical EG free-standing
in vacuum, when $qa = 2$. The
mode $m = 0$ is the only one
that penetrates down to the
center of the cylinder

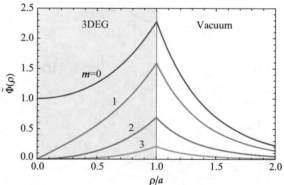

Fig. 2.12, the distribution of induced surface charge of an EG wire is schematically
shown in Fig. 2.13, for $m = 0$, 1, 2, and 3, when $qz = 0$. We note that dipole mode
($m = 1$) can be excited by absorption of electromagnetic radiation. The other modes
can be excited in Raman light scattering experiments.

In the case of a cylindrical insulator in a 3DEG, the corresponding dispersion
relation of SP modes can be found by replacing $\varepsilon_1/\varepsilon_2$ by $\varepsilon_2/\varepsilon_1$ in (2.65). In the case
of a cylindrical channel, i.e., when $\varepsilon_1 = 1$, we obtain

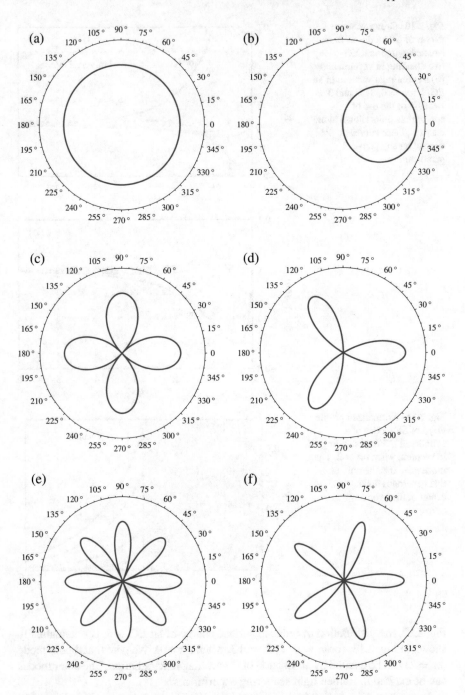

Fig. 2.12 Polar plots of $\cos m\phi$ for the first five values of m, as (**a**) $m = 0$, (**b**) $m = 1$, (**c**) $m = 2$, (**d**) $m = 3$, (**e**) $m = 4$, and (**f**) $m = 5$. Note that $\cos m\phi$ for all value of m, except $m = 0$ that is constant, take on both positive and negative values

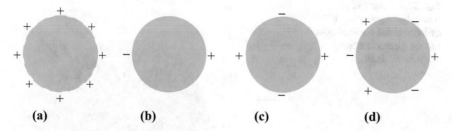

Fig. 2.13 Schematic representation of azimuthal dependence of induced surface charge for the first four SP modes of an isolated cylindrical EG. (**a**) $m = 0$, *monopole mode*, (**b**) $m = 1$, *dipole mode*, (**c**) $m = 2$, *quadrupole mode*, and (**d**) $m = 3$, *octupole mode*

$$\omega^2 = -\omega_p^2 q a I_m(qa) K'_m(qa) . \tag{2.69}$$

To benefit of the reader, panels (b) of Figs. 2.9 and 2.10 show the dispersion curves and corresponding group velocity curves of SP modes of a cylindrical channel for $m = 0, 1, 2$, and 3. It is interesting to note that the squared SP mode frequencies of the EG cylinder, i.e., (2.66) and the cylindrical channel, i.e., (2.69) add up to ω_p^2 for all values of m and q, as was introduced by Apell et al. [31].

2.4.2 Surface Plasmon Resonance

Let an infinitely long and thin EG cylinder (compared with its diameter) of radius a is illuminated by a weak x-polarized plane wave. Because of the smallness of the system with respect to the wavelength of the excitation field, so retardation effect may be neglected [32, 33] and the incident electric field is considered as $\mathbf{E} = E_0 e^{-i\omega t} \mathbf{e}_x$, where E_0 stands for the amplitude of this incident field and ω for its angular frequency. Therefore, we assume that the system is placed in a uniform quasi-static electric field that is normal to its axis. We choose a cylindrical coordinate system $\mathbf{r} = (\rho, \phi, z)$ with the origin taken at the axis of thin wire and the x-axis along the electric field, as shown in Fig. 2.14.

In the following, we work with the electrostatic potential Φ using Laplace's equation $\nabla^2 \Phi(\mathbf{r}) = 0$ for $\rho \neq a$, where Φ and $\mathbf{E}$ must be finite everywhere. The applied electrostatic potential associated with the electric field at far distances can be written as

$$\Phi_a(\rho, \phi) = -E_0 x = -E_0 \rho \cos\phi . \tag{2.70}$$

The above equation constitutes an angle-dependent BC on the potential of the combined scattering problem (EG cylinder + electric field); it introduces ϕ dependence in the potential at arbitrary ρ. However, the electric potentials inside and outside

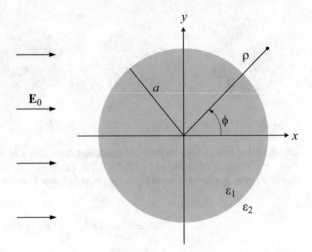

Fig. 2.14 Sketch of an infinitely long and thin EG cylinder of the circular cross section placed into an electrostatic field

the system not to depend on z because the system is infinitely long, and thus the potential will be a function of ρ and ϕ only. Therefore the general solution can be written as [34]

$$\Phi(\rho, \phi) = A_0 + B_0 \ln \rho + \sum_{m=1}^{\infty} \left(A_m \cos m\phi + A'_m \sin m\phi \right) \rho^m$$

$$+ \sum_{m=1}^{\infty} \left(B_m \cos m\phi + B'_m \sin m\phi \right) \rho^{-m} . \qquad (2.71)$$

Because of the symmetry imposed by the polarization of the exciting electric field (x-axis) only $\cos m\phi$ terms should be considered. Also, the ln solution for $m = 0$ should be rejected because it leads to a diverging field at the origin and at infinity. Furthermore, the transmitted potential should be finite as $r \to 0$, and the scattered potential should be zero as $r \to \infty$. Thus, in the presence of the thin EG wire the transmitted and scattered electric field potentials can be written as

$$\Phi_t(\rho, \phi) = \sum_{m=1}^{\infty} A_m \rho^m \cos m\phi , \qquad (2.72)$$

$$\Phi_s(\rho, \phi) = \sum_{m=1}^{\infty} B_m \rho^{-m} \cos m\phi , \qquad (2.73)$$

where A_m and B_m represent the yet unknown transmission and scattering coefficients. To determine the unknown modal coefficients A_m and B_m, the BCs must be enforced on surface of the system $\rho = a$. We have

$$\left(\Phi_a + \Phi_s\right)\big|_{\rho=a} = \Phi_t\big|_{\rho=a} , \tag{2.74a}$$

$$\varepsilon_2 \frac{\partial \left(\Phi_a + \Phi_s\right)}{\partial \rho}\bigg|_{\rho=a} = \varepsilon_1 \frac{\partial \Phi_t}{\partial \rho}\bigg|_{\rho=a} . \tag{2.74b}$$

Here, ε_1 and ε_2 are the relative dielectric constants of the thin EG wire and the surroundings, respectively. In order to evaluate (2.74a) and (2.74b), we use the fact that the functions $\cos m\phi$ are orthogonal. Hence, we obtain

$$- E_0 a + B_1 a^{-1} = A_1 a , \tag{2.75a}$$

$$\varepsilon_2 \left(E_0 + B_1 a^{-2}\right) = -\varepsilon_1 A_1 , \tag{2.75b}$$

and all other $A_m = 0 = B_m$ for $m \geq 2$. Hence,

$$A_1 = -\frac{2\varepsilon_2}{\varepsilon_1 + \varepsilon_2} E_0 , \tag{2.76a}$$

$$B_1 = a^2 \frac{\varepsilon_1 - \varepsilon_2}{\varepsilon_1 + \varepsilon_2} E_0 . \tag{2.76b}$$

With these coefficients the solution for the electric field $\mathbf{E} = -\nabla \Phi$ turns out to be

$$\mathbf{E}_s = E_0 \frac{\varepsilon_1 - \varepsilon_2}{\varepsilon_1 + \varepsilon_2} \frac{a^2}{\rho^2} \left(1 - 2 \sin^2 \phi\right) \mathbf{e}_x + 2 E_0 \frac{\varepsilon_1 - \varepsilon_2}{\varepsilon_1 + \varepsilon_2} \frac{a^2}{\rho^2} \sin \phi \cos \phi \mathbf{e}_y , \tag{2.77a}$$

$$\mathbf{E}_t = E_0 \frac{2\varepsilon_2}{\varepsilon_1 + \varepsilon_2} \mathbf{e}_x , \tag{2.77b}$$

where we reintroduced Cartesian coordinates with the unit vectors $\mathbf{e}_x$, $\mathbf{e}_y$, and $\mathbf{e}_z$. One can see that the direction of the transmitted field $\mathbf{E}_t$ coincides with the direction of the applied field $\mathbf{E}_a$. Furthermore, it is clear that the electric field inside the EG wire is homogeneous that is an unexpected result because the electromagnetic fields decay exponentially into metals. This means that quasi-static approximation is only valid for wires that are smaller in size than the *skin depth* of the EG.

Furthermore, in the quasi-static approximation, the normalized transversal[10] *polarizability* of a thin EG cylinder has been defined by Venermo and Sihvola [35] and Zhu [36]. Using the formula obtained by Venermo and Sihvola, we can write

---

[10]The polarizability of a circular cylinder has two components, known as axial and transversal polarizabilities. In the present case, i.e., when the length-to-diameter ratio of the system goes to infinity, the axial polarizability is equal to $\varepsilon_1 - \varepsilon_2$ [35].

Fig. 2.15 Absolute value of the normalized transversal polarizability α_{pol} of a sub-wavelength EG wire free-standing in vacuum as given by (2.79), with respect to the dimensionless frequency ω/ω_p, when $\gamma = 0.01\omega_p$

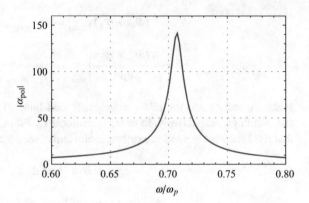

$$\alpha_{pol} = 2\frac{E_2 - E_0}{E_0} , \tag{2.78}$$

where $E_2 = |\mathbf{E}_2|$ is the magnitude of electric field in the region $\rho \geq a$, when $\rho = a$ and $\phi = 0$. From (2.78), we obtain

$$\alpha_{pol} = 2\frac{\varepsilon_1 - \varepsilon_2}{\varepsilon_1 + \varepsilon_2} . \tag{2.79}$$

Equation (2.79) is the central result of this section, the (complex) normalized transversal polarizability of a thin EG wire of sub-wavelength diameter in the electrostatic approximation. Figure 2.15 shows the absolute value of the normalized polarizability α_{pol} of a thin EG wire free-standing in vacuum, with respect to frequency ω (in ω_p units) when ε_1 varying as

$$\varepsilon(\omega) = 1 - \frac{\omega_p^2}{\omega(\omega + i\gamma)} , \tag{2.80}$$

and $\gamma = 0.01\omega_p$. It is clear that one sharp peak is due to the dipole SP of the system under the condition that $\varepsilon_1 + \varepsilon_2$ is a minimum, which for the case of small or slowly varying $\mathrm{Im}\,[\varepsilon_1]$ around the resonance simplifies to $\mathrm{Re}\,[\varepsilon_1] = -\varepsilon_2$. This is the *resonance condition* for the oscillation of dipole SP in an EG wire that is excited by an electric field polarized perpendicular to the wire axis. We note that changes in the dielectric constant of the surrounding medium, i.e., ε_2, lead to shifts of the *resonance frequency* of dipole SP[11] and EG wires are thus ideal platforms for optical sensing of changes in refractive index [34, 37].

---

[11]This resonance frequency red-shifts as ε_2 is increased.

Fig. 2.16 Calculated extinction width (in units of the geometric width) of a sub-wavelength EG wire as given by (2.81b), with respect to the dimensionless frequency ω/ω_p, when $\gamma = 0.01\omega_p$. The different curves refer to (**a**) $\varepsilon_2 = 1$ and (**b**) $\varepsilon_2 = 2$

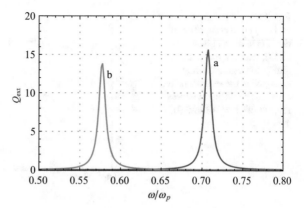

In quasi-static approximation, the widths[12] for *scattering, extinction,*[13] and *absorption* of a thin wire can be calculated via the Poynting vector and the results may be expressed as [38, 39]

$$Q_{sca} = \frac{\pi^2}{16}\varepsilon_2^{3/2}\left(\frac{a\omega}{c}\right)^3|\alpha_{pol}|^2 , \qquad (2.81a)$$

$$Q_{ext} = -\frac{\pi}{2}\varepsilon_2^{1/2}\left(\frac{a\omega}{c}\right)\mathrm{Re}\left[i\alpha_{pol}\right] , \qquad (2.81b)$$

$$Q_{abs} = Q_{ext} - Q_{sca} , \qquad (2.81c)$$

respectively, in units of the geometric width $2a$. Also, c is the light speed in vacuum, as mentioned before. It is clear from (2.81a) that due to the rapid scaling of $Q_{sca} \propto a^3$, it is very difficult to pick out the thin wires of sub-wavelength diameters from a background of larger scatterers and in the extinction formula, i.e., $Q_{ext} = Q_{abs} + Q_{sca}$, the first term that shows the absorption width of the system overwhelms the second term. Figure 2.16 shows the extinction width of a thin wire in the quasi-static approximation calculated using (2.81b) for immersion in two different media.

2.4.3 Effective Permittivity of a Composite of Thin Circular Electron Gas Cylinders

In dipolar theory, the effective dielectric function ε_{eff} of a composite of aligned thin wires embedded in a host matrix with relative dielectric constant ε_2 may be derived, as [40, 41]

---

[12]Here, the optical properties are expressed in terms of widths, which are defined as the cross sections per unit length of the cylinder.

[13]The sum of absorption and scattering is called extinction.

Fig. 2.17 Variation of
$\text{Im}\left[\varepsilon_{eff}\right]$ of a composite of
thin EG wires as given by
(2.83), with respect to the
dimensionless frequency
ω/ω_p, when $\gamma = 0.01\omega_p$ and
$\varepsilon_2 = 1$. The different curves
refer to (**a**) $f = 0.1$ and (**b**)
$f = 0.15$

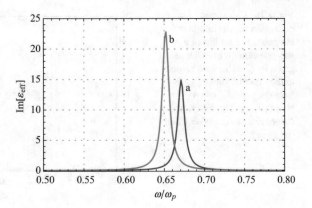

$$\frac{\varepsilon_{eff} - \varepsilon_2}{\varepsilon_{eff} + \varepsilon_2} = \frac{1}{2}f\alpha_{pol} \,, \tag{2.82}$$

where f is the volume fraction of the embedded wires and α_{pol} is the normalized transversal polarizability of a wire, as can be seen in (2.79). By substituting (2.79) into (2.82), we find

$$\frac{\varepsilon_{eff} - \varepsilon_2}{\varepsilon_{eff} + \varepsilon_2} = f\frac{\varepsilon_1 - \varepsilon_2}{\varepsilon_1 + \varepsilon_2} \,. \tag{2.83}$$

This equation is the (complex) effective dielectric function of a composite of aligned thin wires, in the *Maxwell-Garnett* [42] approximation.

In Fig. 2.17 we show imaginary part of the effective dielectric function of the system, as a function of dimensionless frequency ω/ω_p for two different values of volume fraction f, when $\gamma = 0.01\omega_p$ and $\varepsilon_2 = 1$. It is clear that as parameter f increases, the interaction among the wires leads to the resonance frequency at lower values.

2.5   Plasmonic Properties of Circular Electron Gas Tubes

2.5.1   Surface Plasmon Modes

We consider a circular EG tube with internal and external radii denoted by a_1 and a_2, respectively, as a simple model of a metallic tube. The relative dielectric constants of the core $0 < \rho < a_1$ and embedding medium $\rho > a_2$ are ε_1 and ε_3, respectively, while region $a_1 < \rho < a_2$ is filled by an EG characterized by the local dielectric function $\varepsilon_2 = 1 - \omega_p^2/\omega^2$. We consider the propagation of an electrostatic surface wave in the z-direction, so the electric potential is represented in the form of (2.59). For this geometry, the solution of (2.61) has the form

$$\tilde{\Phi}(\rho) = \begin{cases} A_1 I_m(q\rho) \,, & \rho \leq a_1 \,, \\ A_2 I_m(q\rho) + A_3 K_m(q\rho) \,, & a_1 \leq \rho \leq a_2 \,, \\ A_4 K_m(q\rho) \,, & \rho \geq a_2 \,. \end{cases} \tag{2.84}$$

The application of the electrostatic BCs at $\rho = a_1$ and a_2 gives

$$A_1 = \frac{a_2}{a_1} \frac{\varepsilon_2 I'_m(qa_2) K_m(qa_2) - \varepsilon_3 I_m(qa_2) K'_m(qa_2)}{(\varepsilon_2 - \varepsilon_1) I_m(qa_1) I'_m(qa_1)} A_4 \,, \tag{2.85a}$$

$$A_2 = qa_2 \frac{\varepsilon_3 - \varepsilon_2}{\varepsilon_2} K_m(qa_2) K'_m(qa_2) A_4 \,, \tag{2.85b}$$

$$A_3 = \frac{qa_2}{\varepsilon_2} \left[\varepsilon_2 I'_m(qa_2) K_m(qa_2) - \varepsilon_3 I_m(qa_2) K'_m(qa_2) \right] A_4 \,. \tag{2.85c}$$

Furthermore, we obtain the eigenvalue equation for the resonant frequencies of the SP modes of an EG tube suspended in vacuum, with the following two branches of the dispersion relations for each m, as

$$\omega_{\pm}^2(m, q) = \frac{\omega_{wire}^2 + \omega_{chan}^2}{2} \pm \sqrt{\left(\frac{\omega_{wire}^2 - \omega_{chan}^2}{2} \right)^2 + \Delta^2} \,, \tag{2.86}$$

where $\omega_{chan}(m, q)$ is the SP frequency of a cylindrical channel with radius a_1, as shown by (2.69), and $\omega_{wire}(m, q)$ is the SP frequency of an EG wire with radius a_2, as described by (2.66). Also,

$$\Delta = \omega_{wire} \omega_{chan} \left[\frac{I'_m(qa_1)}{I'_m(qa_2)} \frac{K'_m(qa_2)}{K'_m(qa_1)} \right]^{1/2} \tag{2.87}$$

gives the electrostatic coupling between both interfaces [30, 43]. The SP frequencies of a circular EG tube can now be understood as the coupling of the cylindrical channel and wire SPs. In Fig. 2.18, we plot dimensionless frequency ω/ω_p, versus variable qa_1 for several angular modes $m = 0, 1$, and 2, when (a) $a_2/a_1 = 1.2$, and (b) $a_2/a_1 = 1.5$. It is clear that the upper group of SP modes is similar with that of the SP modes of cylindrical channels, when $a_2 \to \infty$ and lower group of SP modes is similar with SP modes in wires, when $a_1 \to 0$. As can be seen, the hybridization of the primitive SPs depends on the thickness of the tube, or the tube aspect ratio a_2/a_1. Also, it is clear that in the limit $qa_1 \to \infty$, all modes tend to the values corresponding to SP frequency of a flat EG-vacuum interface.

Furthermore, we see that the coupling between both interfaces plays an important role on the frequency in the low-q range. In the special case when the longitudinal wavenumber is zero, i.e., $q = 0$, (2.86) can be reduced to the following equation:

Fig. 2.18 Dispersion curves of SP modes of an EG tube in vacuum, as given by (2.86) for $m = 0$, 1, 2, and 3. The dashed horizontal line indicates the asymptotic SP frequency of a flat EG-vacuum interface, i.e., $\omega_s = \omega_p/\sqrt{2}$. The different panels refer to (**a**) $a_2/a_1 = 1.2$ and (**b**) $a_2/a_1 = 1.5$

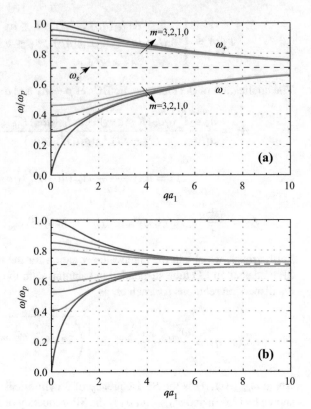

$$\omega_{\pm}(m, q = 0) = \frac{\omega_p}{\sqrt{2}} \left[1 \pm \left(\frac{a_1}{a_2} \right)^m \right]^{1/2} , \quad m \neq 0 . \tag{2.88}$$

At this stage, by using (2.84)–(2.85c), the normalized profiles of electrostatic potentials are presented in Fig. 2.19 for $m = 0$, and 1, when $qa_1 = 2$ and $a_2/a_1 = 1.5$. In this example, we have taken the constant A_4 as independent and equal to one. Also, for the surface charge σ_{ind} induced on the two surfaces of the system, we have

$$\sigma_{ind}(\phi, z)\big|_{\rho=a_1} = \varepsilon_0 \left[\frac{\partial \tilde{\Phi}_1}{\partial \rho} \bigg|_{\rho=a_1} - \frac{\partial \tilde{\Phi}_2}{\partial \rho} \bigg|_{\rho=a_1} \right] \cos qz \, \cos m\phi , \tag{2.89a}$$

$$\sigma_{ind}(\phi, z)\big|_{\rho=a_2} = \varepsilon_0 \left[\frac{\partial \tilde{\Phi}_2}{\partial \rho} \bigg|_{\rho=a_2} - \frac{\partial \tilde{\Phi}_3}{\partial \rho} \bigg|_{\rho=a_2} \right] \cos qz \, \cos m\phi . \tag{2.89b}$$

Therefore, by substituting (2.84) into (2.89a) and (2.89a) and using (2.85a)–(2.85c), for $qz = 0$, we arrive to

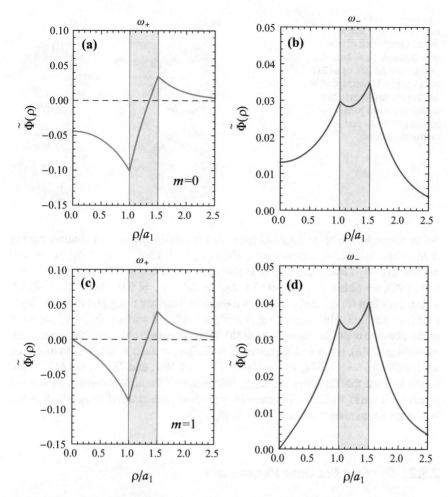

Fig. 2.19 Normalized profile $\tilde{\Phi}(\rho)$ of SP modes of an EG tube in vacuum for $m = 0$ and 1, when $qa_1 = 2$ and $a_2/a_1 = 1.5$. Panels (**a**) and (**c**) show the results for anti-symmetric (odd) modes and panels (**b**) and (**d**) show the results for the symmetric (even) modes

$$\sigma_{ind}(\phi) = \frac{\varepsilon_0}{a_1} A_4 \cos m\phi \begin{cases} \Gamma_1 \,, \; \rho = a_1 \,, \\ \Gamma_2 \,, \; \rho = a_2 \,, \end{cases} \qquad (2.90)$$

where

$$\Gamma_1 = qa_1 \left[\left(\frac{A_1}{A_4} - \frac{A_2}{A_4} \right) I'_m(qa_1) - \frac{A_3}{A_4} K'_m(qa_1) \right] \,, \qquad (2.91)$$

$$\Gamma_2 = qa_1 \left[\frac{A_2}{A_4} I'_m(qa_2) + \left(\frac{A_3}{A_4} - 1 \right) K'_m(qa_2) \right] \,. \qquad (2.92)$$

Fig. 2.20 Schematic representation of surface charge distribution of (**a**) $m = 0$ and (**b**) $m = 1$ modes on the cross section of an EG tube. Up panels correspond to ω_+ that are anti-symmetric modes and down panels correspond to ω_- that are symmetric modes

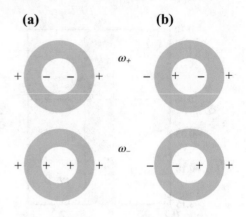

As an example, we choose an EG tube with the inner and outer cylinders having radii a_1 and $a_2 = 1.5a_1$, respectively, when $qa_1 = 2$. Thus, using (2.86), for $m = 0$ we find $\omega_+ = 0.86\omega_p$ and $\omega_- = 0.56\omega_p$. By substituting these values in (2.91) and (2.92), we obtain $\Gamma_{1+} = -0.525$, $\Gamma_{2+} = 0.3$, $\Gamma_{1-} = 0.06$, and $\Gamma_{2-} = 0.117$. This means that for ω_+ and ω_- modes the induced surface charge on the outer layer is always positive, while for ω_+ (ω_-) mode the induced surface charge of the inner surface has the opposite (same) sign of the induced surface charge on the outer layer. Also using (2.88), for $m = 1$ we have $\omega_+ = 0.83\omega_p$ and $\omega_- = 0.57\omega_p$. In this case, we obtain $\Gamma_{1+} = -0.528$, $\Gamma_{2+} = 0.31$, $\Gamma_{1-} = 0.097$, and $\Gamma_{2-} = 0.1142$. These values indicate that the induced charge distribution of the high-frequency modes ω_+ is anti-symmetric, while for low-frequency mode ω_- the induced charge distribution is symmetric, as shown schematically in Fig. 2.20.

2.5.2 Surface Plasmon Resonances

Let us consider the polarizability of a thin EG tube of radius a when it is illuminated by a weak x-polarized plane wave. The relative dielectric constants of the refractive media from inside to outside are ε_1, ε_2, and ε_3, respectively. In this case, we have

$$\Phi(\rho, \phi) = \cos\phi \begin{cases} A_1\rho \,, & \rho \le a_1 \,, \\ A_2\rho + A_3\rho^{-1} \,, & a_1 \le \rho \le a_2 \,, \\ -E_0\rho + A_4\rho^{-1} \,, & \rho \ge a_2 \,. \end{cases} \tag{2.93}$$

We note that A_4 and $A_1 - A_3$ represent the yet unknown scattering and transmission coefficients, respectively. To determine the unknown coefficients $A_1 - A_4$, the BCs must be enforced on surfaces of the system. The BCs yield

$$A_1 = -\frac{4\varepsilon_2\varepsilon_3}{(\varepsilon_1 + \varepsilon_2)(\varepsilon_2 + \varepsilon_3) + \dfrac{a_1^2}{a_2^2}(\varepsilon_1 - \varepsilon_2)(\varepsilon_2 - \varepsilon_3)} E_0 , \qquad (2.94a)$$

$$A_2 = -\frac{2\varepsilon_3(\varepsilon_1 + \varepsilon_2)}{(\varepsilon_1 + \varepsilon_2)(\varepsilon_2 + \varepsilon_3) + \dfrac{a_1^2}{a_2^2}(\varepsilon_1 - \varepsilon_2)(\varepsilon_2 - \varepsilon_3)} E_0 , \qquad (2.94b)$$

$$A_3 = a_1^2\frac{2\varepsilon_3(\varepsilon_1 - \varepsilon_2)}{(\varepsilon_1 + \varepsilon_2)(\varepsilon_2 + \varepsilon_3) + \dfrac{a_1^2}{a_2^2}(\varepsilon_1 - \varepsilon_2)(\varepsilon_2 - \varepsilon_3)} E_0 , \qquad (2.94c)$$

$$A_4 = a_2^2\frac{(\varepsilon_1 + \varepsilon_2)(\varepsilon_2 - \varepsilon_3) + \dfrac{a_1^2}{a_2^2}(\varepsilon_1 - \varepsilon_2)(\varepsilon_2 + \varepsilon_3)}{(\varepsilon_1 + \varepsilon_2)(\varepsilon_2 + \varepsilon_3) + \dfrac{a_1^2}{a_2^2}(\varepsilon_1 - \varepsilon_2)(\varepsilon_2 - \varepsilon_3)} E_0 . \qquad (2.94d)$$

Therefore, from (2.78), we can get the normalized transversal polarizability of the system, as

$$\alpha_{pol} = 2\frac{(\varepsilon_1 + \varepsilon_2)(\varepsilon_2 - \varepsilon_3) + \dfrac{a_1^2}{a_2^2}(\varepsilon_1 - \varepsilon_2)(\varepsilon_2 + \varepsilon_3)}{(\varepsilon_1 + \varepsilon_2)(\varepsilon_2 + \varepsilon_3) + \dfrac{a_1^2}{a_2^2}(\varepsilon_1 - \varepsilon_2)(\varepsilon_2 - \varepsilon_3)} . \qquad (2.95)$$

Figure 2.21 shows the imaginary part of the normalized transversal polarizability α_{pol} of a thin EG tube free-standing in vacuum, with respect to angular frequency ω (in ω_p units) when $\gamma = 0.01\omega_p$ and ε_2 varying according to (2.80). It is clear that two sharp peaks are due to the resonance of dipole SPs of the system. We note that change in the ratio of the inner to outer radius, i.e., a_1/a_2 leads to shift of the resonance frequencies of dipole SPs, as shown in Fig. 2.21. Also, the scattering and extinction widths of sub-wavelength EG tubes may be expressed as

$$Q_{sca} = \frac{\pi^2}{4}\varepsilon_3^{3/2}\left(\frac{a_2\omega}{c}\right)^3 \left|\frac{(\varepsilon_1 + \varepsilon_2)(\varepsilon_2 - \varepsilon_3) + \dfrac{a_1^2}{a_2^2}(\varepsilon_1 - \varepsilon_2)(\varepsilon_2 + \varepsilon_3)}{(\varepsilon_1 + \varepsilon_2)(\varepsilon_2 + \varepsilon_3) + \dfrac{a_1^2}{a_2^2}(\varepsilon_1 - \varepsilon_2)(\varepsilon_2 - \varepsilon_3)}\right|^2 ,$$

$$(2.96)$$

Fig. 2.21 Imaginary part of the normalized transversal polarizability α_{pol} of a sub-wavelength EG tube free-standing in vacuum as given by (2.95), with respect to the dimensionless frequency ω/ω_p, when $\gamma = 0.01\omega_p$. The different curves refer to (**a**) $a_1/a_2 = 0.1$ and (**b**) $a_1/a_2 = 0.15$

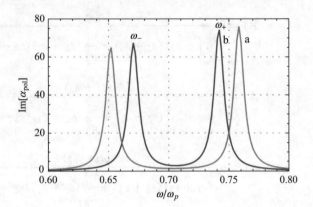

$$
Q_{ext} = -\pi\varepsilon_3^{1/2}\left(\frac{a_2\omega}{c}\right)\mathrm{Re}\left[i\frac{(\varepsilon_1+\varepsilon_2)(\varepsilon_2-\varepsilon_3)+\dfrac{a_1^2}{a_2^2}(\varepsilon_1-\varepsilon_2)(\varepsilon_2+\varepsilon_3)}{(\varepsilon_1+\varepsilon_2)(\varepsilon_2+\varepsilon_3)+\dfrac{a_1^2}{a_2^2}(\varepsilon_1-\varepsilon_2)(\varepsilon_2-\varepsilon_3)}\right].
$$

$$(2.97)$$

Furthermore, the effective dielectric function ε_{eff} of a composite of aligned thin tubes embedded in a host matrix with relative dielectric constant ε_3 may be written as

$$
\frac{\varepsilon_{eff}-\varepsilon_3}{\varepsilon_{eff}+\varepsilon_3} = f\frac{(\varepsilon_1+\varepsilon_2)(\varepsilon_2-\varepsilon_3)+\dfrac{a_1^2}{a_2^2}(\varepsilon_1-\varepsilon_2)(\varepsilon_2+\varepsilon_3)}{(\varepsilon_1+\varepsilon_2)(\varepsilon_2+\varepsilon_3)+\dfrac{a_1^2}{a_2^2}(\varepsilon_1-\varepsilon_2)(\varepsilon_2-\varepsilon_3)},
$$

$$(2.98)$$

where f is the volume fraction of the embedded tubes.

2.6 Surface Plasmon Modes of Cylinder-Cylinder and Cylinder-Plane Electron Gas Systems

To describe the plasmonic response of a system of two thin wires of unequal radii, in [44] the plasmon hybridization theory [45, 46] is used. In this way, a general analytical formalism of SPs dispersion in the matrix form of infinite dimension is obtained. This means that the numerical approach should be chosen in order to obtain convergence for the coupled SP modes of a cylinder-cylinder system. Some of the earlier works in this structure include the works of Schmeits [47], Kottmann and Martin [48], Ditlbacher et al. [49], Manjavacas and Garcia de Abajo [50], Cai et al. [51], Sun et al. [52], Andersen et al. [53], Teperik et al. [54], and Mortensen et al. [55].

Fig. 2.22 Two close EG
cylinders and a bipolar
coordinate system

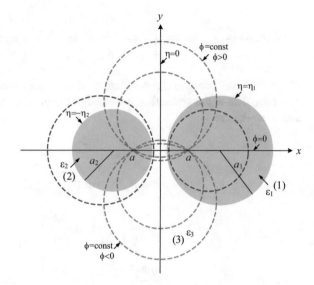

Here, we extend the results in [44] and study the SP modes of two interacting
EG cylinders and an EG cylinder coupled to a semi-infinite EG medium. To do this,
we do not follow the method discussed in [44] and employ an alternative approach,
which utilizes the fact that Laplace's equation is separable in the so-called *bipolar
coordinates*.

The two close cylinders are shown in Fig. 2.22. An EG cylinder of radius a_1
with relative dielectric function ε_1 occupies region (1), while the other cylinder with
relative dielectric function ε_2 occupies region (2), and both spheres are embedded in
region (3) which has relative dielectric function ε_3. In a bipolar coordinate system
[56] two dimensionless coordinates η and ϕ are related to the Cartesian coordinates
as follows:

$$x = \frac{a \sinh \eta}{\cosh \eta - \cos \phi} , \tag{2.99a}$$

$$y = \frac{a \sin \phi}{\cosh \eta - \cos \phi} . \tag{2.99b}$$

The lines $\eta = \pm\text{const}$ are the pairs of circles situated symmetrically in respect to
y-axis (see Fig. 2.22). The electrostatic potential Φ has to obey Laplace's equation
in bipolar coordinates, as [56]

$$\nabla^2 \Phi = \frac{1}{h^2(\eta, \phi)} \left(\frac{\partial^2}{\partial \eta^2} + \frac{\partial^2}{\partial \phi^2} \right) \Phi = 0 , \tag{2.100}$$

where $h(\eta, \phi) = a/(\cosh \eta - \cos \phi)$ is the scaling function. As before, using the
method of separation of variables, the electrostatic potential Φ can be written in the
form

$$\Phi(\eta, \phi) = \sum_{m=0}^{+\infty} \cos m\phi \begin{cases} A_{1m}e^{-m\eta}, & \eta \geq \eta_1, \\ A_{2m}e^{-m\eta} + A_{3m}e^{m\eta}, & -\eta_2 \leq \eta \leq \eta_1, \\ A_{4m}e^{m\eta}, & \eta \leq -\eta_2, \end{cases} \qquad (2.101)$$

where relations between the coefficients $A_{1m} - A_{4m}$ can be determined from the matching BCs at the separation surface. For the present system, the BCs at $\eta = \eta_1$ and $\eta = -\eta_2$ are

$$\Phi_1\big|_{\eta=\eta_1} = \Phi_3\big|_{\eta=\eta_1}, \qquad (2.102a)$$

$$\Phi_2\big|_{\eta=-\eta_2} = \Phi_3\big|_{\eta=-\eta_2}, \qquad (2.102b)$$

$$\varepsilon_1 \frac{\partial \Phi_1}{\partial \eta}\bigg|_{\eta=\eta_1} = \varepsilon_3 \frac{\partial \Phi_3}{\partial \eta}\bigg|_{\eta=\eta_1}, \qquad (2.102c)$$

$$\varepsilon_2 \frac{\partial \Phi_2}{\partial \eta}\bigg|_{\eta=-\eta_2} = \varepsilon_3 \frac{\partial \Phi_3}{\partial \eta}\bigg|_{\eta=-\eta_2}, \qquad (2.102d)$$

where subscript 1 and 2 denote the regions inside the cylinders having radii η_1 and $-\eta_2$, respectively, while subscript 3 denotes the region between the two cylinders. By using the mentioned BCs at the surface of each of the cylinders, after some algebra, the following dispersion relation can be obtained:

$$\frac{\varepsilon_3 - \varepsilon_1}{\varepsilon_3 + \varepsilon_1} \frac{\varepsilon_3 - \varepsilon_2}{\varepsilon_3 + \varepsilon_2} = \exp[2m(\eta_1 + \eta_2)]. \qquad (2.103)$$

This is a general result for cylinder-cylinder and cylinder-plane systems. For every order m there are two branches of SP modes. Several possible applications of the present geometry exist, and we consider four cases illustrated in Figs. 2.23 and 2.24.

In Fig. 2.23, panels (a) and (b), we show the behavior of the SP modes in a pair of cylinders as a function of the η_1 for different values of the parameters η_2, ε_1, ε_1, ε_3, and m. Panel (a) illustrates the case, in which region (1) and region (2) are occupied by same medium with the relative dielectric function $\varepsilon_j = 1 - \omega_{pj}^2/\omega^2$, ($j = 1$ and 2) and region (3) is occupied by vacuum, where $\eta_1 = \eta_2$. One can see for each order m there are two branches of SP modes, and as the separation between the cylinders gets very large, the two modes tend to merge to the SP frequency of an isolated EG cylinder given by $\omega_p/\sqrt{2}$. Panel (b) describes the SP modes of two equal cylinders, when $\omega_{p1} = \omega_p$, $\omega_{p2} = \sqrt{2}\omega_p$ and region (3) is occupied by vacuum. At large η_1 the two branches tend to merge to frequencies given by $\omega_+ = \omega_p$ and $\omega_- = \omega_p/\sqrt{2}$.

In Fig. 2.24, panels (a) and (b), we plot the behavior of the SP modes of a cylinder coupled to a semi-infinite medium ($\eta_2 = 0$) as a function of the η_1 for different values of the parameters ε_1, ε_2, ε_3, and m. One can see that (2.103) exhibits excellent numerical convergence, and is general. In comparison the results of Fig. 2.23, with results of Fig. 2.24, it is easy to find that the two systems [cylinder-cylinder and cylinder-plane] agree only in the large values of η_1.

Fig. 2.23 The dimensionless SP frequencies ω/ω_p of a pair of parallel cylinders, as a function of η_1, for different values of the parameters η_2, ε_1, ε_2, ε_3, and m. (**a**) illustrates different values of the parameter m, when $\omega_{p2} = \omega_p = \omega_{p1}$, $\varepsilon_3 = 1$, and $\eta_1 \neq \eta_2$. (**b**) shows different values of m, when $\omega_{p1} = \omega_p$, $\omega_{p2} = \sqrt{2}\omega_p$, $\varepsilon_3 = 1$, and $\eta_1 = \eta_2$

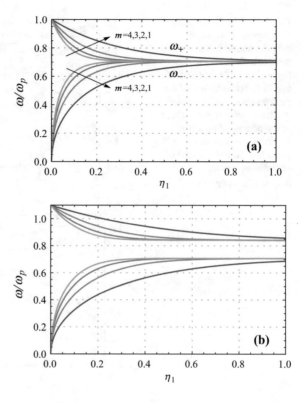

2.7   Plasmonic Properties of Electron Gas Spheres

Problems involving metallic spherical geometries constitute an important class of canonical electrostatic and electromagnetic BVPs that are used to design absorber and scatterers. The field configurations that can be supported by a spherical EG structure can be obtained by analyzing the structure as a BVP. We will concern ourselves here with electrostatic BVPs.

2.7.1   Surface Plasmon Modes

In this section, we derive the *localized* SP frequencies of a spherical EG with a radius a embedded in an insulator. In the spherical coordinates (r, θ, φ), the electrostatic potential of the system is represented in the form

$$\Phi(r, \theta, \varphi) = \tilde{\Phi}(r) Y_{\ell m}(\theta, \varphi) \,, \tag{2.104}$$

Fig. 2.24 The dimensionless SP frequencies ω/ω_p of a cylinder coupled to a semi-infinite medium ($\eta_2 = 0$), as a function of the η_1, for different values of the parameters ε_1, ε_2, ε_3, and m. (**a**) illustrates different values of the parameter m, when $\omega_{p1} = \omega_p = \omega_{p2}$ and $\varepsilon_3 = 0$. (**b**) shows different values of m, when $\omega_{p1} = \omega_p$, $\omega_{p2} = \sqrt{2}\omega_p$, and $\varepsilon_3 = 1$

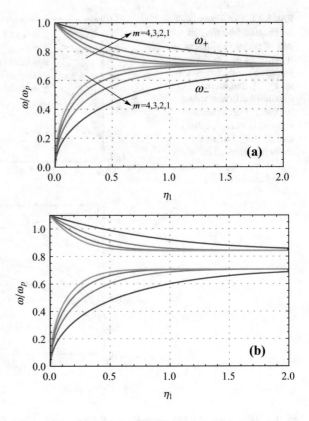

where $Y_{\ell m}$ denote the spherical harmonics. In general, we note that the solution of Laplace's equation in the spherical symmetry is

$$\Phi(r, \theta, \varphi) = \sum_{\ell=0}^{\infty} \sum_{m=-\ell}^{+\ell} \tilde{\Phi}(r) Y_{\ell m}(\theta, \varphi) \,, \qquad (2.105)$$

but, here we consider one of the components which is analogous to taking the Fourier transform in planar geometry. Also, we note that for a BVP possessing azimuthal symmetry $m = 0$, then

$$Y_{\ell 0} = \sqrt{\frac{2\ell + 1}{4\pi}} P_\ell(\cos\theta) \,, \qquad (2.106)$$

Table 2.1 Several Legendre polynomials

ℓ	$P_\ell(\cos\theta)$
0	1
1	$\cos\theta$
2	$\dfrac{1}{2}\left(3\cos^2\theta - 1\right)$
3	$\dfrac{1}{2}\left(5\cos^3\theta - 3\cos\theta\right)$

where $P_\ell(\cos\theta)$ are called the *Legendre polynomials* of order ℓ. Table 2.1 lists the expressions for Legendre polynomials for several values of ℓ. After substitution (2.104) in Laplace's equation $\nabla^2\Phi(\mathbf{r}) = 0$, for $r \neq a$ one finds

$$\frac{d^2\tilde{\Phi}(r)}{dr^2} + \frac{2}{r}\frac{d\tilde{\Phi}(r)}{dr} - \frac{\ell(\ell+1)}{r^2}\tilde{\Phi}(r) = 0 , \qquad r \neq a . \tag{2.107}$$

The solution of (2.107) has the form

$$\tilde{\Phi}(r) = \begin{cases} A_1 r^\ell , & r \leq a , \\ A_2 r^{-(\ell+1)} , & r \geq a , \end{cases} \tag{2.108}$$

where $\ell = 1, 2, \ldots$, and the relations between the coefficients A_1 and A_2 can be determined from the matching BCs at the separation surface. From (1.94) and (1.95), the BCs are

$$\Phi_1|_{r=a} = \Phi_2|_{r=a} , \tag{2.109a}$$

$$\varepsilon_1 \frac{\partial \Phi_1}{\partial r}\bigg|_{r=a} = \varepsilon_2 \frac{\partial \Phi_2}{\partial r}\bigg|_{r=a} . \tag{2.109b}$$

where subscript 1 denotes the region inside the sphere and subscript 2 denotes the region outside the system. The application of the above electrostatic BCs at $r = a$ gives the following dispersion relation

$$\frac{\varepsilon_1}{\varepsilon_2} = -\frac{\ell+1}{\ell} . \tag{2.110}$$

Using (2.7) and $\varepsilon_2 = 1$, we find the Mie plasmons frequencies, as

$$\omega_\ell = \omega_p \sqrt{\frac{\ell}{2\ell+1}} . \tag{2.111}$$

Normalized profile $\tilde{\Phi}(r)$ of SP modes are shown in Fig. 2.25 for $\ell = 1, 2, 3$, and 4. These modes have a cusp behavior at $r = a$, which corresponds to the accumulation of surface charge density at the boundary. This is a characteristic of SPs.

Fig. 2.25 Normalized profile $\tilde{\Phi}(r)$ of SP modes for $\ell = 1$, 2, 3, and 4 of a spherical EG free-standing in vacuum. These modes have maxima at the boundary, $r = a$, corresponding to the accumulation of surface charge density at that place

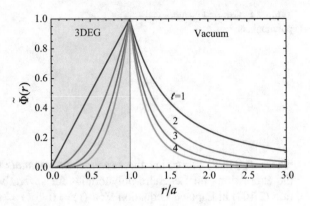

Furthermore, for the surface charge σ_{ind} induced on the surface of the system, we have

$$\sigma_{ind}(\theta)\big|_{r=a} = \varepsilon_0 \left[\frac{\partial \tilde{\Phi}_1}{\partial r}\bigg|_{r=a} - \frac{\partial \tilde{\Phi}_2}{\partial r}\bigg|_{r=a} \right] P_\ell(\cos\theta) . \tag{2.112}$$

Therefore, by substituting (2.104) into (2.112) and using

$$A_1 = -\frac{\varepsilon_2}{\varepsilon_1} \frac{\ell+1}{\ell} a^{-(2\ell+1)} A_2 , \tag{2.113}$$

we obtain

$$\sigma_{ind}(\theta)\big|_{r=a} = \varepsilon_0 (\ell+1) a^{-(\ell+2)} A_2 \left(1 - \frac{\varepsilon_2}{\varepsilon_1}\right) P_\ell(\cos\theta) . \tag{2.114}$$

The above result indicates that, for example, for $\ell = 1$ the induced surface charge is positive in the hemisphere $0 \le \theta \le \pi/2$ and negative in the hemisphere $\pi/2 \le \theta \le \pi$.

In the case of a spherical insulator in a 3DEG, the corresponding localized SP frequencies can be found by replacing $\varepsilon_1/\varepsilon_2$ by $\varepsilon_2/\varepsilon_1$ in (2.110). Thus, in the case of a spherical cavity, we find

$$\omega_\ell = \omega_p \sqrt{\frac{\ell+1}{2\ell+1}} . \tag{2.115}$$

Again, we have $\omega_{sphere}^2 + \omega_{cavity}^2 = \omega_p^2$, which is a property of the SP modes of complementary media [31].

2.7.2 Bulk Plasmon Modes

The bulk modes correspond to electron density oscillations within the volume of an EG sphere with plasma frequency ω_p, therefore we have $\varepsilon(\omega_p) = 0$. These modes are characterized by electron density fluctuations $n_b(r, \theta, \varphi)$ within the volume and should satisfy Poisson's equation, $\nabla^2 \Phi_b = en_b/\varepsilon_0$ [57, 58]. In this case, the electrostatic potential and electron density fluctuations of the system for $r < a$ can be represented in the form

$$\Phi_b(r, \theta, \varphi) = A_1 j_\ell(kr) Y_{\ell m}(\theta, \phi) , \tag{2.116}$$

$$n_b(r, \theta, \varphi) = B_1 j_\ell(kr) Y_{\ell m}(\theta, \phi) , \tag{2.117}$$

where $j_\ell(x)$ is the spherical Bessel function of the first kind and order ℓ. The relation between the coefficients A_1 and B_1 which also depend on k is obtained from Poisson's equation, which yields

$$A_1 = \frac{e}{\varepsilon_0 k^2} B_1 , \tag{2.118}$$

and the BC for these modes is $\Phi_b|_{r=a} = 0$, which yields

$$j_\ell(ka) = 0 . \tag{2.119}$$

Therefore, the values of k are given by $ka = x_{\ell m}$, where $x_{\ell m}$ are the zeros of the spherical Bessel function $j_\ell(x_{\ell m})$, with $m = 1, 2, \cdots$. Panels (a) and (b) of Fig. 2.26 show the characteristics of bulk modes for $\ell = 0$ and 1, respectively, and various values of m. These modes are fully confined within the volume of the sphere. Since the potential Φ_b and induced charge density n_b are locally related by (2.118), the functions shown in Fig. 2.26 illustrate the behavior of both quantities.

2.7.3 Surface Plasmon Resonance

The SP resonance for a small spherical EG of radius a when it is placed in a uniform quasi-static electric field can be found in much the same way as for the thin EG cylinder. Again, we consider the incident electric field as $\mathbf{E} = E_0 \mathbf{e}_z$. The relative dielectric constants of the sphere and the external medium are ε_1 and ε_2, respectively, as shown in Fig. 2.27. We use spherical coordinates with the origin at the center of the sphere. The potentials inside and outside the sphere satisfy Laplace's equation; moreover, the applied electrostatic potential associated with the electric field at far distances is

Fig. 2.26 Normalized profile $\tilde{\Phi}(r)$ of bulk modes of a spherical EG free-standing in vacuum, when $m = 1, 2, 3,$ and 4. Panel (**a**) $\ell = 0$, and panel (**b**) $\ell = 1$. These modes have maxima at the boundary, $r = a$, corresponding to the accumulation of surface charge density at that place

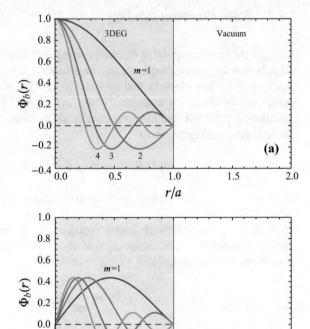

Fig. 2.27 Sketch of a homogeneous spherical EG placed into an electrostatic field

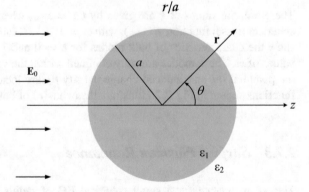

$$\Phi_a(r, \theta) = -E_0 z = -E_0 r \cos\theta . \tag{2.120}$$

The present equation constitutes an angle-dependent BC on the potential of the combined scattering problem (EG sphere + electric field); it introduces θ dependence in the potential at arbitrary r. However, there is still symmetry about the z-axis, and thus the potential will be a function of r and θ only, and therefore the general solution can be written as

$$\Phi(r, \theta) = \sum_{\ell=0}^{\infty} \left[A_\ell r^\ell + B_\ell r^{-(\ell+1)} \right] P_\ell(\cos\theta) . \tag{2.121}$$

Due to the requirement that the transmitted potential should be finite as $r \to 0$, and the scattered potential should be zero as $r \to \infty$, the solutions for the transmitted and scattered electric field potentials can be written as

$$\Phi_t(r, \theta) = \sum_{\ell=0}^{\infty} A_\ell r^\ell P_\ell(\cos\theta) , \tag{2.122}$$

$$\Phi_s(r, \theta) = \sum_{\ell=0}^{\infty} B_\ell r^{-(\ell+1)} P_\ell(\cos\theta) . \tag{2.123}$$

To determine the unknown modal coefficients A_ℓ and B_ℓ, the BCs must be enforced on surface of the system $r = a$. We have

$$\left. (\Phi_a + \Phi_s) \right|_{r=a} = \left. \Phi_t \right|_{r=a} , \tag{2.124a}$$

$$\varepsilon_2 \left. \frac{\partial (\Phi_a + \Phi_s)}{\partial r} \right|_{r=a} = \varepsilon_1 \left. \frac{\partial \Phi_t}{\partial r} \right|_{r=a} . \tag{2.124b}$$

Evaluation of the BCs leads to

$$\Phi_t(r, \theta) = -\frac{3\varepsilon_2}{\varepsilon_1 + \varepsilon_2} E_0 r \cos\theta , \tag{2.125}$$

$$\Phi_s(r, \theta) = \frac{\varepsilon_1 - \varepsilon_2}{\varepsilon_1 + 2\varepsilon_2} \frac{a^3}{r^2} E_0 \cos\theta , \tag{2.126}$$

where Φ_s describes the electrostatic potential of an electric dipole located at the sphere center. To show this, we can rewrite Φ_s by introducing the *dipole moment* $\mathbf{p}$ as

$$\Phi_s(r, \theta) = \frac{\mathbf{p} \cdot \mathbf{r}}{4\pi \varepsilon_0 \varepsilon_2 r^3} , \tag{2.127}$$

where

$$\mathbf{p} = 3\varepsilon_0 \varepsilon_2 V \frac{\varepsilon_1 - \varepsilon_2}{\varepsilon_1 + 2\varepsilon_2} \mathbf{E}_0 , \tag{2.128}$$

Fig. 2.28 Imaginary part of the normalized polarizability α_{pol} of a sub-wavelength EG sphere free-standing in vacuum as given by (2.129) with respect to the dimensionless frequency ω/ω_p, when $\gamma = 0.01\omega_p$

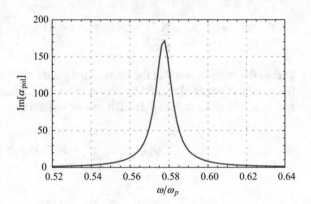

and $V = 4\pi a^3/3$ is the volume of the sphere. This means that the applied electric field induces a dipole moment inside the sphere of magnitude proportional to $|\mathbf{E}_0|$.[14] By introducing the normalized polarizability α_{pol}, defined via $\mathbf{p} = \varepsilon_0\varepsilon_2 V \alpha_{pol}\mathbf{E}_0$, we find

$$\alpha_{pol} = 3\frac{\varepsilon_1 - \varepsilon_2}{\varepsilon_1 + 2\varepsilon_2} . \tag{2.129}$$

Equation (2.129) is the (complex) normalized polarizability of a small sphere of sub-wavelength diameter in the electrostatic approximation. Here, $\text{Re}\,[\varepsilon_1] = -2\varepsilon_2$ is the resonance condition (also called the Fröhlich condition [38]) for the oscillation of dipole SP of an EG sphere that is excited by a quasi-static electric field. Figure 2.28 shows the imaginary part of the normalized polarizability α_{pol} of a small EG sphere in vacuum, with respect to the angular frequency ω (in ω_p units) when ε_1 varying as (2.80) and $\gamma = 0.01\omega_p$.

Also, in quasi-static approximation, the cross sections for scattering, absorption, and extinction of an EG sphere may be expressed as [38]

$$Q_{sca} = \frac{8}{27}\varepsilon_2^2 \left(\frac{a\omega}{c}\right)^4 |\alpha_{pol}|^2 , \tag{2.130a}$$

$$Q_{abs} = \frac{4}{3}\varepsilon_2^{1/2} \left(\frac{a\omega}{c}\right) \text{Im}\,[\alpha_{pol}] , \tag{2.130b}$$

$$Q_{ext} = Q_{sca} + Q_{abs} , \tag{2.130c}$$

in units of the geometric cross section πa^2. Also Im denotes imaginary part and scattering term is small compared with absorption, i.e., we may consider $Q_{ext} = Q_{abs}$. Figure 2.29 shows the extinction cross section of a small sphere in the quasi-

---

[14]The EG sphere has a uniform polarization $\mathbf{P}$ along the z-axis (along the external field); it is equal to $\mathbf{p}/V = \mathbf{p}\,(4\pi a^3/3)^{-1}$.

Fig. 2.29 Extinction cross section (in units of the geometric cross section) of a sub-wavelength EG sphere free-standing in vacuum with respect to the dimensionless frequency ω/ω_p, when $\gamma = 0.01\omega_p$

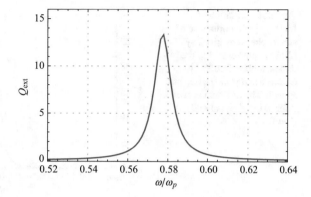

static approximation using (2.130c), when $\gamma = 0.01\omega_p$ and $\varepsilon_2 = 1$. Furthermore, in the absence of the damping effects, the stored electrostatic energy within this polarized EG sphere is

$$U = -\frac{1}{2V} \int \mathbf{p} \cdot \mathbf{E}_0 \, dv = -2\pi a^3 \varepsilon_0 \varepsilon_2 \frac{\varepsilon_1 - \varepsilon_2}{\varepsilon_1 + 2\varepsilon_2} E_0^2 \, . \tag{2.131}$$

This equation indicates the resonance peaks may be appeared in the stored electrostatic energy spectrum.

2.7.4 Effective Permittivity of a Composite of Small Electron Gas Spheres

From the results of previous section, the effective dielectric function ε_{eff} of a composite of small sphere embedded in a host matrix with relative dielectric constant ε_2 can be calculated from a generalized *Clausius–Mossotti* formula [59], as

$$\frac{\varepsilon_{eff} - \varepsilon_2}{\varepsilon_{eff} + 2\varepsilon_2} = \frac{1}{3} f \alpha_{pol} \, , \tag{2.132}$$

where f is the volume fraction of the embedded spheres and α_{pol} is the normalized dipole polarizability of a sphere, as can be seen in (2.129). By substituting (2.129) into (2.132), we find

$$\frac{\varepsilon_{eff} - \varepsilon_2}{\varepsilon_{eff} + 2\varepsilon_2} = f \frac{\varepsilon_1 - \varepsilon_2}{\varepsilon_1 + 2\varepsilon_2} \, . \tag{2.133}$$

This equation is the (complex) effective dielectric function of a composite of small spheres, in the dipolar or electrostatic theory. In Fig. 2.30, we show imaginary part

Fig. 2.30 Variation of $\text{Im}\left[\varepsilon_{eff}\right]$ of a composite of small spheres as given by (2.133) with respect to the dimensionless frequency ω/ω_p, when $\gamma = 0.01\omega_p$ and $\varepsilon_2 = 1$. The different curves refer to (**a**) $f = 0.1$ and (**b**) $f = 0.15$

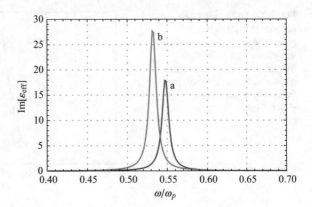

of the effective dielectric function of the system, as a function of dimensionless frequency ω/ω_p for two different values of volume fraction f, when $\gamma = 0.01\omega_p$ and $\varepsilon_2 = 1$. Again, one can see that as parameter f increases, the interaction among the spheres leads to the resonance frequency at lower values.

2.7.5 An Alternative Derivation of Effective Permittivity of a Composite of Small Electron Gas Spheres

The formula of effective dielectric function of a composite of small EG spheres (or aligned thin EG cylinders) can be derived directly from an alternative approach. Referring to panel (a) of Fig. 2.31, we consider a composite consisting of a random distribution of identical EG sphere (1) of radius a_1 and volume fraction f embedded in a background medium (2) with relative permittivity, ε_2. In this approach, we consider an EG sphere plus surrounding insulator shell placed in the effective medium. The effective permittivity of the system ε_{eff} can be determined by requiring that this sphere produces no perturbation of the electric field in the surrounding medium [60, 61].

The system to be studied is shown in panel (b) of Fig. 2.31. The EG sphere, with radius a_1, is enclosed within an insulator sphere of radius a_2, the volume between a_1 and a_2 represents the average envelope per particle and $f = a_1^3/a_2^3$. In this regard, we assume the new system [a coated small spherical EG in panel (b) of Fig. 2.31] is illuminated by a uniform, quasi-static electric field $\mathbf{E} = E_0\mathbf{e}_z$ and the field lines are parallel to the z-direction at sufficient distance from the system. In this case, we have

$$\Phi(r,\theta) = \cos\theta \begin{cases} A_1 r, & r \leq a_1, \\ A_2 r + A_3 r^{-2}, & a_1 \leq r \leq a_2, \\ -E_0 r + A_4 r^{-2}, & r \geq a_2. \end{cases} \tag{2.134}$$

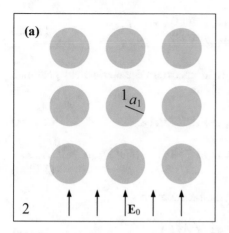
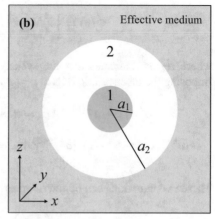

Fig. 2.31 (**a**) EG spheres in an insulator host. (**b**) An EG sphere plus surrounding insulator shell embedded in the effective medium

The coefficients $A_1 - A_4$, in the above equation can be determined from the matching appropriate BCs at the surfaces $r = a_1$ and $r = a_2$. By matching the BCs on the interface of the background and effective medium, i.e., $r = a_2$, we obtain

$$\begin{pmatrix} E_0 \\ A_4 \end{pmatrix} = \frac{a_2^3}{3\varepsilon_{eff}} \begin{pmatrix} -\left[\varepsilon_2 + 2\varepsilon_{eff}\right]a_2^{-3} & 2\left[\varepsilon_2 - \varepsilon_{eff}\right]a_2^{-6} \\ -\left[\varepsilon_2 - \varepsilon_{eff}\right] & \left[2\varepsilon_2 + \varepsilon_{eff}\right]a_2^{-3} \end{pmatrix} \begin{pmatrix} A_2 \\ A_3 \end{pmatrix}. \tag{2.135}$$

By using the condition $A_4 = 0$, for the present system, from (2.135) we obtain

$$\frac{\varepsilon_{eff} - \varepsilon_2}{\varepsilon_{eff} + 2\varepsilon_2} = -\frac{1}{a_2^3}\frac{A_3}{A_2}. \tag{2.136}$$

To determine the unknown coefficients A_2 and A_3, the BCs, i.e., (2.109a) and (2.109b), must be enforced at $r = a_1$. These BCs yield

$$\frac{A_3}{A_2} = -a_1^3 \frac{\varepsilon_1 - \varepsilon_2}{\varepsilon_1 + 2\varepsilon_2}. \tag{2.137}$$

Finally, by substituting (2.137) into (2.136) we obtain (2.133).

2.7.6   Multipolar Response

In order to study the multiple response of an EG sphere, we may consider a general solution of (2.107) as

$$\tilde{\Phi}(r) = \begin{cases} A_{1+}r^{\ell} + A_{1-}r^{-(\ell+1)} \,, & r \leq a \,, \\ A_{2+}r^{\ell} + A_{2-}r^{-(\ell+1)} \,, & r \geq a \,, \end{cases} \tag{2.138}$$

where $A_{2+}r^{\ell}$ indicates that the system is in an external electric field potential. Imposing the electrostatic BCs at $r = a$ leads to

$$a^{\ell}A_{1+} + a^{-(\ell+1)}A_{1-} = a^{\ell}A_{2+} + a^{-(\ell+1)}A_{2-} \,, \tag{2.139a}$$

$$\ell\varepsilon_1 a^{\ell-1}A_{1+} - (\ell+1)\varepsilon_1 a^{-(\ell+2)}A_{1-} = \ell\varepsilon_2 a^{\ell-1}A_{2+} - (\ell+1)\varepsilon_2 a^{-(\ell+2)}A_{2-} \,. \tag{2.139b}$$

These two equations can be rewritten as a matrix equation

$$\mathbf{M}_1 \begin{pmatrix} A_{1+} \\ A_{1-} \end{pmatrix} = \mathbf{M}_2 \begin{pmatrix} A_{2+} \\ A_{2-} \end{pmatrix} \,, \tag{2.140}$$

where

$$\mathbf{M}_j = \begin{pmatrix} a^{\ell} & a^{-(\ell+1)} \\ \ell\varepsilon_j a^{\ell-1} & -(\ell+1)\varepsilon_j a^{-(\ell+2)} \end{pmatrix} \,, \tag{2.141}$$

with $j = 1$ and 2. Then from (2.140), we find

$$\begin{pmatrix} A_{1+} \\ A_{1-} \end{pmatrix} = \mathbf{M} \begin{pmatrix} A_{2+} \\ A_{2-} \end{pmatrix} \,, \tag{2.142}$$

where

$$\mathbf{M} = \mathbf{M}_1^{-1}\mathbf{M}_2 = \begin{pmatrix} M_{11} & M_{12} \\ M_{21} & M_{22} \end{pmatrix} \,, \tag{2.143}$$

and $\mathbf{M}_1^{-1}$ is the inverse matrix of $\mathbf{M}_1$, and

$$M_{11} = \frac{(\ell+1)\varepsilon_1 + \ell\varepsilon_2}{(2\ell+1)\varepsilon_1} \,,$$

$$M_{22} = \frac{\ell\varepsilon_1 + (\ell+1)\varepsilon_2}{(2\ell+1)\varepsilon_1} \,,$$

$$M_{12} = \frac{\ell+1}{2\ell+1} \frac{\varepsilon_1 - \varepsilon_2}{\varepsilon_1} a^{-(2\ell+1)} \,,$$

$$M_{21} = \frac{\ell}{2\ell+1} \frac{\varepsilon_1 - \varepsilon_2}{\varepsilon_1} a^{2\ell+1} \,.$$

Now, we may introduce the normalized multipole polarizability $\alpha_{pol} = -3A_{2-}/a^3 A_{2+}$ for the system, when $A_{1-} = 0$. This yields

$$\alpha_{pol} = \frac{3}{a^3} \frac{M_{21}}{M_{22}} , \tag{2.144}$$

and from (2.143) we obtain

$$\alpha_{pol} = 3\ell a^{2(\ell-1)} \frac{\varepsilon_1 - \varepsilon_2}{\ell \varepsilon_1 + (\ell+1)\varepsilon_2} . \tag{2.145}$$

Equation (2.145) shows the (complex) normalized multipole polarizability of an EG sphere, where ℓ is the multipole order with $\ell = 1$ denoting the dipole mode, $\ell = 2$ the quadrupole mode, and so on. Here, $\ell \varepsilon_1 + (\ell + 1)\varepsilon_2 = 0$ shows the dispersion relation of undamped SP modes, when $\gamma = 0$.

2.8 Plasmonic Properties of Spherical Electron Gas Shells

2.8.1 Surface Plasmon Modes

Consider now a spherical EG shell with internal and external radii denoted by a_1 and a_2, respectively. The relative dielectric constants of the core in the region $0 < r < a_1$ and embedding medium in the region $r > a_2$ are ε_1 and ε_3, respectively, while region $a_1 < r < a_2$ is filled by an EG characterized by the local dielectric function as (2.7). This BVP is similar to that described in Sect. 2.7.1, but slightly more complicated. For this geometry, the solution of (2.104) has the form

$$\tilde{\Phi}(r) = \begin{cases} A_1 r^\ell , & r \leq a_1 , \\ A_2 r^\ell + A_3 r^{-(\ell+1)} , & a_1 \leq r \leq a_2 , \\ A_4 r^{-(\ell+1)} , & r \geq a_2 . \end{cases} \tag{2.146}$$

The application of the electrostatic BCs at $r = a_1$ and a_2 gives

$$A_1 = a_1^{-(2\ell+1)} \frac{\ell \varepsilon_2 + (\ell+1)\varepsilon_3}{\ell(\varepsilon_2 - \varepsilon_1)} A_4 , \tag{2.147a}$$

$$A_2 = a_2^{-(2\ell+1)} \frac{\ell+1}{2\ell+1} \frac{\varepsilon_2 - \varepsilon_3}{\varepsilon_2} A_4 , \tag{2.147b}$$

$$A_3 = \frac{\ell \varepsilon_2 + (\ell+1)\varepsilon_3}{(2\ell+1)\varepsilon_2} A_4 . \tag{2.147c}$$

Furthermore, we obtain the eigenvalue equation for the resonant frequencies of the SP modes of a spherical EG shell free-standing in vacuum, with the following two branches of dispersion relations for each ℓ, as

Fig. 2.32 Dispersion curves of SP modes of a spherical EG shell in vacuum versus the variable a_1/a_2, as given by (2.148) for $\ell = 1, 2, 3,$ and 4

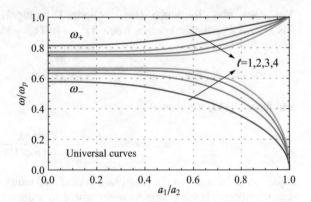

$$\omega_\pm^2 = \frac{\omega_p^2}{2} \left[1 \pm \frac{1}{2\ell+1} \sqrt{1 + 4\ell(\ell+1)\left(\frac{a_1}{a_2}\right)^{2\ell+1}} \right] . \tag{2.148}$$

In Fig. 2.32, we plot dimensionless frequency ω/ω_p, versus variable a_1/a_2 for several angular modes $\ell = 1, 2,$ and 3. Also, by using (2.146)–(2.147c), the normalized profiles of electrostatic potentials are presented in Fig. 2.33 for $\ell = 1$ and 2, when $a_2/a_1 = 1.5$. In this example, we have taken the constant A_4 as independent and equal to one. Furthermore, due to the azimuthal symmetry of the problem, for the surface charge σ_{ind} induced on the two surfaces of the system, we have

$$\sigma_{ind}(\theta)\big|_{r=a_1} = \varepsilon_0 \left[\frac{\partial \tilde{\Phi}_1}{\partial r}\bigg|_{r=a_1} - \frac{\partial \tilde{\Phi}_2}{\partial r}\bigg|_{r=a_1} \right] P_\ell(\cos\theta) , \tag{2.149a}$$

$$\sigma_{ind}(\theta)\big|_{r=a_2} = \varepsilon_0 \left[\frac{\partial \tilde{\Phi}_2}{\partial r}\bigg|_{r=a_2} - \frac{\partial \tilde{\Phi}_3}{\partial r}\bigg|_{r=a_2} \right] P_\ell(\cos\theta) . \tag{2.149b}$$

Therefore, by substituting (2.146) into above equations and using (2.147a)–(2.147c), we arrive to

$$\sigma_{ind}(\theta) = \varepsilon_0 a_1^{-(\ell+2)} A_4 P_\ell(\cos\theta) \begin{cases} \Gamma_1 , & r = a_1 , \\ \Gamma_2 , & r = a_2 , \end{cases} \tag{2.150}$$

where

$$\Gamma_1 = \frac{\ell\varepsilon_2 + (\ell+1)\varepsilon_3}{\varepsilon_2 - \varepsilon_1} + \frac{\ell+1}{2\ell+1} \left(\frac{\varepsilon_3}{\varepsilon_2} + \ell \frac{\varepsilon_2+\varepsilon_3}{\varepsilon_2} \left[1 - \frac{\varepsilon_2-\varepsilon_3}{\varepsilon_2+\varepsilon_3} \left(\frac{a_1}{a_2}\right)^{2\ell+1} \right] \right) , \tag{2.151}$$

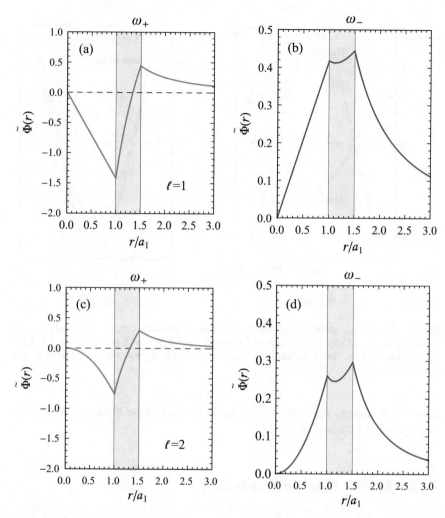

Fig. 2.33 Normalized profile $\tilde{\Phi}(r)$ of SP modes of a spherical EG shell free-standing in vacuum for $\ell = 1$, and 2, when $a_2/a_1 = 1.5$. Panels (**a**) and (**c**) show the results for anti-symmetric (odd) modes and panels (**b**) and (**d**) show the results for the symmetric (even) modes

$$\Gamma_2 = (\ell + 1) \left(\frac{a_1}{a_2} \right)^{\ell+2} \left(1 - \frac{\varepsilon_3}{\varepsilon_2} \right) . \tag{2.152}$$

The normalized induced surface charge in units of $\varepsilon_0 a_1^{-(\ell+1)} A_4$ is shown in Fig. 2.34 as a function of θ for $\ell = 1$, when $a_2/a_1 = 1.5$. It is clear that the induced surface charge on the outer surface is positive for $0 \leq \theta \leq \pi/2$ and negative for $\pi/2 \leq \theta \leq \pi$. Also, surface charge distribution of the system shows that for ω_- (ω_+) mode, the induced surface charge of the inner surface has the same (opposite) sign of the induced surface charge on the outer surface. Therefore ω_- mode is a symmetric

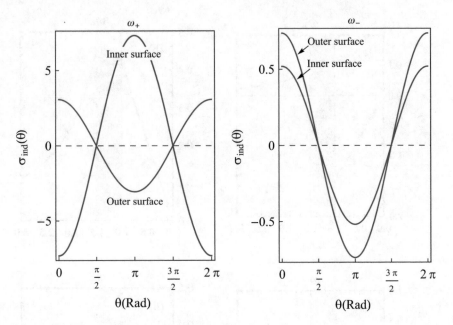

Fig. 2.34 Normalized surface charge distribution induced on the two surfaces of a spherical EG shell, as given by (2.150) with respect to the function θ, when $\ell = 1$ and $a_2/a_1 = 1.5$. Panel (**a**) corresponds to ω_- that is a symmetric mode and panel (**b**) corresponds to ω_+ that is an anti-symmetric mode

mode, while the ω_+ mode is anti-symmetric mode, that are quite similar with the corresponding results of EG tubes.

2.8.2 Surface Plasmon Resonances

We consider a small spherical EG shell with internal and external radii denoted by a_1 and a_2, respectively. The relative dielectric constants of media from inside to outside are ε_1, ε_2, and ε_3, respectively (see Fig. 2.35). The plasmon resonance of the present system in the electrostatic approximation can be found in much the same way as for the coaxial cylinders. In this case, we have

$$\Phi(r, \theta) = \cos\theta \begin{cases} A_1 r, & r \leq a_1, \\ A_2 r + A_3 r^{-2}, & a_1 \leq r \leq a_2, \\ -E_0 r + A_4 r^{-2}, & r \geq a_2. \end{cases} \tag{2.153}$$

We note that A_4 and A_1–A_3 represent the yet unknown scattering and transmission coefficients, respectively. To determine the unknown coefficients $A_1 - A_4$, the BCs must be enforced on surfaces of the system. The BCs yield

Fig. 2.35 Cross section of a small spherical EG shell placed into an electrostatic field. The internal and external radii denoted by a_1 and a_2, respectively

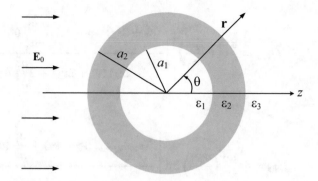

$$A_1 = -\frac{9\varepsilon_2\varepsilon_3}{(\varepsilon_1 + 2\varepsilon_2)(\varepsilon_2 + 2\varepsilon_3) + 2\dfrac{a_1^3}{a_2^3}(\varepsilon_1 - \varepsilon_2)(\varepsilon_2 - \varepsilon_3)}E_0 \,, \qquad (2.154a)$$

$$A_2 = -\frac{3\varepsilon_3(\varepsilon_1 + 2\varepsilon_2)}{(\varepsilon_1 + 2\varepsilon_2)(\varepsilon_2 + 2\varepsilon_3) + 2\dfrac{a_1^3}{a_2^3}(\varepsilon_1 - \varepsilon_2)(\varepsilon_2 - \varepsilon_3)}E_0 \,, \qquad (2.154b)$$

$$A_3 = a_1^3\frac{3\varepsilon_3(\varepsilon_1 - \varepsilon_2)}{(\varepsilon_1 + 2\varepsilon_2)(\varepsilon_2 + 2\varepsilon_3) + 2\dfrac{a_1^3}{a_2^3}(\varepsilon_1 - \varepsilon_2)(\varepsilon_2 - \varepsilon_3)}E_0 \,, \qquad (2.154c)$$

$$A_4 = a_2^3\frac{(\varepsilon_1 + 2\varepsilon_2)(\varepsilon_2 - \varepsilon_3) + \dfrac{a_1^3}{a_2^3}(\varepsilon_1 - \varepsilon_2)(2\varepsilon_2 + \varepsilon_3)}{(\varepsilon_1 + 2\varepsilon_2)(\varepsilon_2 + 2\varepsilon_3) + 2\dfrac{a_1^3}{a_2^3}(\varepsilon_1 - \varepsilon_2)(\varepsilon_2 - \varepsilon_3)}E_0 \,. \qquad (2.154d)$$

Therefore, the normalized polarizability of the system may be written as

$$\alpha_{pol} = 3\frac{(\varepsilon_1 + 2\varepsilon_2)(\varepsilon_2 - \varepsilon_3) + \dfrac{a_1^3}{a_2^3}(\varepsilon_1 - \varepsilon_2)(2\varepsilon_2 + \varepsilon_3)}{(\varepsilon_1 + 2\varepsilon_2)(\varepsilon_2 + 2\varepsilon_3) + 2\dfrac{a_1^3}{a_2^3}(\varepsilon_1 - \varepsilon_2)(\varepsilon_2 - \varepsilon_3)} \,. \qquad (2.155)$$

Also, the cross sections of scattering and extinction Q_{sca} and Q_{abs} of sub-wavelength spherical EG shells may be expressed as

$$Q_{sca} = \frac{8}{3}\varepsilon_2^2 \left(\frac{a_2\omega}{c}\right)^4 \left|\frac{(\varepsilon_1 + 2\varepsilon_2)(\varepsilon_2 - \varepsilon_3) + \dfrac{a_1^3}{a_2^3}(\varepsilon_1 - \varepsilon_2)(2\varepsilon_2 + \varepsilon_3)}{(\varepsilon_1 + 2\varepsilon_2)(\varepsilon_2 + 2\varepsilon_3) + 2\dfrac{a_1^3}{a_2^3}(\varepsilon_1 - \varepsilon_2)(\varepsilon_2 - \varepsilon_3)}\right|^2 ,$$

(2.156a)

$$Q_{abs} = 4\varepsilon_2^{1/2}\left(\frac{a_2\omega}{c}\right)\mathrm{Im}\left[\frac{(\varepsilon_1 + 2\varepsilon_2)(\varepsilon_2 - \varepsilon_3) + \dfrac{a_1^3}{a_2^3}(\varepsilon_1 - \varepsilon_2)(2\varepsilon_2 + \varepsilon_3)}{(\varepsilon_1 + 2\varepsilon_2)(\varepsilon_2 + 2\varepsilon_3) + 2\dfrac{a_1^3}{a_2^3}(\varepsilon_1 - \varepsilon_2)(\varepsilon_2 - \varepsilon_3)}\right] .$$

(2.156b)

Furthermore, the effective dielectric function ε_{eff} of a composite of small EG shells embedded in a host matrix with relative dielectric constant ε_3 may be written as

$$\frac{\varepsilon_{eff} - \varepsilon_3}{\varepsilon_{eff} + \varepsilon_3} = f\frac{(\varepsilon_1 + 2\varepsilon_2)(\varepsilon_2 - \varepsilon_3) + \dfrac{a_1^3}{a_2^3}(\varepsilon_1 - \varepsilon_2)(2\varepsilon_2 + \varepsilon_3)}{(\varepsilon_1 + 2\varepsilon_2)(\varepsilon_2 + 2\varepsilon_3) + 2\dfrac{a_1^3}{a_2^3}(\varepsilon_1 - \varepsilon_2)(\varepsilon_2 - \varepsilon_3)} ,$$

(2.157)

where f is the volume fraction of the embedded shells.

2.9   Surface Plasmon Modes of Sphere-Sphere and Sphere-Plane Electron Gas Systems

The physical behavior of SP modes of two spherical nanoparticles has been well known for a long time [62–70]; however, some problems remain partially open and attract much interest of today researchers [70]. In this section, we wish to obtain a general dispersion relation for SP frequencies of sphere-sphere and sphere-plane EG systems that exhibits excellent numerical convergence. Let the two EG spheres with radii a_1, a_2 and relative dielectric functions ε_1, ε_2 separated by a small distance δ and immersed in a uniform medium with a relative dielectric function ε_3. In order to study the sphere-sphere and sphere-plane systems, as shown in Fig. 2.36, it is convenient to make use of *bispherical coordinates* $(-\infty < \eta < \infty, 0 \leq \theta < \pi, 0 \leq \psi < 2\pi)$, where (η, θ, ψ) are related to the rectangular coordinate system (x, y, z) as described by Morse and Feshbach [71]

Fig. 2.36 Geometry of the problem, i.e., two close EG spheres. The parameter c relates the Cartesian coordinates (x, z) to the bispherical coordinates (η, θ) when azimuthal symmetry about z is assumed. Also a_1, a_2, and δ uniquely define c, η_1, and η_2

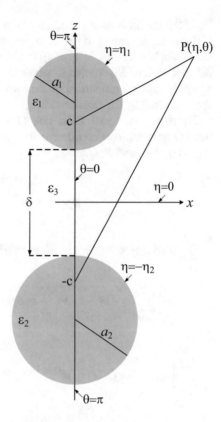

$$x = \frac{a \sin \theta \cos \psi}{\cosh \eta - \cos \theta} \, , \tag{2.158a}$$

$$y = \frac{a \sin \theta \sin \psi}{\cosh \eta - \cos \theta} \, , \tag{2.158b}$$

$$z = \frac{a \sinh \eta}{\cosh \eta - \cos \theta} \, . \tag{2.158c}$$

For the case of azimuthal symmetry about the z, the electrostatic potential Φ can be written in the form

$$\Phi_i(\eta, \theta) = (\cosh \eta - \cos \theta)^{1/2} \sum_{\ell=0}^{+\infty} \left\{ A_{i\ell} e^{-\left(\ell + \frac{1}{2}\right)(\eta - \eta_1)} + B_{i\ell} e^{\left(\ell + \frac{1}{2}\right)(\eta - \eta_1)} \right\} P_\ell(\cos \theta) \, . \tag{2.159}$$

Here the coefficients $A_{i\ell}$ and $B_{i\ell}$ take on different values in the three different regions corresponding to $i = 1$, in the inside the sphere 1 ($\eta > \eta_1$); $i = 3$, between the two spheres ($-\eta_2 < \eta < \eta_1$); and $i = 2$, inside the sphere 2 ($\eta < -\eta_2$). In the first region, $B_{1\ell}$ vanishes, while in the second region $A_{2\ell}$ must vanish. Also, again the optical properties of the two EG spheres and background medium are approximated by the Drude model.

By using the usual BCs, i.e., (1.94) and (1.95), at the surface of each of the EG spheres, one can derive the following equations for all the coefficients; $A_{1\ell}$, $A_{3\ell}$, $B_{3\ell}$, and $B_{2\ell}$, as

$$A_{1\ell} = A_{3\ell} + B_{3\ell} , \qquad (2.160)$$

$$B_{2\ell} = A_{3\ell} e^{(2\ell+1)(\eta_1+\eta_2)} + B_{3\ell} , \qquad (2.161)$$

$$\sum_{\ell=0}^{+\infty} \{\varepsilon_1 A_{1\ell} - \varepsilon_3 A_{3\ell} - \varepsilon_3 \Gamma(\ell, \eta_1) B_{3\ell}\} [\sinh \eta_1 - (2\ell + 1) \cosh \eta_1] P_\ell(\cos\theta)$$

$$+ \sum_{\ell=0}^{+\infty} (\varepsilon_1 A_{1\ell} - \varepsilon_3 A_{3\ell} + \varepsilon_3 B_{3\ell}) (2\ell + 1) \cos\theta\, P_\ell(\cos\theta) = 0 , \qquad (2.162)$$

$$\sum_{\ell=0}^{+\infty} \left\{\varepsilon_2 B_{2\ell} - \varepsilon_3 \Gamma(\ell, \eta_2) e^{(2\ell+1)(\eta_1+\eta_2)} A_{3\ell} - \varepsilon_3 B_{3\ell}\right\}$$

$$\times e^{-\left(\ell+\frac{1}{2}\right)(\eta_1+\eta_2)} [\sinh \eta_2 - (2\ell + 1) \cosh \eta_2] P_\ell(\cos\theta)$$

$$+ \sum_{\ell=0}^{+\infty} \left\{\varepsilon_2 B_{2\ell} + \varepsilon_3 A_{3\ell} e^{(2\ell+1)(\eta_1+\eta_2)} - \varepsilon_3 B_{3\ell}\right\}$$

$$\times e^{-\left(\ell+\frac{1}{2}\right)(\eta_1+\eta_2)} (2\ell + 1) \cos\theta\, P_\ell(\cos\theta) = 0 , \qquad (2.163)$$

where

$$\Gamma(\ell, \eta_i) = \frac{\sinh \eta_i + (2\ell + 1) \cosh \eta_i}{\sinh \eta_i - (2\ell + 1) \cosh \eta_i} . \qquad (2.164)$$

To avoid secular terms in (2.162) and (2.163), we eliminate contributions proportional to $\cos\theta\, P_\ell(\cos\theta)$ on the second terms of them. The conditions for the elimination of these secular terms are

$$\varepsilon_1 A_{1\ell} - \varepsilon_3 A_{3\ell} + \varepsilon_3 B_{3\ell} = 0 , \qquad (2.165)$$

$$\varepsilon_2 B_{2\ell} + \varepsilon_3 A_{3\ell} e^{(2\ell+1)(\eta_1+\eta_2)} - \varepsilon_3 B_{3\ell} = 0 . \qquad (2.166)$$

Substituting (2.165) and (2.166) into (2.162) and (2.163), we obtain the following expressions for the coefficients $A_{1\ell}$, $A_{3\ell}$, $B_{3\ell}$, and $B_{2\ell}$, as

$$\varepsilon_1 A_{1\ell} - \varepsilon_3 A_{3\ell} - \varepsilon_3 \Gamma(\ell, \eta_1) B_{3\ell} = 0 \,, \tag{2.167}$$

$$\varepsilon_2 B_{2\ell} - \varepsilon_3 \Gamma(\ell, \eta_2) e^{(2\ell+1)(\eta_1+\eta_2)} A_{3\ell} - \varepsilon_3 B_{3\ell} = 0 \,. \tag{2.168}$$

By combining (2.165) and (2.166) with (2.160) and (2.161), we can get the dispersion relation obtained by Batson in [64]. This means that there is a major point in the previous work [64], where (2.165) and (2.166) are not satisfied. Here, we improve the theory by returning to the full set of equations. After doing some algebra, we find

$$\chi_1(\ell, \eta_1)\chi_2(\ell, \eta_2) = \exp[(2\ell + 1)(\eta_1 + \eta_2)] \,, \tag{2.169}$$

where

$$\chi_i(\ell, \eta_i) = \frac{\varepsilon_3 - \varepsilon_i}{\varepsilon_3 \Gamma(\ell, \eta_i) - \varepsilon_i} \,. \tag{2.170}$$

Equation (2.169) shows a general result for sphere-sphere and sphere-plane systems. For each order ℓ there are two branches of SP modes similar to that of the resonant frequencies of SP modes of parallel EG cylinder systems. Let us note that (2.169) reproduces equation (11) of work by Rendell et al. [72], when $\varepsilon_1 = -\infty$ and $\eta_2 = 0$.

Note that at large $\eta_1 + \eta_2$ the upper and lower branches tend to merge to the frequencies given by $\omega_{\pm} = \sqrt{\ell/(2\ell + 1)}\omega_p$, as shown in panel (a) of Fig. 2.37, for the case $\eta_2 = 2\eta_1$. Also, panel (b) for a sphere-plane ($\eta_2 = 0$) system indicates that in the limit of large η_1 the all upper modes approach the SP frequency of a flat EG-vacuum interface, i.e., $\omega_+ = \omega_p/\sqrt{2}$ and each lower mode with order ℓ approaches the SP frequency of an isolated EG sphere, i.e., $\omega_- = \sqrt{\ell/(2\ell + 1)}\omega_p$. Thus the expression that has been obtained in (2.169) correctly reproduces expected results in the limit of large $\eta_1 + \eta_2$, in addition to explaining the behavior in the low separation limit as $\eta_1 + \eta_2 \to 0$.

Fig. 2.37 Panel (**a**) represents dispersion curves of SP modes of a sphere-sphere system when $\eta_2 = 2\eta_1$, for $\ell = 1, 2, 3,$ and 4. Panel (**b**) shows the results for a sphere-plane ($\eta_2 = 0$) system. Here $\omega_{p2} = \omega_p = \omega_{p1}$ and $\varepsilon_3 = 1$

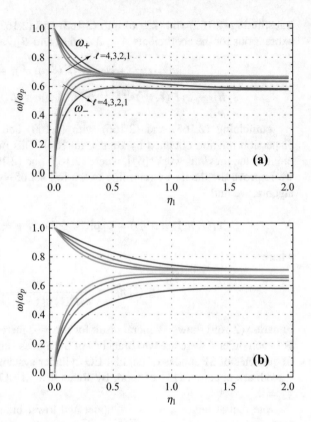

References

1. H.J. Lee, Electrostatic surface waves in a magnetized two-fluid plasma. Plasma Phys. Control. Fusion **37**, 755–762 (1995)
2. M. Marklund, G. Brodin, L. Stenflo, C.S. Liu, New quantum limits in plasmonic devices. Europhys. Lett. **84**, 17006 (2008)
3. A. Moradi, Bohm potential and inequality of group and energy transport velocities of plasmonic waves on metal-insulator waveguides. Phys. Plasmas **24**, 072104 (2017)
4. J.D. Jackson, *Classical Electrodynamics* (Wiley, New York, 1999)
5. R.H. Ritchie, Plasma losses by fast electrons in thin films. Phys. Rev. **106**, 874–881 (1957)
6. J. Nkoma, R. Loudon, D.R. Tilley, Elementary properties of surface polaritons. J. Phys. C **7**, 3547–3559 (1974)
7. S.S. Gupta, N.C. Srivastava, Power flow and energy distribution of magnetostatic bulk waves-in dielectric-layered structure. J. Appl. Phys. **50**, 6697–6699 (1979)
8. M.S. Kushwaha, Interface polaritons in layered structures with metallized surfaces. J. Appl. Phys. **59**, 2136–2141 (1986)
9. K.L. Ngai, E.N. Economou, M.H. Cohen, Theory of inelastic electron-surface-plasmon interactions in metal-semiconductor tunnel junctions. Phys. Rev. Lett. **22**, 1375–1378 (1969)
10. B. Huttner, S.M. Barnett, Quantization of the electromagnetic field in dielectrics. Phys. Rev. A **46**, 4306–4322 (1992)
11. M. Bertolotti, C. Sibilia, A. Guzman, *Evanescent Waves in Optics* (Springer, Basel, 2017)

12. J.J. Brion, R.F. Wallis, Theory of pseudosurface polaritons in semiconductors in magnetic fields. Phys. Rev. B **10**, 3140–3143 (1974)
13. R.F. Wallis, J.J. Brion, E. Burstein, A. Hartstein, Theory of surface polaritons in anisotropic dielectric media with application to surface magnetoplasmons in semiconductors. Phys. Rev. B **9**, 3424–37 (1974)
14. B. Hu, Y. Zhang, Q. Wang, Surface magneto plasmons and their applications in the infrared frequencies. Nanophotonics **4**, 383 (2015)
15. A. Moradi, Comment on: propagation of surface waves on a semi-bounded quantum magnetized collisional plasma. Phys. Plasmas **23**, 044701 (2016)
16. A. Moradi, Comment on: propagation of a TE surface mode in a relativistic electron beam-quantum plasma system. Phys. Lett. A **380**, 2580–2581 (2016)
17. K.W. Chiu, J.J. Quinn, Magnetoplasma surface waves in metals. Phys. Rev. B **5**, 4707–4709 (1972)
18. J.J. Brion, R.F. Wallis, A. Hartstein, E. Burstein, Theory of surface magnetoplasmons in semiconductors. Phys. Rev. Lett. **28**, 1455–1458 (1972)
19. A. Moradi, Electrostatic surface waves on a magnetized quantum plasma half-space. Phys. Plasmas **23**, 034501 (2016)
20. A. Moradi, Comment on: surface electromagnetic wave equations in a warm magnetized quantum plasma. Phys. Plasmas **23**, 074701 (2016)
21. A.F. Alexandrov, L.S. Bogdankevich, A.A. Rukhadze, *Principles of Plasma Electrodynamics* (Springer, Berlin, 1984)
22. G.C. Aers, A.D. Boardman, E.D. Issac, High-frequency electrostatic surface waves on a warm gaseous magnetoplasma half-space. IEEE Trans. Plasma Sci. **PS-5**, 123–130 (1977)
23. A. Moradi, Energy density and energy flow of magnetoplasmonic waves on graphene. Solid State Commun. **253**, 63 (2017)
24. A. Moradi, Energy density and energy flow of surface waves in a strongly magnetized graphene. J. Appl. Phys **123**, 04310 (2018)
25. E. Prodan, C. Radloff, N.J. Halas, P. Nordlander, A hybridization model for the plasmon response of complex nanostructures. Science **302**, 419–422 (2003)
26. E.N. Economou, Surface plasmons in thin films. Phys. Rev. **182**, 539–554 (1969)
27. J.J. Burke, G.I. Stegeman, T. Tamir, Surface-polariton-like waves guided by thin, lossy metal films. Phys. Rev. B **33**, 5186–5201 (1986)
28. S.A. Maiera, H.A. Atwater, Plasmonics: localization and guiding of electromagnetic energy in metal/dielectric structures. J. Appl. Phys. **98**, 011101 (2005)
29. B.E. Sernelius, *Surface Modes in Physics* (Wiley-VCH, Berlin, 2001)
30. A. Moradi, Plasmon hybridization in metallic nanotubes. J. Phys. Chem. Solids **69**, 2936–2938 (2008)
31. S.P. Apell, P.M. Echenique, R.H. Ritchie, Sum rules for surface plasmon frequencies, Ultramicroscopy **65**, 53–60 (1995)
32. N. Daneshfar, K. Bazyari, Optical and spectral tunability of multilayer spherical and cylindrical nanoshells. Appl. Phys. A Mater. Sci. Process. **116**, 611–620 (2014)
33. N. Daneshfar, H. Foroughi, Optical bistability in plasmonic nanoparticles: effect of size, shape and embedding medium. Phys. E **83**, 268–274 (2016)
34. L. Novotny, B. Hecht, *Principles of Nano-Optics* (Cambridge University Press, Cambridge, 2006)
35. J. Venermo, A. Sihvola, Dielectric polarizability of circular cylinder. J. Electrostat. **63**, 101–117 (2005)
36. J. Zhu, Surface plasmon resonance from bimetallic interface in Au-Ag core-shell structure nanowires. Nanoscale. Res. Lett. **4**, 977–981 (2009)
37. S.A. Maier, *Plasmonics: Fundamentals and Applications* (Springer, Berlin, 2007)
38. C.F. Bohren, D.R. Huffman, *Absorption and Scattering of Light by Small Particles* (Wiley, New York, 1983)
39. A. Moradi, Extinction properties of metallic nanowires: quantum diffraction and retardation effects. Phys. Lett. A **379**, 2379–2383 (2015)

40. R. Ruppin, Evaluation of extended Maxwell-Garnett theories. Opt. Commun. **182**, 273–279 (2000)
41. Y. Battie, A. Resano-Garcia, N. Chaoui, A. En Naciri, Optical properties of plasmonic nanoparticles distributed in size determined from a modified Maxwell-Garnett-Mie theory. Phys. Status Solidi C **12**, 142–146 (2015)
42. J.C. Maxwell-Garnett, Colours in metal glasses and in metallic films. Philos. Trans. R. Soc. Lond. Ser. B **203**, 385–420 (1904)
43. A. Moradi, Plasmon hybridization in tubular metallic nanostructures. Phys. B **405**, 2466–2469 (2010)
44. A. Moradi, Plasmon hybridization in parallel nano-wire systems. Phys. Plasmas **18**, 064508 (2011)
45. A. Moradi, Plasmon hybridization in coated metallic nanowires. J. Opt. Soc. Am. B **29**, 625–629 (2012)
46. A. Moradi, Plasma wave propagation in a pair of carbon nanotubes. JETP Lett. **88**, 795–798 (2008)
47. M. Schmeits, Surface-plasmon coupling in cylindrical pores. Phys. Rev. B **39**, 7567–7577 (1989)
48. J.P. Kottmann, O.J.F. Martin, Plasmon resonant coupling in metallic nanowires. Opt. Express **8**, 655–663 (2001)
49. H. Ditlbacher, A. Hohenau, D. Wagner, U. Kreibig, M. Rogers, F. Hofer, F.R. Aussenegg, J.R. Krenn, Silver nanowires as surface plasmon resonators. Phys. Rev. Lett. **95**, 257403 (2005)
50. A. Manjavacas, F.J. Garcia de Abajo, Robust plasmon waveguides in strongly interacting nanowire Arrays. Nano Lett. **9**, 1285–1289 (2009)
51. W. Cai, L. Wang, X. Zhang, J. Xu, F.J. Garcia de Abajo, Controllable excitation of gap plasmons by electron beams in metallic nanowire pairs. Phys. Rev. B **82**, 125454 (2010)
52. S. Sun, H.-T Chen, W.-J. Zheng, G.-Y Guo, Dispersion relation, propagation length and mode conversion of surface plasmon polaritons in silver double-nanowire systems. Opt. Express **21**, 14591–14605 (2013)
53. K. Andersen, K.L. Jensen, N.A. Mortensen, K.S. Thygesen, Visualizing hybridized quantum plasmons in coupled nanowires: from classical to tunneling regime. Phys. Rev. B **87**, 235433 (2013)
54. T.V. Teperik, P. Nordlander, J. Aizpurua, A.G. Borisov, Quantum effects and nonlocality in strongly coupled plasmonic nanowire dimers. Opt. Express **21**, 27306–27325 (2013)
55. N.A. Mortensen, S. Raza, M. Wubs, T. Sondergaard, S. I. Bozhevolnyi, A generalized non-local optical response theory for plasmonic nanostructures. Nat. Commun. **5**, 3809 (2014)
56. P. Moon, D.E. Spencer, *Field Theory Handbook: Including Coordinate Systems, Differential Equations and Their Solutions* (Springer, New York, 1988)
57. J.L. Gervasoni, N.R. Arista, Plasmon excitations in cylindrical wires by external charged particles. Phys. Rev. B **68**, 235302 (2003)
58. J.L. Gervasoni, S. Segui, N. Arista, Dispersion relation due to plasmon excitation in nanostructures by charged particles. Vacuum **84**, 262–265 (2007)
59. U. Kreibig, M. Vollmer, *Optical Properties of Metal Clusters* (Springer, Berlin, 1995)
60. G.B. Smith, Dielectric constants for mixed media. J. Phys. D: Appl. Phys. **10**, L39–L42 (1977)
61. A. Moradi, Maxwell-Garnett effective medium theory: quantum nonlocal effects. Phys. Plasmas **22**, 042105 (2015)
62. A.A. Lucas, A. Ronveaux, M. Schmeits, F. Delanaye, van der Waals energy between voids in dielectrics. Phys. Rev. B **12**, 5372–5380 (1975)
63. R. Ruppin, Surface modes of two spheres. Phys. Rev. B **26**, 3440–3444 (1982)
64. P.E. Batson, Surface plasmon coupling in clusters of small spheres. Phys. Rev. Lett. **49**, 936–940 (1982)
65. P. Clippe, R. Evrard, A.A. Lucas, Aggregation effect on the infrared absorption spectrum of small ionic crystals. Phys. Rev. B **14**, 1715–1721 (1976)
66. R. Brako, M. Sunjic, V. Sips, Dispersion interaction between small crystals. Solid State Commun. **19**, 161–164 (1975)

67. I. Olivares, R. Rojas, F. Claro, Surface modes of a pair of unequal spheres. Phys. Rev. B **35**, 2453–2455 (1987)
68. M. Schmeits, L. Dambly, Fast-electron scattering by bispherical surface-plasmon modes. Phys. Rev. B **44**, 12706–12712 (1991)
69. J.S. Nkoma, Surface modes of two spheres embedded into a third medium. Surf. Sci. **245**, 207–212 (1991)
70. P. Nordlander, C. Oubre, E. Prodan, K. Li, M.I. Stockman, Plasmon hybridization in nanoparticle dimers. Nano Lett. **4**, 899–903 (2004)
71. P.M. Morse, H. Feshbach, *Methods of Theoretical Physics* (McGraw-Hill, New York, 1953)
72. R.W. Rendell, D.J. Scalapino, B. Muhlschlegel, Role of local plasmon modes in light emission from small-particle tunnel junctions. Phys. Rev. Lett. **41**, 1746–1750 (1978)

Chapter 3
Problems in Electromagnetic Theory

Abstract In this chapter, some electromagnetic boundary-value problems involving bounded electron gases with different geometries are studied. By considering the local Drude model, explicit plasmonic results are given in planar, cylindrical, and spherical geometries. Comparisons are made with the results of Chap. 2. For brevity, in many sections of this chapter the $\exp(-i\omega t)$ time factor is suppressed. Furthermore, all media under consideration are nonmagnetic and attention is only confined to the linear phenomena.

3.1 Total Reflection of a Plane Wave on a Semi-Infinite Electron Gas

Let us consider an incident plane electromagnetic wave coming from below $z < 0$ and propagating in an insulator medium with a relative real permittivity ε_1, reflected by a semi-infinite EG interface located at $z = 0$, as indicated in Fig. 3.1. The EG being characterized by a relative permittivity ε_2, as shown by (2.7). Without loss of generality, we choose a Cartesian coordinate system such that the interface between two media is perpendicular to the z-axis and the plane of incidence[1] is perpendicular to the y-axis, as shown in Fig. 3.1. The incident plane wave has the electric field given by $\mathbf{E} = \mathbf{E}_0 \exp\left[i\left(k_x x + k_z z\right)\right]$ and magnetic field given by $\mathbf{H} = \mathbf{H}_0 \exp\left[i\left(k_x x + k_z z\right)\right]$, where $\mathbf{E}_0$ and $\mathbf{H}_0$ are constant vectors (by assumption, the x and z dependences of the fields are $\exp(ik_x x)$ and $\exp(ik_z z)$, respectively, and $\partial \mathbf{E}/\partial y = 0 = \partial \mathbf{H}/\partial y$). The similar equations hold for reflected and transmitted waves.

Hence, from Maxwell's curl equations, i.e., (1.33c) and (1.33d), for the present system we arrive at the following set of coupled equations as

---

The original version of this chapter was revised. The correction to this chapter is available at https://doi.org/10.1007/978-3-030-43836-4_11

[1]The plane of incidence is the plane determined by the normal to the interface and the direction of the propagation of the electromagnetic wave.

© The Editor(s) (if applicable) and The Author(s), under exclusive license to Springer Nature Switzerland AG 2020, corrected publication 2020
A. Moradi, *Canonical Problems in the Theory of Plasmonics*, Springer Series in Optical Sciences 230, https://doi.org/10.1007/978-3-030-43836-4_3

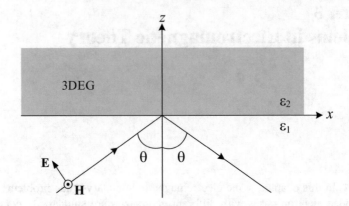

Fig. 3.1 Reflection of a p-polarized plane wave from a flat interface between an insulator and a semi-infinite EG

$$\frac{\partial E_x}{\partial z} - \frac{\partial E_z}{\partial x} = i\omega\mu_0 H_y \,, \qquad (3.1a)$$

$$\frac{\partial H_y}{\partial z} = i\omega\varepsilon_0\varepsilon E_x \,, \qquad (3.1b)$$

$$-\frac{\partial H_y}{\partial x} = i\omega\varepsilon_0\varepsilon E_z \,, \qquad (3.1c)$$

and

$$\frac{\partial H_z}{\partial x} - \frac{\partial H_x}{\partial z} = i\omega\varepsilon_0\varepsilon E_y \,, \qquad (3.1d)$$

$$-\frac{\partial E_y}{\partial z} = i\omega\mu_0 H_x \,, \qquad (3.1e)$$

$$\frac{\partial E_y}{\partial x} = i\omega\mu_0 H_z \,. \qquad (3.1f)$$

The above equations indicate that the calculation of the reflection and transmission coefficients at the flat boundary $z = 0$ can be discussed in two special wave polarizations.

The first case is that in which the vector of the magnetic field, **H**, of the incident wave is parallel to the interface, as shown in Fig. 3.1. This case is known as the *transverse magnetic* (TM) wave (or p-polarized[2] wave), where only the field components E_x, E_z, and H_y are non-zero.

---

[2] p-polarized means parallel polarized.

The second case that is called the *transverse electric* (TE) wave (that is also known as *s*-polarized[3] wave), with only H_x, H_z, and E_y being non-zero, is that in which the vector of the electric field, **E**, of the incident wave is parallel to the plane $z = 0$. We first ponder on the *p*-polarized case, the results for *s*-polarized will then directly follow from the duality principle.

3.1.1   *p*-Polarized Wave

Here, we consider a *p*-polarized wave, which has E_x, E_z, and H_y components, as shown in Fig. 3.1. Note that, the total magnetic field H_y in the insulator space (i.e., $z < 0$) is taken to be in the form of a superposition of incident and reflected waves. Thus

$$H_y = -H_0 \left[e^{ik_z z} + r_p e^{-ik_z z} \right] e^{ik_x x} , \tag{3.2}$$

where r_p is, by definition, the *Fresnel reflection coefficient* for *p*-polarized waves, H_0 is the amplitude of the incident magnetic field H_y, and $k_z^2 = \varepsilon_1 k_0^2 - k_x^2$ ($k_0 = \omega/c$ is the wavenumber of the propagating wave in vacuum). Also, using (3.1b) the total electric field along the *x*-direction in the insulator space has the form

$$E_x = \frac{-k_z}{\omega \varepsilon_0 \varepsilon_1} H_0 \left[e^{ik_z z} - r_p e^{-ik_z z} \right] e^{ik_x x} . \tag{3.3}$$

In the EG, the magnetic field can be written as

$$H_y = -t_p H_0 e^{-\kappa z} e^{ik_x x} , \tag{3.4}$$

where t_p is, by definition, the *Fresnel transmission coefficient* for *p*-polarized waves and $\kappa^2 = k_x^2 - \varepsilon_2 k_0^2$. Again, using (3.1b) the electric field along the *x*-direction in the EG space has the form

$$E_x = \frac{-i\kappa}{\omega \varepsilon_0 \varepsilon_2} t_p H_0 e^{-\kappa z} e^{ik_x x} . \tag{3.5}$$

Let us note that in the above equations, we have used $k_{rx} = k_x = k_{tx}$, because of the phase-matching at $z = 0$. In other words, $k_{rx} = k_x = k_{tx}$ means that the continuity of the tangential components of both electric and magnetic fields must be satisfied at any point *x* of the interface. Also, we have considered $k_{rz} = -k_z$, because from the dispersion relation of the electromagnetic wave [see (1.50)] in the insulator medium, we have $k_x^2 + k_z^2 = k_x^2 + k_{rz}^2$. As the *x*-component of the wavevector is preserved, we find that $k_{rz} = -k_z$ [1].

---

[3] *s*-polarized means senkrecht polarized.

The magnetic field H_y and the x-component of the electric field, i.e., E_x, are continuous at $z = 0$, so that we find

$$1 + r_p = t_p \, , \tag{3.6a}$$

$$1 - r_p = \frac{\varepsilon_1}{\varepsilon_2} \frac{i\kappa}{k_z} t_p \, . \tag{3.6b}$$

Now, we can solve the system of (3.6a) and (3.6b) to obtain

$$r_p = \frac{\varepsilon_2 k_z - i\varepsilon_1 \kappa}{\varepsilon_2 k_z + i\varepsilon_1 \kappa} \, . \tag{3.7}$$

The *reflectance coefficient* is given by the ratio of the power flow densities in two media. Using the relations between E_x and H_y, we can express the time-averaged power flow densities perpendicular to the surface, given by (1.64), as

$$S_i = \frac{1}{2} \text{Re} \left[E_{ix} H_{iy}^* \right] = \frac{1}{2} \frac{\text{Re}\,[k_z]}{\omega \varepsilon_0 \varepsilon_1} H_0^2 \, , \tag{3.8a}$$

$$S_r = \frac{1}{2} \text{Re} \left[E_{rx} H_{ry}^* \right] = \frac{1}{2} \frac{\text{Re}\,[k_z]}{\omega \varepsilon_0 \varepsilon_1} |r_p|^2 H_0^2 \, , \tag{3.8b}$$

where H_0 has been defined to be real. Therefore, the reflectance coefficient is

$$R_p = |r_p|^2 = 1 \, , \tag{3.9}$$

since r_p is the ratio of two complex numbers with the same absolute value. Let us note that in the present calculations, we have used $k_{tz} = i\kappa$ that is purely imaginary. This means that no energy is transferred through the interface because the Poynting vector is proportional to the real part of k_{tz}. This result is in agreement with our result $R_p = 1$, that the energy of the incident wave is totally reflected back to the first medium.

3.1.2   s-Polarized Wave

The preceding analysis refers to p-polarized electromagnetic waves where the electrical field is in the plane of incidence. The complementary problem concerns the case of a s-polarized wave so that the electric field is perpendicular to the plane of incidence and parallel to the plane interface. Thus, the total electric field in the dielectric space, which has a y-component only, is given by

$$E_y = E_0 \left[e^{ik_z z} + r_s e^{-ik_z z} \right] e^{ik_x x} \, , \tag{3.10}$$

where r_s is, by definition, the Fresnel reflection coefficient for s-polarized waves, and E_0 is the amplitude of the incident electric field E_y. Using (3.1e) the total magnetic field along the x-direction in the insulator space has the form

$$H_x = \frac{-k_z}{\omega\mu_0} E_0 \left[e^{ik_z z} - r_s e^{-ik_z z} \right] e^{ik_x x} . \tag{3.11}$$

In this case, the magnetic and electric fields in the EG can be written as

$$E_y = t_s E_0 e^{-\kappa z} e^{ik_x x} , \tag{3.12}$$

$$H_x = \frac{-i\kappa}{\omega\mu_0} t_s E_0 e^{-\kappa z} e^{ik_x x} , \tag{3.13}$$

where t_s is, by definition, the Fresnel transmission coefficient for s-polarized waves. Following the same procedure used for a p-polarized wave, we apply the same BCs for the s-polarized waves case, too. It may be verified that the BCs are satisfied, if

$$r_s = \frac{k_z - i\kappa}{k_z + i\kappa} , \tag{3.14}$$

and the reflectance coefficient is given by

$$R_s = |r_s|^2 = 1 . \tag{3.15}$$

3.2 Plasmonic Properties of Semi-Infinite Electron Gases

3.2.1 Dispersion Relation

Here, we wish to obtain the dispersion relation of a surface electromagnetic wave, known as a *surface plasmon polariton* (SPP), that propagates in a wave like fashion along the planar interface between an EG and an insulator medium, often vacuum, and whose amplitude decays exponentially with increasing distance into each medium from the interface [2].

We assume that a SPP propagates along a semi-infinite EG surface (in the half-space $z > 0$). Thus in the EG region and also the insulator medium, this wave has the electric field given by $\mathbf{E} \exp\left(i\mathbf{k}_\parallel \cdot \mathbf{r}_\parallel\right)$ and magnetic field given by $\mathbf{H} \exp\left(i\mathbf{k}_\parallel \cdot \mathbf{r}_\parallel\right)$, where $\mathbf{k}_\parallel = k_x \mathbf{e}_x + k_y \mathbf{e}_y$, $\mathbf{r}_\parallel = x\mathbf{e}_x + y\mathbf{e}_y$, $\mathbf{E} = \left(E_x, E_y, E_z\right)$, and E_x, E_y, and E_z are functions of z but not x and y (since the x-dependence of the field is $\exp(ik_x x)$ by assumption and without loss of generality, we consider $\partial\mathbf{E}/\partial y = 0$). The same assumptions hold for $\mathbf{H}$.

Again, from Maxwell's curl equations, we arrive to a set of coupled equations as (3.1a)–(3.1f) and it can easily be shown that the present system allows two sets of self-consistent solutions with different polarization properties of the propagating waves. The first set is a *p*-polarized (or TM) wave [see (3.1a)–(3.1c)], where only the field components E_x, E_z, and H_y are non-zero, and the second set is a *s*-polarized (or TE) wave [see (3.1d)–(3.1f)], with only H_x, H_z, and E_y being non-zero.

In the following, we consider the propagation of a SPP with *p*-polarized polarization, since SPPs with the *s*-polarized polarization cannot propagate on the surface of the present system [2]. Substituting (3.1c) into (3.1a), after some manipulation, gives

$$H_y = \frac{i\omega\varepsilon_0\varepsilon}{\kappa^2}\frac{\partial E_x}{\partial z} \, , \tag{3.16a}$$

$$E_z = -\frac{ik_x}{\kappa^2}\frac{\partial E_x}{\partial z} \, , \tag{3.16b}$$

where $\kappa^2 = k_x^2 - \varepsilon k_0^2$. Thus H_y and E_z can be determined if the non-zero longitudinal component E_x is known. Substituting (3.16a) and (3.16b) into (3.1b), one can obtain the following equation for the *x*-component E_x of the expanding coefficient **E**, as

$$\left[\frac{d^2}{dz^2} - \kappa^2\right] E_x(z) = 0 \, , \quad z \neq 0 \, , \tag{3.17}$$

that is the so-called *Helmholtz equation*. For the present system, the solution of (3.17) has the form

$$E_x(z) = \begin{cases} A_1 e^{+\kappa_1 z} \, , \, z \leq 0 \, , \\ A_2 e^{-\kappa_2 z} \, , \, z \geq 0 \, , \end{cases} \tag{3.18}$$

where $\kappa_j^2 = k_x^2 - \varepsilon_j k_0^2$, and the index j describes the media: $j = 1$ at $z < 0$, and $j = 2$ at $z > 0$. Also, the relations between the coefficients A_1 and A_2 can be determined from the matching BCs at the separation surface, i.e., (1.92a) and (1.92b). On applying the electromagnetic BCs at $z = 0$, we find

$$\frac{\varepsilon_1}{\kappa_1} + \frac{\varepsilon_2}{\kappa_2} = 0 \, . \tag{3.19}$$

From (3.19) we obtain an explicit expression for the SPPs wavenumber (or propagation constant) k_x, as

$$k_x = k_0\sqrt{\frac{\varepsilon_1\varepsilon_2}{\varepsilon_1 + \varepsilon_2}} \, , \tag{3.20}$$

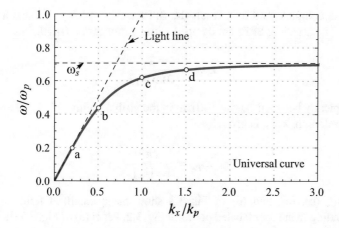

Fig. 3.2 Dispersion curve of SPPs at a flat EG-vacuum interface, as obtained from (3.21). Here $k_p = \omega_p/c$ and the field profiles of the four labeled points a, b, c, and d, are shown in Fig. 3.3, panels (a), (b), (c), and (d), respectively. Because of the use of appropriate units shown along the axes, these curves do not depend on any physical parameters

a relation that is valid even if ε_2 is complex. When the EG is characterized by (2.7), and $\varepsilon_1 = 1$, the frequency of a SPP is given by

$$\omega^2 = \frac{\omega_p^2}{2} + c^2 k_x^2 - \sqrt{\frac{\omega_p^4}{4} + c^4 k_x^4} \,, \tag{3.21}$$

which in the retarded region (where $k_x < \omega_s/c$) couples with the free electromagnetic field, while for the resonance condition or in the nonretarded (electrostatic) limit[4] ($k_x \to \infty$), we obtain the classical non-dispersive SP frequency (or resonance condition) $\omega_s = \omega_p/\sqrt{2}$ which is the maximum SP frequency.

Figure 3.2 displays typical dispersion curve for the propagation of SPPs at an EG-vacuum interface as obtained from the numerical solution of (3.21). For comparison, the electrostatic solution has been shown by broken line. The light line for propagation in the bulk free space in Fig. 3.2 appears as a straight line through the origin with a slope = 1. For small values of k_x, the dispersion curve approaches the light line. However, the SPPs dispersion curve is always to the right of the light line. This type of non-radiative wave, associated with $\varepsilon_2 < 0$, is commonly referred to as a *Fano wave*.

Note that the present discussions of Fig. 3.2 have assumed an ideal semi-infinite metallic medium with Im $[\varepsilon_2] = 0$. Excitations of the conduction electrons of real metals however suffer both from free electron and interband damping. For a real

---

[4]In the resonance region, the retardation effects are unimportant and an electrostatic approximation to the field equations is adequate.

case, ε_2 is complex, and as a result the SPPs wavenumber k_x is also a complex quantity. The traveling SPPs are damped with a *propagation length*,[5] as

$$L_{spp} = \frac{1}{2\text{Im}\,[k_x]}\,, \tag{3.22}$$

that is typically between 10 and 100 μm in the visible regime [3, 4]. Therefore, the SPPs wavelength, λ_{spp}, is given by

$$\lambda_{spp} = \frac{2\pi}{\text{Re}\,[k_x]}\,. \tag{3.23}$$

Panels (a), (b), (c), and (d) of Fig. 3.3 show the normalized field profiles E_x corresponding to the four labeled points in Fig. 3.2. Panel (a) of Fig. 3.3 shows a SPP that is very close to the light line; therefore, the evanescent tail in the free space is relatively long. In panels (b), (c), and (d) of Fig. 3.3, the SPPs have relatively smaller phase velocities $v_p = \omega/k_x$ and shorter evanescent tails in the vacuum. It should also be noted that as $\omega \to \omega_s$ asymptotically, the group velocity $v_g = d\omega/dk_x \to 0$.

3.2.2 Electric Field Lines

We can also at this point discuss the *electric field lines* associated with a SPP on a flat EG-insulator interface. Although the development in Sect. 3.2.1 was written in a complex notation, we now take the real parts of the complex fields as the physical fields. Keeping the phase factor $\exp\,[i\,(k_x x - \omega t)]$ in mind, we find for the insulator medium

$$E_x = A_1 e^{+\kappa_1 z} \cos\,(k_x x - \omega t)\,, \tag{3.24a}$$

$$E_z = A_1 \frac{k_x}{\kappa_1} e^{+\kappa_1 z} \sin\,(k_x x - \omega t)\,, \tag{3.24b}$$

while in the EG medium, we have

$$E_x = A_2 e^{-\kappa_2 z} \cos\,(k_x x - \omega t)\,, \tag{3.25a}$$

$$E_z = -A_2 \frac{k_x}{\kappa_2} e^{-\kappa_2 z} \sin\,(k_x x - \omega t)\,. \tag{3.25b}$$

We note that in each medium, the field components E_z and H_y are 90° out of phase with E_x. The electric field lines or streamlines represent the direction of the **E** field

---

[5]The distance over which the power/intensity of the wave falls to $1/e$ of its initial value.

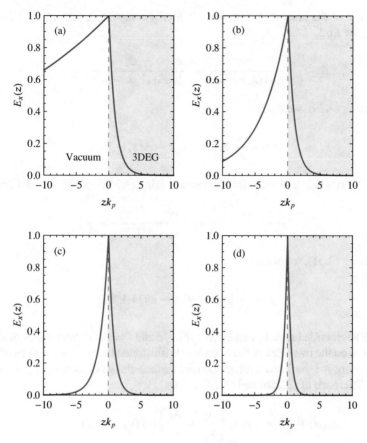

Fig. 3.3 Normalized profile of the tangential component of the electric field of a SPP on a flat EG-vacuum interface for four labeled points a, b, c, and d, that are shown in Fig. 3.2

in space. We set

$$\mathbf{dl} = \mathscr{C}\mathbf{E} \, , \tag{3.26}$$

where $\mathscr{C}$ is a constant. In Cartesian coordinates, (3.26) becomes

$$\mathbf{e}_x \, dx + \mathbf{e}_y \, dy + \mathbf{e}_z \, dz = \mathscr{C} \left(\mathbf{e}_x E_x + \mathbf{e}_y E_y + \mathbf{e}_z E_z \right) \, , \tag{3.27}$$

which can be written

$$\frac{dx}{E_x} = \frac{dy}{E_y} = \frac{dz}{E_z} \, . \tag{3.28}$$

For a SPP of the present system there is no E_y component, and using (3.24a) and (3.24b) we find

$$\frac{dx}{\cos(k_x x - \omega t)} = \frac{\kappa_1}{k_x} \frac{dz}{\sin(k_x x - \omega t)} . \qquad (3.29)$$

Integrating (3.29), we obtain

$$z = -\frac{1}{\kappa_1} \ln|\cos(k_x x - \omega t)| - \mathscr{C}_E , \qquad (3.30)$$

where $\mathscr{C}_E$ is a constant. In a similar manner, using (3.25a) and (3.25b) we find

$$\frac{dx}{\cos(k_x x - \omega t)} = -\frac{\kappa_2}{k_x} \frac{dz}{\sin(k_x x - \omega t)} . \qquad (3.31)$$

Integrating (3.31), we obtain

$$z = \frac{1}{\kappa_2} \ln|\cos(k_x x - \omega t)| + \mathscr{C}_E . \qquad (3.32)$$

For the nonretarded region, i.e., $k_x \gg \sqrt{\varepsilon_j}\omega/c$ and $t = 0$, the vector plot of electric field lines on the two sides of the boundary is illustrated in Fig. 3.4. This electric field arrangement is associated with an induced surface charge density σ_{ind} proportional to the difference of (3.25b) and (3.24b). Thus

$$\sigma_{ind}(x)\big|_{z=0} = -\varepsilon_0 k_x \left[\frac{A_2}{\kappa_2} + \frac{A_1}{\kappa_1}\right] \sin(k_x x - \omega t) . \qquad (3.33)$$

The normal surface wave which has emerged from this analysis possesses the following three important characteristics:

1. it is a wave which travels parallel to the surface.
2. it is associated with a surface charge.
3. the electromagnetic fields of this wave decay exponentially from the surface. Thus this normal surface wave is appropriately identified as a SPP.

3.2.3 Penetration Depth

Now, we look at the spatial extension of the electromagnetic fields associated with the SPPs. Substituting (3.20) into equation $\kappa_j^2 = k_x^2 - \varepsilon_j k_0^2$, we obtain the SPPs decay constant κ_j perpendicular to the interface, as

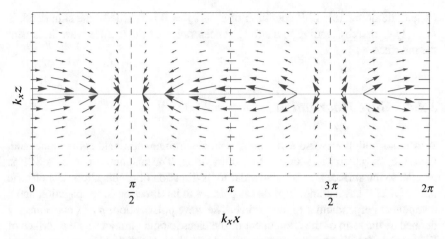

Fig. 3.4 Electric field pattern of a SPP in the zx plane when $k_x \gg \sqrt{\varepsilon_j}\omega/c$ and $t = 0$. The field magnitudes are oscillatory in the x-direction but decrease exponentially away from the boundary

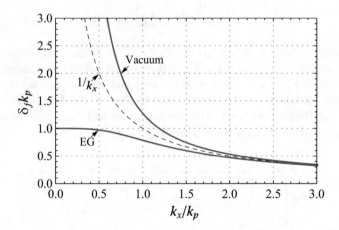

Fig. 3.5 Dimensionless penetration depth $\delta_j k_p$ of SPPs of a flat EG-vacuum interface, versus k_x/k_p, as obtained from (3.34). Here $k_p = \omega_p/c$ and the dotted line represents the large k_x limit of both δ_1 and δ_2, i.e., $1/k_x$

$$\kappa_j = \frac{\omega}{c}\sqrt{\frac{-\varepsilon_j^2}{\varepsilon_1 + \varepsilon_2}} \, , \qquad (3.34)$$

which allows to define the *penetration depth* $\delta_j = 1/\kappa_j$ at which the electromagnetic field falls to $1/e$ [5]. Figure 3.5 shows dimensionless penetration depth $\delta_j k_p$ of SPPs in an EG-vacuum interface as a function of dimensionless variable k_x/k_p, where $k_p = \omega_p/c$. In the vacuum side of the interface, the penetration depth is over the wavelength involved ($\delta_1 > 1/k_x$), whereas the penetration depth into the EG (metal) is determined at long wavelengths ($k_x \to 0$) by the skin depth. At large k_x

(where the nonretarded SP condition of $\varepsilon_1 + \varepsilon_2 = 0$ is fulfilled), the skin depth is $\ell_1 \sim 1/k_x$ thereby leading to a strong concentration of the surface wave field near the interface.

3.2.4 Wave Polarization

A SPP has both transverse and longitudinal electromagnetic field components and, therefore, in general has an *elliptic polarization*. The magnetic field of a SPP is parallel to the surface and perpendicular to its direction of propagation. The electric field of a SPP has a component that is parallel to its direction of propagation and a component perpendicular to the surface. The wave polarization, p_w, is conveniently defined as the ratio of the component of the electric field parallel to the interface of EG-insulator to the component perpendicular to the system, that is

$$p_w = \left| \frac{i E_x}{E_z} \right| = \frac{1}{k_x} \begin{cases} \kappa_1 \,, z \leq 0 \,, \\ \kappa_2 \,, z \geq 0 \,. \end{cases} \tag{3.35}$$

Figure 3.6 shows wave polarization of SPPs of an EG-vacuum interface as a function of dimensionless variable k_x/k_p. The transverse component is dominant in the electric field of SPPs with small wavenumbers and low frequencies (photon-like excitations) for which the dispersion curve is close to the free space light line. Only for very large wavenumbers are the transverse and longitudinal components comparable. Indeed, they are equal at the frequency given by ω_s.

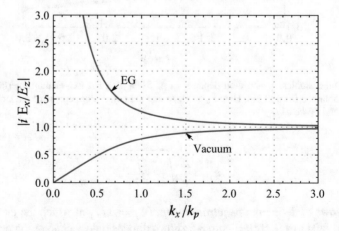

Fig. 3.6 Wave polarization $p_w = \left| i E_x/E_z \right|$ of SPPs of a flat EG-vacuum interface, versus k_x/k_p, as obtained from (3.35)

3.2.5 Power Flow

The power flow density associated with the SPPs is determined by the Poynting vectors in the two media, we have,

$$S = \begin{cases} \mathbf{E}_1 \times \mathbf{H}_1 \, , \, z \leq 0 \, , \\ \mathbf{E}_2 \times \mathbf{H}_2 \, , \, z \geq 0 \, , \end{cases} \tag{3.36}$$

where subscript 1 denotes the regions outside the EG, while subscript 2 denotes the region inside the EG. Both vectors have components in the x- and z-directions, but their z-components vanish on averaging over a cycle of oscillation of the fields [6–8]. The cycle-averaged x-components are

$$S_x = -\frac{1}{2} \begin{cases} \mathrm{Re}\left[E_{1z}H_{1y}^*\right] = -\dfrac{\varepsilon_0\varepsilon_1 k_x \omega}{\kappa_1^2} A_1^2 e^{+2\kappa_1 z} \, , \, z \leq 0 \, , \\ \mathrm{Re}\left[E_{2z}H_{2y}^*\right] = -\dfrac{\varepsilon_0\varepsilon_2 k_x \omega}{\kappa_2^2} A_2^2 e^{-2\kappa_2 z} \, , \, z \geq 0 \, , \end{cases} \tag{3.37}$$

where A_1 and A_2 are real. It is seen that the power flow density for $z < 0$ is positive, while it is negative for $z > 0$. The power flow densities are largest at the boundary, and away from $z = 0$ they fall off more rapidly in the EG than in the insulator. The total power flow associated with the SPPs is determined by an integration over z. If the integrated Poynting vectors are denoted by angle brackets, the power flow through an area in the yz plane of infinite length in the z-direction and unit width in the y-direction is

$$\langle S_x \rangle = \frac{\varepsilon_0 k_x \omega}{4} \left(\frac{\varepsilon_1}{\kappa_1^3} + \frac{\varepsilon_2}{\kappa_2^3} \right) A_1^2 \, . \tag{3.38}$$

This total power flow density (per unit width) is positive when k_x is positive; the magnitude of the power flow density in the insulator is larger than that in the EG medium. Note that for the resonance condition or the nonretarded (electrostatic) limit ($k_x \to \infty$), the total power flow density vanishes, as shown in Sect. 2.1.2.

3.2.6 Energy Distribution

We now consider the energy density distribution in the transverse direction. As described in Sect. 1.4.5, the energy density of the electromagnetic wave in a lossless medium can be expressed by (1.67). For the cycle-averaged of energy density distribution associated with the SPPs of a semi-infinite EG, we have,

$$U = \frac{1}{4} \begin{cases} \varepsilon_0 \varepsilon_1 |\mathbf{E}_1|^2 + \mu_0 |\mathbf{H}_1|^2 \,, & z \le 0 \,, \\ \varepsilon_0 \left(\varepsilon_2 + \omega \dfrac{d\varepsilon_2}{d\omega} \right) |\mathbf{E}_2|^2 + \mu_0 |\mathbf{H}_2|^2 \,, & z \ge 0 \,, \end{cases} \tag{3.39}$$

that yields,

$$U = \frac{\varepsilon_0}{2} A_1^2 \begin{cases} \varepsilon_1 \dfrac{k_x^2}{\kappa_1^2} e^{+2\kappa_1 z} \,, & z \le 0 \,, \\ \left(\varepsilon_2 \dfrac{k_x^2}{\kappa_2^2} + \omega \dfrac{d\varepsilon_2}{d\omega} \dfrac{k_x^2 + \kappa_2^2}{2\kappa_2^2} \right) e^{-2\kappa_2 z} \,, & z \ge 0 \,, \end{cases} \tag{3.40}$$

where both contributions to the energy density are positive. The total energy density associated with the SPPs is again determined by integration over z, the energy per unit surface area being

$$\langle U \rangle = \frac{\varepsilon_0 k_x^2}{4} \left(\frac{\varepsilon_1}{\kappa_1^3} + \frac{\varepsilon_2}{\kappa_2^3} + \omega \frac{d\varepsilon_2}{d\omega} \frac{k_x^2 + \kappa_2^2}{2k_x^2 \kappa_2^3} \right) A_1^2 \,. \tag{3.41}$$

Also, the position of the center of energy along the z-axis can be written as

$$\langle z \rangle \equiv \frac{1}{\langle U \rangle} \int_{-\infty}^{+\infty} z\, U\, dz = \frac{1}{2} \frac{-\dfrac{\varepsilon_1}{\kappa_1^4} + \dfrac{\varepsilon_2}{\kappa_2^4} + \omega \dfrac{d\varepsilon_2}{d\omega} \dfrac{k_x^2 + \kappa_2^2}{2k_x^2 \kappa_2^4}}{\dfrac{\varepsilon_1}{\kappa_1^3} + \dfrac{\varepsilon_2}{\kappa_2^3} + \omega \dfrac{d\varepsilon_2}{d\omega} \dfrac{k_x^2 + \kappa_2^2}{2k_x^2 \kappa_2^3}} \,. \tag{3.42}$$

Dependence of the energy center position of SPPs of an EG-vacuum interface with respect to the SPP frequency is shown in Fig. 3.7. The numerical result shows that for $\omega < 0.6369\omega_p$ the center of energy is located in the vacuum, i.e., $\langle z \rangle < 0$, while for $0.6369\omega_p < \omega < \omega_p/\sqrt{2}$ it moves into the EG region, i.e., $\langle z \rangle > 0$ [9].

3.2.7 Energy Velocity

The energy velocity of the SPPs is given as the ratio of the total power flow density (per unit width) and total energy density (per unit area), such as

$$v_e = \frac{\omega}{k_x} \frac{\dfrac{\varepsilon_1}{\kappa_1^3} + \dfrac{\varepsilon_2}{\kappa_2^3}}{\dfrac{\varepsilon_1}{\kappa_1^3} + \dfrac{\varepsilon_2}{\kappa_2^3} + \omega \dfrac{d\varepsilon_2}{d\omega} \dfrac{k_x^2 + \kappa_2^2}{2k_x^2 \kappa_2^3}} \,. \tag{3.43}$$

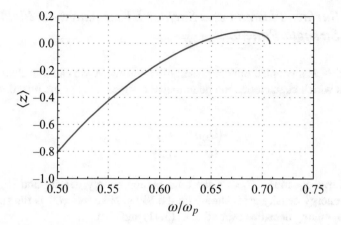

Fig. 3.7 Normalized variation of energy center position of SPPs of a flat EG-vacuum interface versus ω/ω_p as obtained from (3.42)

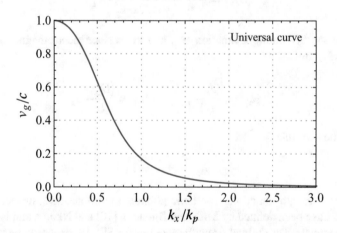

Fig. 3.8 Group (energy) velocity curve of SPPs of an EG-vacuum system, as given by (3.43). Because of the use of appropriate units shown along the axes, these curves do not depend on any physical parameters

The expression on the right-hand side is precisely that obtained from the usual definition of the SPPs group velocity, i.e., $v_g = d\omega/dk_x$. According to (3.43), the group velocity or energy velocity of SPPs of an EG-vacuum system is plotted in Fig. 3.8.

3.2.8 Surface Plasmon and Surface Electromagnetic Field Strength Functions

Since a SPP is a coupled optical plasmon-photon wave, we can introduce strength functions which characterize the quantitative compositions of the mixed wave. We have

$$\frac{\langle U_{sp} \rangle}{\langle U \rangle} + \frac{\langle U_{ph} \rangle}{\langle U \rangle} = 1 \,. \tag{3.44}$$

In the above equation, $\langle U_{ph} \rangle$ is the total photon energy density and $\langle U_{sp} \rangle$ is the total SP energy density associated with the SPPs. Note that $\langle U \rangle$ is the sum of the integrated energy densities over all z, as (3.41) and also

$$\langle U_{sp} \rangle = \frac{\varepsilon_0 k_x^2 \omega}{4} \frac{d\varepsilon_2}{d\omega} \frac{k_x^2 + \kappa_2^2}{2 k_x^2 \kappa_2^3} A_1^2 \,. \tag{3.45}$$

We define the SPs strength function Θ_{sp} and the surface electromagnetic strength function Θ_{ph} by

$$\Theta_{sp} = \frac{\langle U_{sp} \rangle}{\langle U \rangle} \,, \qquad \Theta_{ph} = \frac{\langle U_{ph} \rangle}{\langle U \rangle} \,, \tag{3.46}$$

obeying the sum rule

$$\Theta_{sp} + \Theta_{ph} = 1 \,. \tag{3.47}$$

Analogous strength functions for bulk phonon polaritons and surface phonon polaritons have been defined by Mills and Burstein [10] and Nkoma and Rwaboona [11], respectively. The defined strength functions for SPs can explicitly be evaluated, and we obtain

$$\Theta_{sp} = \frac{\omega \dfrac{d\varepsilon_2}{d\omega} \dfrac{k_x^2 + \kappa_2^2}{2 k_x^2 \kappa_2^3}}{\dfrac{\varepsilon_1}{\kappa_1^3} + \dfrac{\varepsilon_2}{\kappa_2^3} + \omega \dfrac{d\varepsilon_2}{d\omega} \dfrac{k_x^2 + \kappa_2^2}{2 k_x^2 \kappa_2^3}} \,, \tag{3.48}$$

and by using (3.19)–(3.21) in the above equation, we obtain

$$\Theta_{sp} = \frac{\varepsilon_1 \omega^2 \omega_s^2}{\left(\omega^2 - \omega_p^2 \right) \left(\omega^2 - \omega_s^2 \right) + \varepsilon_1 \omega^2 \omega_s^2} \,, \tag{3.49}$$

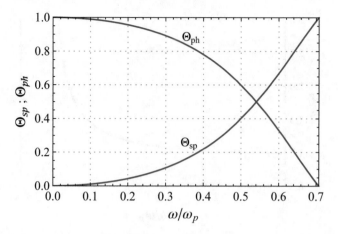

Fig. 3.9 Frequency dependence of the strength functions Θ_{sp} and Θ_{ph} of SPPs of an EG-vacuum system, as given by (3.47) and (3.49). Because of the use of appropriate units shown along the axes, these curves do not depend on any physical parameters

where ω_s defines the maximum frequency for SPPs, as given by (2.8). It is interesting to note the behavior of the strength functions at the two limiting frequencies, as

$$\Theta_{sp}(0) = 0 , \qquad \Theta_{ph}(0) = 1 , \tag{3.50a}$$

$$\Theta_{sp}(\omega_s) = 1 , \qquad \Theta_{ph}(\omega_s) = 0 . \tag{3.50b}$$

Figure 3.9 shows the variation of Θ_{sp} and Θ_{ph} with respect to the frequency when $\varepsilon_1 = 1$. Note that the sum rule given in (3.47) is satisfied throughout the frequency range in consideration.

3.2.9   Normalized Propagation Constant

Let a TM SPP propagates along the x-direction with the wavenumber k_x. Therefore, the normalized *propagation constant* (also called the effective refractive index of a SPP) is given by

$$\beta_{TM} = \frac{k_x}{k_0} , \tag{3.51}$$

assumes a real and positive value in a lossless EG medium. For a SPP that propagates along a semi-infinite EG surface, from (3.20) we find [12]

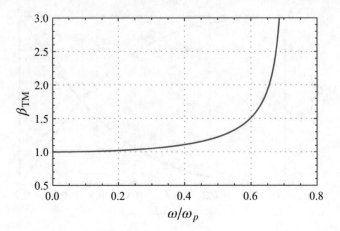

Fig. 3.10 Frequency dependence of the normalized propagation constant β_{TM} of SPPs of an EG-vacuum system, as given by (3.52)

$$\beta_{TM} = \sqrt{\frac{\varepsilon_1 \varepsilon_2}{\varepsilon_1 + \varepsilon_2}} \, . \tag{3.52}$$

Figure 3.10 shows the variation of normalized propagation constant of SPPs of a semi-infinite EG surface with respect to the frequency when $\varepsilon_1 = 1$.

3.2.10   Wave Impedance

Again, we assume that a SPP propagates along the x-direction. The *wave impedance*, η, is given by

$$\eta = \frac{E_\perp}{H_\perp} \, , \tag{3.53}$$

where $E_\perp$ and $H_\perp$ are the components of the electric and magnetic fields normal to the x-direction, respectively. For a SPP that propagates along a semi-infinite EG surface, we obtain

$$\eta = \left| \frac{E_z}{H_y} \right| = \eta_0 \begin{cases} 1/\varepsilon_1 \, , \ z \leq 0 \, , \\ 1/\varepsilon_2 \, , \ z \geq 0 \, , \end{cases} \tag{3.54}$$

where $\eta_0 = \sqrt{\mu_0/\varepsilon_0} \approx 377\,\Omega$ is the *vacuum impedance*.

3.2.11 Quantization of Surface Plasmon Polariton Fields

The electric and magnetic fields associated with the SPPs have so far been treated as classical variables. They can also be treated as quantum-mechanical observables represented by operators. Again, the canonical procedure for quantization of the SPPs can be used here [13–17], but we leave it for the reader as a BVP and follow an alternative approach [6].

To do this, it is useful to introduce and work with them instead of with the fields. Let $\mathbf{A}$ and Φ denote the vector and scalar potentials. Therefore (1.14b) is fulfilled if we let

$$\mu_0 \mathbf{H} = \nabla \times \mathbf{A} . \tag{3.55}$$

Then, substituting (3.55) into (1.14c) gives

$$\nabla \times \left(\mathbf{E} + \frac{\partial \mathbf{A}}{\partial t} \right) = 0 , \tag{3.56}$$

that is fulfilled if what is inside the parentheses is the gradient of a scalar. Thus, let

$$\mathbf{E} + \frac{\partial \mathbf{A}}{\partial t} = -\nabla \Phi . \tag{3.57}$$

Here, it is convenient to work in the gauge $\Phi = 0$, where no scalar potential is required to specify the radiation field. To derive the electric and magnetic fields from the *vector potential* $\mathbf{A}$, it should be represented by different operators in media 1 and 2. The vector potential operator associated with an excitation of wavevector $\mathbf{k}_{\parallel}$ can be written in the form

$$\hat{\mathbf{A}}_{\mathbf{k}_{\parallel}} = \left[\mathbf{e}_{\mathbf{k}_{\parallel}} \hat{a}_{\mathbf{k}_{\parallel}} e^{i\left(\mathbf{k}_{\parallel} \cdot \mathbf{r}_{\parallel} - \omega t\right)} + \mathbf{e}^*_{\mathbf{k}_{\parallel}} \hat{a}^{\dagger}_{\mathbf{k}_{\parallel}} e^{-i\left(\mathbf{k}^*_{\parallel} \cdot \mathbf{r}_{\parallel} - \omega t\right)} \right] \begin{cases} f_{\mathbf{k}_{\parallel}} e^{+\kappa_1 z} , & z \leq 0 , \\[2mm] g_{\mathbf{k}_{\parallel}} e^{-\kappa_2 z} , & z \geq 0 , \end{cases} \tag{3.58}$$

where $\hat{\mathbf{A}}_{\mathbf{k}_{\parallel}}$ is now a quantum-mechanical operator, $f_{\mathbf{k}_{\parallel}}$ and $g_{\mathbf{k}_{\parallel}}$ are unknown factors to be determined, $\hat{a}_{\mathbf{k}_{\parallel}}$ and $\hat{a}^{\dagger}_{\mathbf{k}_{\parallel}}$ are annihilation and creation operators for SPP quanta and obey (2.19) and (2.20). Also, $\mathbf{k}^*_{\parallel}$ is the complex conjugate of $\mathbf{k}_{\parallel}$ and the complex unit vector $\mathbf{e}_{\mathbf{k}_{\parallel}}$ is parallel to the complex classical electric field $\mathbf{E} \exp\left[i \left(\mathbf{k}_{\parallel} \cdot \mathbf{r}_{\parallel} - \omega t \right) \right]$ and satisfies $\mathbf{e}_{\mathbf{k}_{\parallel}} \cdot \mathbf{e}^*_{\mathbf{k}_{\parallel}} = 1$. With the complex conjugations shown in (3.58), $\hat{\mathbf{A}}_{\mathbf{k}_{\parallel}}$ is a Hermitian operator.

Now, the correspondence between the SPPs energy given by the usual quantum-harmonic-oscillator expression and the classical expression obtained from (3.41) is

$$A \langle U \rangle = \left(n_{\mathbf{k}_\parallel} + \frac{1}{2} \right) \hbar \omega , \tag{3.59}$$

where $|n_{\mathbf{k}_\parallel}\rangle$ is a state in which $n_{\mathbf{k}_\parallel}$ quanta of SPPs are excited, and A is the area of the boundary between the two media. Also, in making the transition from classical to quantum mechanics, we must have

$$\frac{1}{2}|\mathbf{E}|^2 \to \langle n_{\mathbf{k}_\parallel} | \hat{\mathbf{E}}^2 | n_{\mathbf{k}_\parallel} \rangle = 2\omega^2 \left(n_{\mathbf{k}_\parallel} + \frac{1}{2} \right) \begin{cases} f_{\mathbf{k}_\parallel}^2 e^{+2\kappa_1 z} , z \le 0 , \\ g_{\mathbf{k}_\parallel}^2 e^{-2\kappa_2 z} , z \ge 0 , \end{cases} \tag{3.60}$$

In the final step it has been assumed that the evaluation of the expectation value uses the electric field operator obtained from $\hat{\mathbf{E}} = -\partial \hat{\mathbf{A}}/\partial t$ and (3.58) together with the usual properties of the shift operators. Then (3.41) becomes

$$\langle U \rangle = \varepsilon_0 k_x^2 \omega^2 \left(n_{\mathbf{k}_\parallel} + \frac{1}{2} \right) \left(\frac{\varepsilon_1}{\kappa_1^3} + \frac{\varepsilon_2}{\kappa_2^3} + \omega \frac{d\varepsilon_2}{d\omega} \frac{k_x^2 + \kappa_2^2}{2k_x^2 \kappa_2^3} \right) \begin{cases} \dfrac{\kappa_1^2}{k_x^2 + \kappa_1^2} f_{\mathbf{k}_\parallel}^2 , z \le 0 , \\ \dfrac{\kappa_2^2}{k_x^2 + \kappa_2^2} g_{\mathbf{k}_\parallel}^2 , z \ge 0 , \end{cases} \tag{3.61}$$

Comparison of (3.59) and (3.61) now determines the unknown factors in (3.58) as

$$\begin{pmatrix} f_{\mathbf{k}_\parallel} \\ g_{\mathbf{k}_\parallel} \end{pmatrix} = \left(\frac{\hbar}{\varepsilon_0 A \omega} \right)^{1/2} \left(\frac{\varepsilon_1}{\kappa_1^3} + \frac{\varepsilon_2}{\kappa_2^3} + \omega \frac{d\varepsilon_2}{d\omega} \frac{k_x^2 + \kappa_2^2}{2k_x^2 \kappa_2^3} \right)^{-1/2} \begin{pmatrix} \dfrac{[k_x^2 + \kappa_1^2]^{1/2}}{k_x \kappa_1} \\ \dfrac{[k_x^2 + \kappa_2^2]^{1/2}}{k_x \kappa_2} \end{pmatrix} , \tag{3.62}$$

where ω can be found from (3.19). The vector potential operators in both medium 1 and 2 are thus fully determined. Now, the total vector potential operators in the two media are obtained by

$$\hat{\mathbf{A}}(\mathbf{r}, t) = \sum_{\mathbf{k}_\parallel} \left[\mathbf{e}_{\mathbf{k}_\parallel} \hat{a}_{\mathbf{k}_\parallel} e^{i(\mathbf{k}_\parallel \cdot \mathbf{r}_\parallel - \omega t)} + \mathbf{e}_{\mathbf{k}_\parallel}^* \hat{a}_{\mathbf{k}_\parallel}^\dagger e^{-i\left(\mathbf{k}_\parallel^* \cdot \mathbf{r}_\parallel - \omega t\right)} \right] \begin{cases} f_{\mathbf{k}_\parallel} e^{+\kappa_1 z} , z \le 0 , \\ g_{\mathbf{k}_\parallel} e^{-\kappa_2 z} , z \ge 0 , \end{cases} \tag{3.63}$$

which shows the vector potential field in terms of $\hat{a}_{\mathbf{k}_\parallel}$ and $\hat{a}_{\mathbf{k}_\parallel}^\dagger$.

3.3 Excitation of Surface Plasmon Polaritons on Semi-Infinite Electron Gases

In general, a SPP is a p-polarized surface wave, traveling along the interface of two different media. In the previous section, we assumed one of the media to be an EG and the other to be an insulator. This SPP propagates parallel to the interface (along the x-direction, for example) and is exponentially attenuated in the normal direction in both the EG and insulator. The wavenumber k_x of a SPP on a semi-infinite EG bounded by vacuum is given by

$$k_x = \frac{\omega}{c}\sqrt{\frac{\varepsilon(\omega)}{1+\varepsilon(\omega)}} \ . \tag{3.64}$$

However, for most metals $\varepsilon(\omega)$ is less than -1 in the visible region of the spectrum, and we see from (3.64) that k_x is greater than the wavenumber of an electromagnetic wave in vacuum at same ω, as can be seen in Fig. 3.11. In the quantum picture, this means that a SPP possesses a larger momentum than photons of the same energy. For this reason, free space light beams in vacuum (or in the real case, air with $\varepsilon_{air} \approx 1$) cannot simply couple to a SPP due to the fact that it is not possible in the frequency range of its dispersion to satisfy simultaneously the energy and momentum conservation laws. Thus, a SPP may be excited only with the evanescent wave and is therefore classified as non-radiative [18–20].

The decrease in reflectivity from aluminum films in the total internal reflection was first observed by Turbader [21]; however, he did not attribute this phenomenon to a SPP resonance. The excitation of a SPP in silver films was first recognized by Otto [22], using the method of *attenuated total reflection* (ATR) where the

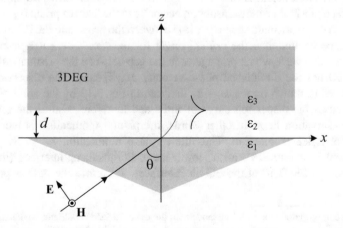

Fig. 3.11 Experimental setup for excitation and observation of SPPs, known as Otto's configuration

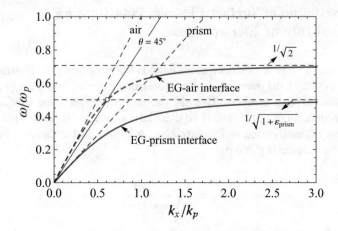

Fig. 3.12 Prism coupling and dispersion curves of SPPs at the interfaces of a semi-infinite EG bounded by air ($\varepsilon_{air} \approx 1$) and a prism ($\varepsilon_{prism} = 3$). Also plotted are the air and prism light lines (dashed lines) and the corresponding SP frequencies (dashed lines). The dashed portion of SPP at the EG-air interface, lying to the right of the vacuum light line while inside the prism light cone, is accessible. Therefore the dashed portion of the SPPs dispersion at the EG-air interface, possessing wavevectors between the respective light lines of air and the prism, can be excited by the ATR method (the solid black line which crosses the dispersion curve of SPPs at the EG-air interface, and, therefore, ensures the fulfillment of the conservation laws)

evanescent wave is either coupled from the prism into the metal through a thin air spacing.

Now, by the ATR method we relate the amplitude of a SPP to the exciting incident radiation. Figure 3.11 gives the experimental setup of the ATR method. This configuration is called *Otto's configuration*[6] or prism-air-medium configuration.[7] The reflection coefficient is measured of an electromagnetic wave impinging upon the interface of a higher-index insulator, normally in the form of prism ($\varepsilon_1 = \varepsilon_{prism}$) and a lower-index insulator gap layer (ε_2) between the prism and the EG at an angle of incidence greater than the *critical angle* for the *total internal reflection*. The electromagnetic wave which propagates in the gap between the prism and the semi-infinite EG has the diminished phase velocity $c/\sqrt{\varepsilon_1}\sin\theta$. The dispersion curve of this wave is shown on the $\omega - k_x$ diagram in Fig. 3.12 by the solid black line which crosses the dispersion curve of SPPs and therefore ensures the fulfillment of the conservation laws. In other words, the photon momentum is increased in the optically denser medium. Thus the in-plane momentum $k_x = k_0\sqrt{\varepsilon_1}\sin\theta$ is sufficient to excite SPPs at the lower-index EG-insulator interface (in current case shown in Fig. 3.12 at the EG-air interface). Therefore the dashed portion of

---

[6]The Otto configuration is a useful method when the metal surface should not be damaged, but it is difficult to keep the constant distance between the metal and the prism [23].

[7]Another setup is *Kretschmann's configuration* [24] or prism-medium-air configuration. In this case, the thickness of the metal film must be smaller than the skin depth of the evanescent field.

the SPPs dispersion at the EG-air interface, possessing wavevectors between the respective light lines of air and the prism, can be excited (see Fig. 3.12). In this case the total internal reflection becomes frustrated, the absorption of light occurs and the reflection coefficient R becomes less than unity. The positions of the minima appearing in the reflection spectrum $R(\omega)$ correspond to the frequency of a SPP, if certain conditions are satisfied, see below.

As shown in Fig. 3.11, a monochromatic p-polarized wave in a prism is incident on a thin insulator spacing at the hypotenuse face of the prism and at an angle of incidence greater than the critical angle for the total internal reflection. In the prism medium we have an incident and reflected propagating wave of the total form

$$H_y = -\left[A_1 e^{ik_{1z}z} + A_2 e^{-ik_{1z}z}\right] e^{ik_x x}, \quad z \leq 0, \tag{3.65}$$

where $k_{1z} = k_0\sqrt{\varepsilon_1}\cos\theta$ and the corresponding total electric field along the x-direction can be derived, using (3.1b). In the thin insulator spacing we write the total magnetic field as

$$H_y = -\left[A_3 e^{\kappa_2 z} + A_4 e^{-\kappa_2 z}\right] e^{ik_x x}, \quad 0 \leq z \leq d, \tag{3.66}$$

where $\kappa_2^2 = k_x^2 - \varepsilon_2 k_0^2$ and the transmitted wave in the EG is a evanescent wave, as

$$H_y = -A_5 e^{-\kappa_3 z} e^{ik_x x}, \quad z \geq d, \tag{3.67}$$

where $\kappa_3^2 = k_x^2 - \varepsilon_3 k_0^2$ and $\varepsilon_3 = \varepsilon_{3r} + i\varepsilon_{3i}$ is the complex dielectric function of the EG. Again, one can express E_x in term of the corresponding magnetic field component H_y, using (3.1b). The four unknown amplitudes A_2–A_5 can be related to the incident amplitude A_1 through the BCs. Continuity of the tangential components of $\mathbf{E}$ and $\mathbf{H}$ at the $z = 0$ and $z = d$ boundaries lead to a linear system of four equations, which may be casted as a standard $\mathbf{Mb} = \mathbf{c}$ matrix equation, namely

$$\mathbf{M}\begin{pmatrix} A_2 \\ A_3 \\ A_4 \\ A_5 \end{pmatrix} = \begin{pmatrix} A_1 \\ A_1 \\ 0 \\ 0 \end{pmatrix}, \tag{3.68}$$

where the matrix $\mathbf{M}$ is given by

$$\mathbf{M} = \begin{pmatrix} -1 & 1 & 1 & 0 \\ 1 & -i\dfrac{\varepsilon_1}{\varepsilon_2}\dfrac{\kappa_2}{k_{1z}} & i\dfrac{\varepsilon_1}{\varepsilon_2}\dfrac{\kappa_2}{k_{1z}} & 0 \\ 0 & e^{\kappa_2 d} & e^{-\kappa_2 d} & -e^{-\kappa_3 d} \\ 0 & e^{\kappa_2 d} & -e^{-\kappa_2 d} & \dfrac{\varepsilon_2}{\varepsilon_3}\dfrac{\kappa_3}{\kappa_2}e^{-\kappa_3 d} \end{pmatrix}. \tag{3.69}$$

Now, using Cramer's rule, we can find the Fresnel reflection coefficient from the following quotient of determinants:

$$r_p = \frac{A_2}{A_1} = \frac{1}{A_1} \frac{\det \mathbf{m}}{\det \mathbf{M}} , \tag{3.70}$$

where the matrix $\mathbf{m}$ is built by replacing the first column of $\mathbf{M}$ by the column vector $(A_1, A_1, 0, 0)^T$, that is, the vector on the right-hand side of (3.68). After some algebra one arrives to the following expression for the reflection coefficient, as

$$r_p = \frac{r_{12} + r_{23} \exp[-2\kappa_2 d]}{1 + r_{12}r_{23} \exp[-2\kappa_2 d]} , \tag{3.71}$$

where the Fresnel reflection coefficients, with the 12 and 23 subscripts for glass-insulator and EG-insulator boundaries, respectively, are given by

$$r_{12} = \frac{\varepsilon_2 k_{1z} - i\varepsilon_1 \kappa_2}{\varepsilon_2 k_{1z} + i\varepsilon_1 \kappa_2} , \tag{3.72a}$$

$$r_{23} = \frac{\varepsilon_3 \kappa_2 - \varepsilon_2 \kappa_3}{\varepsilon_3 \kappa_2 + \varepsilon_2 \kappa_3} . \tag{3.72b}$$

According to the (3.9), the reflectance coefficient is just the modulus squared of the Fresnel reflection coefficient. The effect of the SPP resonance on the linear reflectivity can be observed only for the case of non-zero absorption given by $\varepsilon_{3i} \neq 0$. In this case, the energy of a SPP is absorbed in the EG medium when this SPP is excited. The absorption of the energy causes a decrease of the reflection coefficient. If $\varepsilon_{3i} = 0$, the reflectance coefficient calculated from $R_p = |r_p|^2$ remains unity for all angles of incidence including the SPP resonance.

The calculated ATR spectra are shown in Fig. 3.13, for the incident angle $\theta = 45°$, when $d = 7c/\omega_p$ and $\gamma = 0.01\omega_p$. One can see a sharp dip in the reflectance coefficient when the frequency of the incident wave coincides with the frequency of SPPs.

3.4 Surface Magneto Plasmon Polaritons of Semi-Infinite Magnetized Electron Gases: Voigt Configuration

Here, we study the propagation of a *surface magneto plasmon polariton* (SMPP) along the interface between an insulator and a semi-infinite magnetized EG in the presence of a static magnetic field $\mathbf{B}_0$ in the Voigt configuration [25–28]. We note that the SMPP remains *p*-polarized wave only in this geometry [27]. For other directions of the static magnetic field $\mathbf{B}_0$, a SMPP acquires both *p*-polarized and *s*-

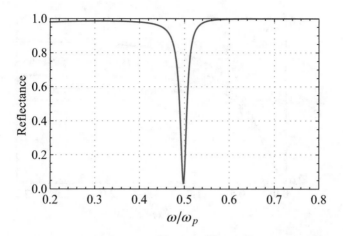

Fig. 3.13 Frequency dependence of the reflectance coefficient for the value of the incident angle $\theta = 45°$. A decrease in the reflectance coefficient means that the incident energy is transferred, through the air slab, to the excitation of the SPPs. The position of the minimum in $R(\omega)$ agrees with the frequency of the labeled point in Fig. 3.12. The parameters used in this calculation are $d = 7c/\omega_p$ and $\gamma = 0.01\omega_p$

polarized components due to the cyclotron motion of electrons. Again, we consider a semi-infinite EG occupying the half-space $z > 0$ and assume that a SMPP is propagating in the x-direction. In this case, the dielectric tensor of the system is represented in the form of (2.36) and (3.1b) and (3.1c) must be rewrite as

$$\frac{\partial H_y}{\partial z} = i\omega\varepsilon_0 \left(\varepsilon_{xx}E_x + \varepsilon_{xz}E_z\right) , \tag{3.73a}$$

$$-\frac{\partial H_y}{\partial x} = i\omega\varepsilon_0 \left(\varepsilon_{zx}E_x + \varepsilon_{zz}E_z\right) , \tag{3.73b}$$

Substituting (3.73b) into (3.1a), after some manipulation, gives

$$H_y = \frac{i\omega\varepsilon_0\varepsilon_{zz}}{k_x^2 - \varepsilon_{zz}k_0^2} \left(\frac{\partial}{\partial z} + ik_x\frac{\varepsilon_{zx}}{\varepsilon_{zz}}\right) E_x , \tag{3.74a}$$

$$E_z = \frac{1}{k_x^2 - \varepsilon_{zz}k_0^2} \left(-ik_x\frac{\partial}{\partial z} + \varepsilon_{zx}k_0^2\right) E_x , \tag{3.74b}$$

Then, after substituting (3.74a) and (3.74b) into (3.73a), one can obtain the following equation for the x-component E_x of the expanding coefficient **E**, as

$$\left[\frac{d^2}{dz^2} - \kappa_2^2\right] E_x(z) = 0 , \quad z \geq 0 , \tag{3.75}$$

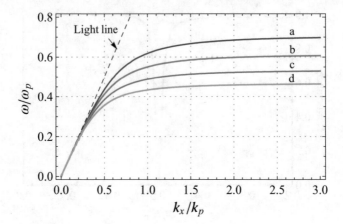

Fig. 3.14 Dispersion curves of SMPPs of a flat EG-vacuum interface in the Voigt configuration, as obtained from (3.76). Here $k_p = \omega_p/c$ and the different curves refer to (**a**) $\omega_c/\omega_p = 0$, (**b**) $\omega_c/\omega_p = 0.2$, (**c**) $\omega_c/\omega_p = 0.4$, and (**d**) $\omega_c/\omega_p = 0.6$

where $\kappa_2^2 = k_x^2 - \varepsilon_v k_0^2$ and $\varepsilon_v = \varepsilon_{zz} + \varepsilon_{xz}^2/\varepsilon_{zz}$ is the so-called *Voigt dielectric constant*. Therefore, the solutions of (3.17) for $z \leq 0$ and (3.75) for $z \geq 0$ have the form of (3.18). On applying the electromagnetic BCs at $z = 0$, we find

$$\varepsilon_V \kappa_1 + \varepsilon_1 \left(\kappa_2 + i k_x \frac{\varepsilon_{zx}}{\varepsilon_{zz}} \right) = 0 . \tag{3.76}$$

From (3.76) one notes immediately that the dispersion equations for waves propagating along the positive and negative x-directions are different from each other, i.e., SMPP behavior is nonreciprocal. In Fig. 3.14, we plot dimensionless frequency ω/ω_p versus dimensionless variable k_x/k_p for forward-going SMPPs with several values of ω_c/ω_p when $\varepsilon_1 = 1$. One can see that the increasing of ω_c/ω_p red-shifts the SMPP frequency.

3.5 Surface Plasmon Polaritons of Planar Electron Gas Slabs

Up to now, we investigated the main features of SPPs in a semi-infinite EG. We note the SPPs on the surfaces of EG films possess interesting properties that SPPs on the surface of a semi-infinite metal do not possess. Therefore, for a planar EG slab, we expect new physical behavior of the dispersion relation, in comparison with those obtained for a semi-infinite EG. The aim of the present section is to study the main features of SPPs of planar EG slabs. Figure 2.2 depicts schematic view of a planar EG slab of thickness d and relative dielectric constant ε_2 along the z-direction and infinitely wide in the x- and y-directions. In the present structure, a substrate with relative dielectric constant ε_1 is supposed to occupy the region $z < 0$ underneath the EG film, whereas the region $z > d$ is assumed to be a semi-infinite insulator with relative dielectric constant ε_3. Now, we consider the propagation of a p-polarized SPP in the x-direction, so the solution of (3.17) has the form

$$E_x(z) = \begin{cases} A_1 e^{+\kappa_1 z}, & z \leq 0, \\ A_2 e^{+\kappa_2 z} + A_3 e^{-\kappa_2 z}, & 0 \leq z \leq d, \\ A_4 e^{-\kappa_3 z}, & z \geq d, \end{cases} \tag{3.77}$$

where the relations between the coefficients A_1–A_4 can be determined from the matching BCs at the separation surfaces. Using the electromagnetic BCs at $z = 0$ and $z = d$, i.e., (1.92a) and (1.92b), one arrives at the following homogeneous linear system of equations

$$\begin{pmatrix} -1 & 1 & 1 & 0 \\ -1 & \dfrac{\varepsilon_2}{\varepsilon_1} \dfrac{\kappa_1}{\kappa_2} & -\dfrac{\varepsilon_2}{\varepsilon_1} \dfrac{\kappa_1}{\kappa_2} & 0 \\ 0 & e^{\kappa_2 d} & e^{-\kappa_2 d} & -e^{-\kappa_3 d} \\ 0 & e^{\kappa_2 d} & -e^{-\kappa_2 d} & \dfrac{\varepsilon_3}{\varepsilon_2} \dfrac{\kappa_2}{\kappa_3} e^{-\kappa_3 d} \end{pmatrix} \begin{pmatrix} A_1 \\ A_2 \\ A_3 \\ A_4 \end{pmatrix} = 0. \tag{3.78}$$

The dispersion relation of SPP modes of a planar EG slab is then obtained by solving the condition that the determinant of the matrix in (3.78) must be zero in order to exist a non-trivial solution, yielding

$$\left(1 + \frac{\varepsilon_2}{\varepsilon_1} \frac{\kappa_1}{\kappa_2}\right)\left(1 + \frac{\varepsilon_2}{\varepsilon_3} \frac{\kappa_3}{\kappa_2}\right) = \left(1 - \frac{\varepsilon_2}{\varepsilon_1} \frac{\kappa_1}{\kappa_2}\right)\left(1 - \frac{\varepsilon_2}{\varepsilon_3} \frac{\kappa_3}{\kappa_2}\right) e^{-2\kappa_2 d}. \tag{3.79}$$

This expression must be solved numerically. It is interesting to consider the particular case of a symmetrical environment, in which the both superstrate and substrate of the planar EG slab are the same, i.e., $\varepsilon_1 = \varepsilon_3$ and $\kappa_1 = \kappa_3$. In this case (3.79) can be further simplified, and splits into two modes delivered by the following pair of equations:

$$\tanh\left(\frac{\kappa_2 d}{2}\right) = -\frac{\varepsilon_2}{\varepsilon_1} \frac{\kappa_1}{\kappa_2}, \tag{3.80a}$$

$$\coth\left(\frac{\kappa_2 d}{2}\right) = -\frac{\varepsilon_2}{\varepsilon_1} \frac{\kappa_1}{\kappa_2}, \tag{3.80b}$$

corresponding to a high-frequency mode and a low-frequency mode, respectively.

Figure 3.15 shows the dispersion behavior of SPP modes of a planar EG slab bounded by vacuum, computed using (3.80a) and (3.80b) for several different widths. The dimensionless frequencies ω_+/ω_p and ω_-/ω_p are plotted versus the dimensionless wavenumber k_x/k_p, where $k_p = \omega_p/c$ as remarked before. It

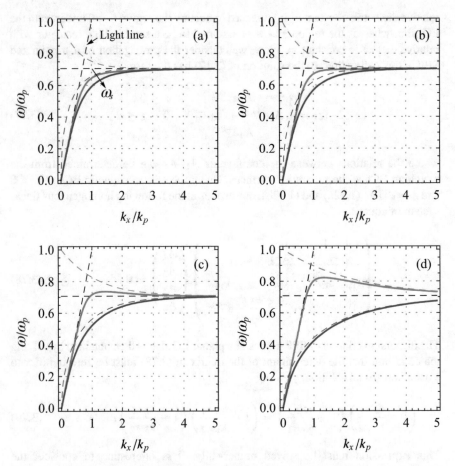

Fig. 3.15 Dispersion curves of SPP modes of a planar EG slab bounded by vacuum, showing the hybridized, coupled modes for various EG layer thicknesses d. Here the different panels refer to (**a**) $d = 2/k_p$, (**b**) $d = 1.5/k_p$, (**c**) $d = 1/k_p$, and (**d**) $d = 0.5/k_p$. Also $k_p = \omega_p/c$ and the solid red and blue curves represent the calculation with the account of retardation for ω_+ and ω_-, respectively, and the dashed thin red and blue curves, without retardation effect

can be seen that the high-frequency mode ω_+ is particularly strongly affected by retardation. In addition, the high- and low-frequency modes can be classified as an odd (anti-symmetric) and even (symmetric) modes with respect to the tangential component of the electric field. Note that for odd mode, the tangential component of the electric field (E_x) is anti-symmetric with respect to the plane $z = d/2$, while for an even mode, the tangential component of the electric field is symmetric relative to the $z = d/2$ plane, as shown in Fig. 3.16.

Finally, to obtain the dispersion relation of SPP modes of an insulator slab gap (with relative dielectric constants ε_1) in an EG with transverse dielectric function ε_2, we reverse the EG slab geometry and find a relation between ω and k_x, as

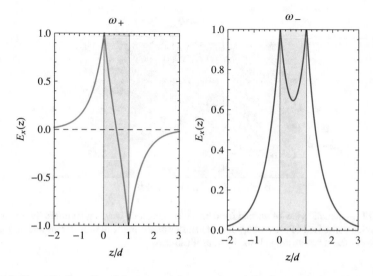

Fig. 3.16 Normalized profile of the tangential component of the electric field of SPP modes of a planar EG slab bounded by vacuum. Left: anti-symmetric (odd) mode. Right: symmetric (even) mode. Parameters: $d = 0.5/k_p$ and $k_x/k_p = 4$

$$\coth\left(\frac{\kappa_1 d}{2}\right) = -\frac{\varepsilon_1}{\varepsilon_2}\frac{\kappa_2}{\kappa_1} \, , \qquad (3.81a)$$

$$\tanh\left(\frac{\kappa_1 d}{2}\right) = -\frac{\varepsilon_1}{\varepsilon_2}\frac{\kappa_2}{\kappa_1} \, , \qquad (3.81b)$$

for odd and even modes, respectively.

3.6 Electromagnetic Waves Propagation in One-Dimensional Superlattices of Alternating Electron Gas Layers

Figure 3.17 shows an infinite 1D superlattice composed of alternating thin EG and insulator material, where media EG and insulator are of thickness a and b, respectively, and the corresponding relative dielectric constants are ε_1 and ε_2. The unit cells of the structure are designated by the index n and $L = a+b$ is the length of a unit cell, as illustrated in the figure. Here we wish to obtain the dispersion relation of normal incident electromagnetic waves on the structure which propagate in the z-direction and have angular frequency ω. Our starting point is the 1D Helmholtz equation given by

$$\left[\frac{\mathrm{d}^2}{\mathrm{d}z^2} - \varepsilon(z)k_0^2\right] E_x(z) = 0 \, , \qquad (3.82)$$

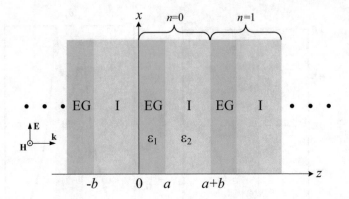

Fig. 3.17 Schematic view of normal incident electromagnetic wave on an infinitely extended 1D superlattice composed of alternating thin EG and insulator material of different thicknesses. The unit cells of the structure are indexed by n, as illustrated

where

$$\varepsilon(z) = \begin{cases} \varepsilon_1 \,, \ nL \leq z \leq nL + a \,, \\ \varepsilon_2 \,, \ nL + a \leq z \leq (n+1)L \,, \end{cases} \tag{3.83}$$

and ε_1 is characterized by (2.7) and $k_0 = \omega/c$ as mentioned before. For the present BVP when $z \neq nL$ and $z \neq nL + a$, the solutions of electric and magnetic fields, respectively, have the form

$$E_x(z) = \begin{cases} A_+ e^{+ik_{1z}(z-nL)} + A_- e^{-ik_{1z}(z-nL)} \,, & nL \leq z \leq nL + a \,, \\ B_+ e^{+ik_{2z}(z-nL-a)} + B_- e^{-ik_{2z}(z-nL-a)} \,, & nL + a \leq z \leq (n+1)L \,, \end{cases} \tag{3.84}$$

$$H_y(z) = \frac{1}{\omega\mu_0} \begin{cases} k_{1z}\left[A_+ e^{+ik_{1z}(z-nL)} - A_- e^{-ik_{1z}(z-nL)} \right] \,, & nL \leq z \leq nL + a \,, \\ k_{2z}\left[B_+ e^{+ik_{2z}(z-nL-a)} - B_- e^{-ik_{2z}(z-nL-a)} \right] \,, & nL+a \leq z \leq (n+1)L \,, \end{cases} \tag{3.85}$$

for $\varepsilon_1 > 0$, and

$$E_x(z) = \begin{cases} C_+ e^{-\kappa_{1z}(z-nL)} + C_- e^{+\kappa_{1z}(z-nL)} \,, & nL \leq z \leq nL + a \,, \\ D_+ e^{+ik_{2z}(z-nL-a)} + D_- e^{-ik_{2z}(z-nL-a)} \,, & nL + a \leq z \leq (n+1)L \,, \end{cases} \tag{3.86}$$

$$H_y(z) = \frac{1}{\omega\mu_0} \begin{cases} i\kappa_{1z}\left[C_+ e^{-\kappa_{1z}(z-nL)} - C_- e^{+\kappa_{1z}(z-nL)} \right] \,, & nL \leq z \leq nL + a \,, \\ k_{2z}\left[D_+ e^{+ik_{2z}(z-nL-a)} - D_- e^{-ik_{2z}(z-nL-a)} \right] \,, & nL + a \leq z \leq (n+1)L \,, \end{cases} \tag{3.87}$$

for $\varepsilon_1 < 0$, where $k_{jz} = \sqrt{\varepsilon_1}k_0$, $k_{1z} = i\kappa_{1z}$ and the relations between the coefficients $A_\pm$, $B_\pm$, $C_\pm$, and $D_\pm$ can be determined from the matching BCs at the separation surfaces. The BCs at $z = nL + a$ are

$$E_{x1}\big|_{z=nL+a} = E_{x2}\big|_{z=nL+a} , \tag{3.88a}$$

$$H_{y1}\big|_{z=nL+a} = H_{y2}\big|_{z=nL+a} , \tag{3.88b}$$

where E_{x1} (H_{y1}) and E_{x2} (H_{y2}) denote the electric (magnetic) fields in the media 1 and 2, respectively. Another appropriate BCs are

$$E_{x1}\big|_{z=nL} = e^{-iq_z L} E_{x2}\big|_{z=(n+1)L} , \tag{3.89a}$$

$$H_{y1}\big|_{z=nL} = e^{-iq_z L} H_{y2}\big|_{z=(n+1)L} , \tag{3.89b}$$

where the constant q_z is the Bloch wavenumber in the *Bloch theorem*.[8] Now, for the case in which $\varepsilon_1 > 0$, from (3.88) and (3.89), we obtain the dispersion relation of electromagnetic waves propagation in a 1D superlattice of alternating EG layers, as

$$\cos q_z L = \cos k_{1z}a \, \cos k_{2z}b - \frac{k_{1z}^2 + k_{2z}^2}{2k_{1z}k_{2z}} \sin k_{1z}a \, \sin k_{2z}b , \tag{3.90}$$

and for $\varepsilon_1 < 0$, as

$$\cos q_z L = \cosh \kappa_{1z}a \, \cos k_{2z}b + \frac{\kappa_{1z}^2 - k_{2z}^2}{2\kappa_{1z}k_{2z}} \sinh k_{1z}a \, \sin k_{2z}b . \tag{3.91}$$

We note that there are three selective parameters, $a\omega_p/c$, b/a, and ε_2, in the numerical calculations [29, 30]. Figure 3.18 is a plot of $\omega a/c$ versus $q_z L$, when $a = 2c/\omega_p$, $b/a = 2$, and $\varepsilon_2 = 1$. One can see that the dispersion curve becomes a *band structure* with frequency gaps.

3.7 Plasmonic Properties of Circular Electron Gas Cylinders

Up to this point, only SPPs in Cartesian coordinates have been considered due to the geometry of the considered structures. However, the propagation of SPPs in cylindrical coordinates is also one of the important BVPs in the plasmonics. This is because some of the important MNSs including metallic *nanowire*, metallic tubes, and cylindrical dielectric core have cylindrical like geometry. Therefore, plasmonic properties of circular EG cylinders will be examined here in some detail.

---

[8]Since the structure in Fig. 3.17 is periodic in the z-direction, according to the Bloch theorem, the solution of $E_x(z)$ is of the form $E_x(z) = e^{iq_z z}u_{q_z}(z)$, where for any integer n, $u_{q_z}(z)$ is periodic with a period L, i.e., $u_{q_z}(z+L) = u_{q_z}(z)$. This periodic condition for the Bloch wave is considered in (3.89).

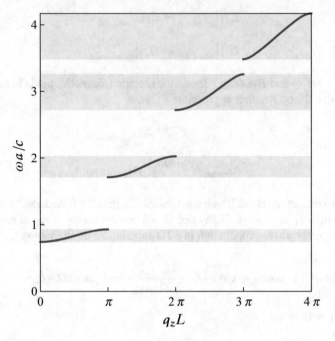

Fig. 3.18 Dispersion curves of electromagnetic waves propagation in an infinitely extended 1D superlattice of alternating EG layers, as obtained from (3.90) or (3.91) when $a = 2c/\omega_p$, $b/a = 2$, and $\varepsilon_2 = 1$

3.7.1   Surface Plasmon Polariton Modes

Consider a circular EG cylinder of radius a and relative dielectric function ε_1 embedded in an insulator having a relative dielectric constant ε_2. We use cylindrical coordinates (ρ, ϕ, z). The axis of the cylinder lies along the z-axis. To obtain the SPP modes of the system we have to solve Maxwell's curl equations for electric and magnetic fields, i.e., (1.33c) and (1.33d), as

$$\nabla \times \mathbf{E} = i\mu_0\omega\mathbf{H} , \tag{3.92a}$$

$$\nabla \times \mathbf{H} = -i\omega\mathbf{D} . \tag{3.92b}$$

Note that $D = \varepsilon_0\varepsilon_1\mathbf{E}$ inside the cylinder and $D = \varepsilon_0\varepsilon_2\mathbf{E}$ outside it. We now look for transverse solutions of (3.92a) and (3.92b), i.e., modes for which $\nabla \cdot \mathbf{E} = 0$. Operating on (3.92a) and (3.92b) with the curl operator, using second Maxwell's equation, i.e., (1.33b), and eliminating either $\mathbf{E}$ or $\mathbf{H}$, we find that both fields have to satisfy Helmholtz equation

$$\nabla^2 \begin{pmatrix} \mathbf{E} \\ \mathbf{H} \end{pmatrix} + k^2 \begin{pmatrix} \mathbf{E} \\ \mathbf{H} \end{pmatrix} = 0 , \tag{3.93}$$

where k is given by $k_1 = \sqrt{\varepsilon_1} k_0$ (with $k_0 = \omega/c$) inside the EG cylinder and by $k_2 = \sqrt{\varepsilon_2} k_0$ outside it. The above vector wave functions can be generated from the cylindrical solutions of the corresponding scalar wave equation $\nabla^2 \psi + k^2 \psi = 0$, which in cylindrical polar coordinates ρ, ϕ, and z is

$$\frac{1}{\rho} \frac{\partial}{\partial \rho} \left(\rho \frac{\partial \psi}{\partial \rho} \right) + \frac{1}{\rho^2} \frac{\partial^2 \psi}{\partial \phi^2} + \frac{\partial^2 \psi}{\partial z^2} + k^2 \psi = 0 . \tag{3.94}$$

Separable solutions to (3.94) that are single-valued functions of ϕ are of the form

$$\psi_m = Z_m(\kappa \rho) \exp(im\phi) \exp(iqz) , \tag{3.95}$$

where $\kappa = \sqrt{k^2 - q^2}$ and Z_m is a solution to the *Bessel equation*

$$\frac{1}{\rho} \frac{d}{d\rho} \left(\rho \frac{d}{d\rho} Z_m \right) + \left(\kappa^2 - \frac{m^2}{\rho^2} \right) Z_m = 0 . \tag{3.96}$$

Here Z_m is a cylindrical Bessel or Hankel function of the first kind,[9] J_m and H_m of order m. Two sets of solutions of the vector Helmholtz equation are given by Bohren and Huffman [31]

$$\mathbf{M}_m = \nabla \times (\mathbf{e}_z \psi_m) , \tag{3.97a}$$

$$\mathbf{N}_m = \frac{\nabla \times \mathbf{M}_m}{k} , \tag{3.97b}$$

where $\mathbf{e}_z$ is unit vector in the z-direction. In component form these vector harmonics are

$$\mathbf{M}_m = \kappa \left(im \frac{Z_m(\kappa \rho)}{\kappa \rho} \mathbf{e}_\rho - Z'_m(\kappa \rho) \mathbf{e}_\phi \right) e^{i(qz + m\phi)} , \tag{3.98a}$$

$$\mathbf{N}_m = \frac{\kappa}{k} \left(iq Z'_m(\kappa \rho) \mathbf{e}_\rho - qm \frac{Z_m(\kappa \rho)}{\kappa \rho} \mathbf{e}_\phi + \kappa Z_m(\kappa \rho) \mathbf{e}_z \right) e^{i(qz + m\phi)} , \tag{3.98b}$$

where as remarked before, $\mathbf{e}_\rho$ and $\mathbf{e}_\phi$ are unit vectors in ρ and ϕ-directions, respectively, and $Z'_m(x) = dZ_m(x)/dx$. Unlike the planar case in the previous

---

[9]The Hankel function of the first kind H_m (also called a Bessel function of the third kind) can be written, in terms of Bessel function of the first and second kind, J_m and Y_m (occasionally the notation N_m is also used for the Bessel functions of the second type Y_m and called the Neumann function), as $H_m = J_m + iY_m$.

sections, pure TE or TM modes cannot in general exist independently [32, 37, 38]. Mixed modes have to be used in order to satisfy the BCs at the surface of the EG cylinder. We expand the internal and external fields of a SPP of type m, q in the form

$$\mathbf{E} = A_m \mathbf{M}_m + B_m \mathbf{N}_m , \tag{3.99a}$$

$$\mathbf{H} = -\frac{ik}{\mu_0 \omega} (A_m \mathbf{N}_m + B_m \mathbf{M}_m) . \tag{3.99b}$$

The internal and external radial functions are $J_m(\kappa_1 \rho)$ and $H_m(\kappa_2 \rho)$, where $\kappa_j = \sqrt{k_j^2 - q^2}$ (with $j = 1, 2$). From now, the factor $\exp(im\phi) \exp(iqz)$ which is common to all the fields will be omitted for brevity. The tangential field components inside and outside the EG cylinder are given by

$$H_z = \frac{-i}{\mu_0 \omega} \begin{cases} A_{m1} \kappa_1^2 J_m(\kappa_1 \rho) , & \rho \leq a , \\ A_{m2} \kappa_2^2 H_m(\kappa_2 \rho) , & \rho \geq a , \end{cases} \tag{3.100}$$

$$E_z = \begin{cases} B_{m1} \dfrac{\kappa_1^2}{k_1} J_m(\kappa_1 \rho) , & \rho \leq a , \\ B_{m2} \dfrac{\kappa_2^2}{k_2} H_m(\kappa_2 \rho) , & \rho \geq a , \end{cases} \tag{3.101}$$

$$H_\phi = \frac{i}{\mu_0 \omega} \begin{cases} A_{m1} \dfrac{qm}{\rho} J_m(\kappa_1 \rho) + B_{m1} k_1 \kappa_1 J'_m(\kappa_1 \rho) , & \rho \leq a , \\ A_{m2} \dfrac{qm}{\rho} H_m(\kappa_2 \rho) + B_{m2} k_2 \kappa_2 H'_m(\kappa_2 \rho) , & \rho \geq a , \end{cases} \tag{3.102}$$

$$E_\phi = - \begin{cases} A_{m1} \kappa_1 J'_m(\kappa_1 \rho) + B_{m1} \dfrac{qm}{k_1 \rho} J_m(\kappa_1 \rho) , & \rho \leq a , \\ A_{m2} \kappa_2 H'_m(\kappa_2 \rho) + B_{m2} \dfrac{qm}{k_2 \rho} H_m(\kappa_2 \rho) , & \rho \geq a , \end{cases} \tag{3.103}$$

where the relations between the coefficients A_{m1}, A_{m2}, B_{m1}, and B_{m2} can be determined from the matching BCs at the separation surface. We note that the tangential components E_z, E_ϕ, H_z, and H_ϕ have to be continuous at $\rho = a$. This requirement yields the following homogeneous linear system of equations:

$$\begin{pmatrix} \kappa_1^2 J_m(\kappa_1 a) & -\kappa_2^2 H_m(\kappa_2 a) & 0 & 0 \\ 0 & 0 & \dfrac{\kappa_1^2}{k_1} J_m(\kappa_1 a) & -\dfrac{\kappa_2^2}{k_2} H_m(\kappa_2 a) \\ \dfrac{qm}{a} J_m(\kappa_1 a) & -\dfrac{qm}{a} H_m(\kappa_2 a) & k_1 \kappa_1 J'_m(\kappa_1 a) & -k_2 \kappa_2 H'_m(\kappa_2 a) \\ \kappa_1 J'_m(\kappa_1 a) & -\kappa_2 H'_m(\kappa_2 a) & \dfrac{qm}{ak_1} J_m(\kappa_1 a) & \dfrac{qm}{ak_2} H_m(\kappa_2 a) \end{pmatrix} \begin{pmatrix} A_{m1} \\ A_{m2} \\ B_{m1} \\ B_{m2} \end{pmatrix} = 0 . \tag{3.104}$$

The dispersion relation of SPPs of an EG cylinder is then obtained by solving the condition that the determinant of the matrix in (3.104) must be zero in order to exist a non-trivial solution, yielding

$$
\left[\frac{1}{\kappa_2} \frac{H'_m(\kappa_2 a)}{H_m(\kappa_2 a)} - \frac{1}{\kappa_1} \frac{J'_m(\kappa_1 a)}{J_m(\kappa_1 a)} \right] \left[\frac{k_2^2}{\kappa_2} \frac{H'_m(\kappa_2 a)}{H_m(\kappa_2 a)} - \frac{k_1^2}{\kappa_1} \frac{J'_m(\kappa_1 a)}{J_m(\kappa_1 a)} \right]
$$
$$
= \frac{m^2 q^2}{a^2} \left(\frac{1}{\kappa_1^2} - \frac{1}{\kappa_2^2} \right)^2 . \tag{3.105}
$$

Equation (3.105) is a transcendental equation and determines the dependence of ω on m and q with a as a parameter. This expression must be solved numerically. Partly because of the appearance of the azimuthal quantum number m, the mode spectrum is much richer than in the analogous case of the SPP modes of an EG slab. The above equation describes modes in which the field amplitudes are maximum at the interface $\rho = a$ and decrease with distance from this interface. Therefore, $q^2 > k_j^2$ and κ_j has to be imaginary. Also, it should be noticed that (3.105) can be written in terms of the modified Bessel functions of the first kind $I_m(x)$ and second kind $K_m(x)$, as [32]

$$
\left[\frac{1}{\kappa_2} \frac{K'_m(\kappa_2 a)}{K_m(\kappa_2 a)} - \frac{1}{\kappa_1} \frac{I'_m(\kappa_1 a)}{I_m(\kappa_1 a)} \right] \left[\frac{k_2^2}{\kappa_2} \frac{K'_m(\kappa_2 a)}{K_m(\kappa_2 a)} - \frac{k_1^2}{\kappa_1} \frac{I'_m(\kappa_1 a)}{I_m(\kappa_1 a)} \right]
$$
$$
= \frac{m^2 q^2}{a^2} \left(\frac{1}{\kappa_1^2} - \frac{1}{\kappa_2^2} \right)^2 , \tag{3.106}
$$

if we replace $\kappa_j = \sqrt{k_j^2 - q^2}$ by $\kappa_j = \sqrt{q^2 - k_j^2}$.

An important special case is $q = 0$, for which pure TE^z and TM^{z}[10] modes can exist. In this case (3.105) simplified and yields the equations

$$
k_2 \frac{H'_m(k_2 a)}{H_m(k_2 a)} - k_1 \frac{J'_m(k_1 a)}{J_m(k_1 a)} = 0 , \tag{3.107a}
$$

$$
\frac{1}{k_2} \frac{H'_m(k_2 a)}{H_m(k_2 a)} - \frac{1}{k_1} \frac{J'_m(k_1 a)}{J_m(k_1 a)} = 0 , \tag{3.107b}
$$

for the TM^z and TE^z modes, respectively. Another case for which pure TE^z and TM^z modes exist is $m = 0$ and q arbitrary. In this case (3.106) simplified and yields

---

[10]Here, TE^z (TM^z) means that the electric (magnetic) field of the wave does not have a z-component, so the field configuration is referred to as transverse electric (magnetic) to z.

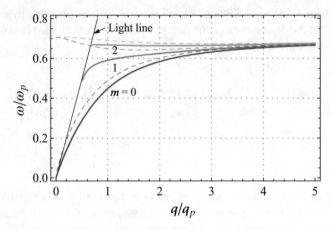

Fig. 3.19 Dispersion curves of the first three non-radiative SPP modes of an EG cylinder with radius $a = c/\omega_p$. Here $q_p = \omega_p/c$ and the dispersion curves of SP modes are also plotted with dashed lines

the equations

$$\frac{k_2^2}{\kappa_2} \frac{K'_0(\kappa_2 a)}{K_0(\kappa_2 a)} - \frac{k_1^2}{\kappa_1} \frac{I'_0(\kappa_1 a)}{I_0(\kappa_1 a)} = 0 , \qquad (3.108a)$$

$$\frac{1}{\kappa_2} \frac{K'_0(\kappa_2 a)}{K_0(\kappa_2 a)} - \frac{1}{\kappa_1} \frac{I'_0(\kappa_1 a)}{I_0(\kappa_1 a)} = 0 , \qquad (3.108b)$$

for the TM^z and TE^z modes, respectively. As an example, we show in Fig. 3.19 the first three non-radiative SPP modes of an EG cylinder with radius $a = c/\omega_p$.

3.7.2   Extinction Property

The theory of light scattering and absorption by an infinitely long cylinder was first studied by Ignatowski [33] in 1905 and then by Seitz [34] in 1906. The BVP of scattering of a plane wave at normal incidence by a dielectric infinite cylinder was later independently solved by Lord Rayleigh [35] in 1918. Also, by Wait in 1955 [36], a solution has been given for the problem of a plane wave incident obliquely on a circular cylinder of infinite length. A very good summary of the light scattering and absorption solutions by a cylinder has been given by Bohren and Huffman [31].

Now, let an infinitely long EG cylinder (compared with its diameter) is illuminated by a plane wave at oblique incidence. This means that the projection of the propagation vector k_2 on the xz plane makes an angle ζ with the z-axis, as shown in Fig. 3.20. We define TE^z and TM^z polarizations of the incident wave as follows: (a)

Fig. 3.20 An infinitely long
EG cylinder and a cylindrical
polar coordinate system. The
z-axis lies along the axis of
the EG cylinder. The system
is illuminated by
electromagnetic plane wave
with TM^z polarization at the
incidence angle ζ. The z-axis
coincides with the axis of the
cylinder and the xz plane
shows the incidence plane

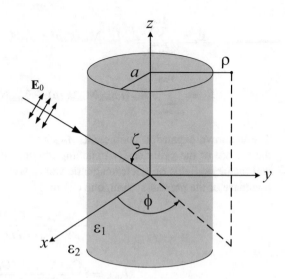

The case in which the incident electric field is perpendicular to the incident plane,
i.e., the plane that contains the z-axis and the direction of propagation of the incident
wave (xz plane), is referred to as TE^z polarization. (b) The case in which the incident
electric field is parallel to the incident plane is referred to as TM^z polarization.

Again, the vector cylindrical harmonics functions can be defined according
to the previous section. Here k and κ are given as in the previous section and
$q = -k_2 \cos \zeta$. Also, $Z_n(\kappa \rho)$ represents a cylindrical Bessel or Hankel function,
and is chosen as follows. Inside the cylinder $J_m(\kappa_1 \rho)$ is used to indicate that the
transmitted wave is a wave traveling in the inward radial direction. Outside the
cylinder $J_m(\kappa_2 \rho)$ and $H_m(\kappa_2 \rho)$ are used for the incident and scattered waves,
respectively. The Hankel function is chosen to indicate that the scattered field is
a wave traveling in the outward radial direction.

3.7.2.1 TE^z Polarization

For TE^z polarization, the incident electric field $\mathbf{E}_i$ can be expanded as [31]

$$\mathbf{E}_i = -i \sum_{m=-\infty}^{+\infty} E_m \mathbf{M}_m(\kappa_2 \rho) , \tag{3.109}$$

where $E_m = E_0(-i)^m/(k_2 \sin \zeta)$. The transmitted and scattered electric fields can
be represented as

$$\mathbf{E}_t = \sum_{m=-\infty}^{+\infty} E_m \left[g_m \mathbf{M}_m(\kappa_1 \rho) + f_m \mathbf{N}_m(\kappa_1 \rho) \right] , \tag{3.110}$$

$$\mathbf{E}_s = \sum_{m=-\infty}^{+\infty} E_m \left[i a_{m\perp} \mathbf{M}_m(\kappa_2 \rho) + b_{m\perp} \mathbf{N}_m(\kappa_2 \rho) \right] . \tag{3.111}$$

The unknown expansion coefficients $a_{m\perp}$ and $b_{m\perp}$ are determined by the BCs at the surface of the cylinder. By matching the tangential components of the electric and magnetic fields of electromagnetic waves with the angular frequency ω at the boundary of the present system, one can find

$$a_{m\perp} = -\frac{A_m V_m - i C_m D_m}{W_m V_m + i D_m^2} , \tag{3.112a}$$

$$b_{m\perp} = i \frac{C_m W_m + A_m D_m}{W_m V_m + i D_m^2} , \tag{3.112b}$$

where

$$A_m = i v \left[v J_m(v) J'_m(u) - u J_m(u) J'_m(v) \right] ,$$

$$C_m = u J_m(v) J_m(u) \frac{qm}{k_2} \left(\frac{v^2}{u^2} - 1 \right) ,$$

$$D_m = u H_m(v) J_m(u) \frac{qm}{k_2} \left(\frac{v^2}{u^2} - 1 \right) ,$$

$$V_m = v \left[\frac{k_1^2}{k_2^2} v H_m(v) J'_m(u) - u J_m(u) H'_m(v) \right] ,$$

$$W_m = i v \left[u J_m(u) H'_m(v) - v H_m(v) J'_m(u) \right] ,$$

with $v = a\sqrt{k_2^2 - q^2}$ and $u = a\sqrt{k_1^2 - q^2}$. In this case the extinction, scattering, and absorption widths can be expressed as

$$Q_{ext\perp} = \frac{2}{k_2 a} \mathrm{Re} \left[a_{0\perp} + 2 \sum_{m=1}^{+\infty} a_{m\perp} \right] , \tag{3.113}$$

$$Q_{sca\perp} = \frac{2}{k_2 a} \left[|a_{0\perp}|^2 + 2 \sum_{m=1}^{+\infty} \left(|a_{m\perp}|^2 + |b_{m\perp}|^2 \right) \right] , \tag{3.114}$$

$$Q_{abs\perp} = Q_{ext\perp} - Q_{sca\perp} , \tag{3.115}$$

respectively, in units of the geometric width $2a$. Let us note that the denominators of $a_{m\perp}$ and $b_{m\perp}$ vanish at the frequencies of the mixed TE^z and TM^z SPP modes and from the equation $W_m V_m + i D_m^2 = 0$, the well-known dispersion relation associated with SPP modes propagating along the z-axis having the factor $\exp(iqz + im\phi)$ can be obtained as (3.105).

3.7.2.2 TM^z Polarization

For TM^z polarization, the incident electric field $\mathbf{E}_i$ can be expanded as

$$\mathbf{E}_i = \sum_{m=-\infty}^{+\infty} E_m \mathbf{N}_m(\kappa_2\rho) . \tag{3.116}$$

The transmitted and scattered electric fields can be represented as

$$\mathbf{E}_t = \sum_{m=-\infty}^{+\infty} E_m \left[p_m \mathbf{M}_m(\kappa_1\rho) + q_m \mathbf{N}_m(\kappa_1\rho) \right] , \tag{3.117}$$

$$\mathbf{E}_s = -\sum_{m=-\infty}^{+\infty} E_m \left[i a_{m\|} \mathbf{M}_m(\kappa_2\rho) + b_{m\|} \mathbf{N}_m(\kappa_2\rho) \right] . \tag{3.118}$$

The coefficients of the scattered field can be written in the form

$$a_{m\|} = -\frac{C_m V_m - B_m D_m}{W_m V_m + i D_m^2} , \tag{3.119a}$$

$$b_{m\|} = \frac{B_m W_m + i D_m C_m}{W_m V_m + i D_m^2} , \tag{3.119b}$$

where D_m, C_m, and so on were defined in the preceding section and

$$B_m = v \left[\frac{k_1^2}{k_2^2} v J_m(v) J'_m(u) - u J_m(u) J'_m(v) \right] .$$

In this case, the extinction, scattering, and absorption widths can be expressed as

$$Q_{ext\|} = \frac{2}{k_2 a} \mathrm{Re} \left[b_{0\|} + 2 \sum_{m=1}^{+\infty} b_{m\|} \right] , \tag{3.120}$$

$$Q_{sca\parallel} = \frac{2}{k_2 a} \left[|b_{0\parallel}|^2 + 2 \sum_{m=1}^{+\infty} \left(|b_{m\parallel}|^2 + |a_{m\parallel}|^2 \right) \right] , \tag{3.121}$$

$$Q_{abs\parallel} = Q_{ext\parallel} - Q_{sca\parallel} , \tag{3.122}$$

respectively, in units of the geometric width $2a$. In Fig. 3.21, the extinction spectra of an EG cylinder in vacuum with various parameter ζ are plotted for both TE^z and TM^z polarizations when $a = c/\omega_p$. The simple case $q = 0$, which corresponds to normal incidence of the irradiating plane wave, is shown in panel (a). When the incident electric field is parallel to the cylinder axis, i.e., TM^z polarization, no SPP resonance peak is found in the far-field spectra because the polarization direction along the axis of EG cylinder cannot excite the collective motions of the conduction electrons [39–41]. However, for the TE^z polarization, i.e., $\mathbf{E} \perp \mathbf{e}_z$, one can observe that for a very small cylinder, $a = c/\omega_p$ the spectrum is dominated by the SPP modes peaks in the region between $\omega < \omega_p/\sqrt{2}$. As shown in panels (b)–(d) of Fig. 3.21, we find for the incidence angle $\zeta \neq \pi/2$, the incident polarization is not along the axis of the system and the SP resonance peak can be found in the far-field spectra of TM^z polarization. Also, by decreasing the angle of incidence wave with the system, the extinction peaks decrease.

Moreover, in Fig. 3.22 we use both the full electromagnetic formula, i.e., (3.113) when $\zeta = \pi/2$ and the electrostatic approximation formula, i.e., (2.81b) to calculate and compare the results of extinction cross section of a thin EG cylinder in vacuum for different values of a. It is clear that for $a \approx 0.03c/\omega_p$ (about 1nm for a sodium cylinder) the electrostatic result is quite equal with the result of the full electromagnetic theory. Therefore, to use the electrostatic approximation formula, the radius of a thin EG cylinder should be about a few *nanometers*.

In Fig. 3.23, we have investigated the dependence of the extinction spectra on the parameter a for both TE^z and TM^z polarizations when $\zeta = \pi/4$. As shown in panels (a)–(d), when a increases, the band of surface modes becomes richer in structure and shifts to lower frequencies. Physically, for a *nano* cylinder, only the mode $m = 1$ is observable [42, 43]. When the radius increases the contribution of the other modes is also noticeable.

3.8   Surface Plasmon Polaritons of Curved Semi-Infinite Electron Gases

As mentioned before, a p-polarized SPP cannot propagate on the surface of a planar EG-insulator interface. However, the situation is quite different if the EG-insulator interface is not planar but curved [44, 45]. Here we investigate the propagation of p-polarized and s-polarized electromagnetic waves circumferentially around a portion of a circular interface between an insulator and a semi-infinite EG when the EG is concave toward the insulator. We note that the electromagnetic fields in this case are not required to be single valued [44–46].

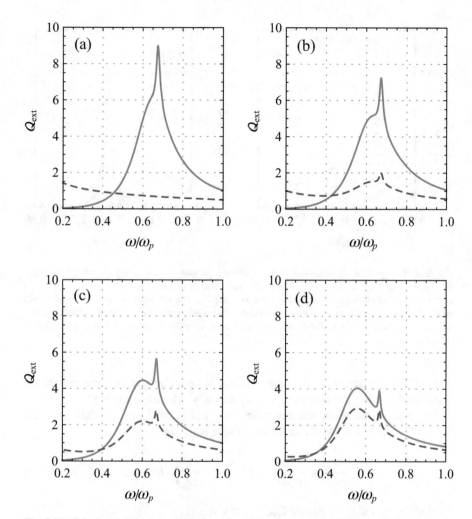

Fig. 3.21 Calculated extinction width (in units of the geometric width) of an EG cylinder in vacuum with respect to the dimensionless frequency ω/ω_p for both TE^z and TM^z polarizations when $a = c/\omega_p$ and $\gamma = 0.01\omega_p$. The red solid curves are for TE^z (or $\mathbf{E} \perp \mathbf{e}_z$) and the blue dashed ones for TM^z (or $\mathbf{E} \parallel \mathbf{e}_z$). The different panels refer to (**a**) $\zeta = \pi/2$, (**b**) $\zeta = \pi/3$, (**c**) $\zeta = \pi/4$, and (**d**) $\zeta = \pi/6$

3.8.1 p-Polarized Wave

Figure 3.24 shows a portion of a circular interface between an insulator and a semi-infinite EG with relative dielectric constants ε_1 and ε_2, respectively. The insulator occupies the region corresponding to the range $\rho < a$ in the usual cylindrical coordinate system (ρ, ϕ, z). We assume the z-direction to be coincident with the axis of the curved system. Also, all electromagnetic field components contain a common factor $\exp(im\phi - i\omega t)$, where ω is the angular frequency of the wave. Let

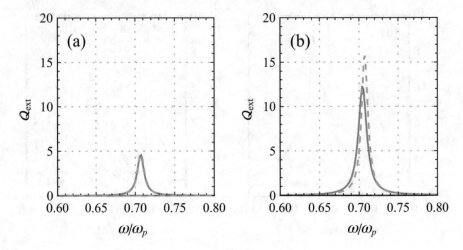

Fig. 3.22 Comparison of extinction width spectrum (in units of the geometric width) of a thin EG cylinder in vacuum from (3.113) for TE^z polarization (when $\zeta = \pi/2$ and $\gamma = 0.01\omega_p$) with the result of electrostatic approximation, i.e., (2.81b). The red solid curve is the result of full electromagnetic formula and the green dashed one shows the result of electrostatic approximation. The different panels refer to (**a**) $a = 0.03c/\omega_p$ (about 1nm for a sodium cylinder) and (**b**) $a = 0.1c/\omega_p$

us note that the parameter m is not required to be an integer, because we are not considering a complete cylinder, but only a locally cylindrical surface.

For p-polarized wave, the magnetic field **H** is parallel to the z-axis and from Maxwell's equations, we obtain the Helmholtz equation

$$\frac{d^2 H_z}{d\rho^2} + \frac{1}{\rho}\frac{d H_z}{d\rho} + \left(\varepsilon_1 \frac{\omega^2}{c^2} - \frac{m^2}{\rho^2}\right) H_z = 0 , \qquad (3.123)$$

outside the curved semi-infinite EG, i.e., $\rho < a$ and

$$\frac{d^2 H_z}{d\rho^2} + \frac{1}{\rho}\frac{d H_z}{d\rho} - \left(|\varepsilon_2|\frac{\omega^2}{c^2} + \frac{m^2}{\rho^2}\right) H_z = 0, \qquad (3.124)$$

inside the EG, i.e., $\rho > a$. Therefore, for the present BVP, the solution of (3.123) and (3.124) may be written as

$$H_z = H_0 \begin{cases} \dfrac{J_m\left(\sqrt{\varepsilon_1}\dfrac{\omega}{c}\rho\right)}{J_m\left(\sqrt{\varepsilon_1}\dfrac{\omega}{c}a\right)} , & \rho \leq a , \\[4mm] \dfrac{K_m\left(\sqrt{|\varepsilon_2|}\dfrac{\omega}{c}\rho\right)}{K_m\left(\sqrt{|\varepsilon_2|}\dfrac{\omega}{c}a\right)} , & \rho \geq a , \end{cases} \qquad (3.125)$$

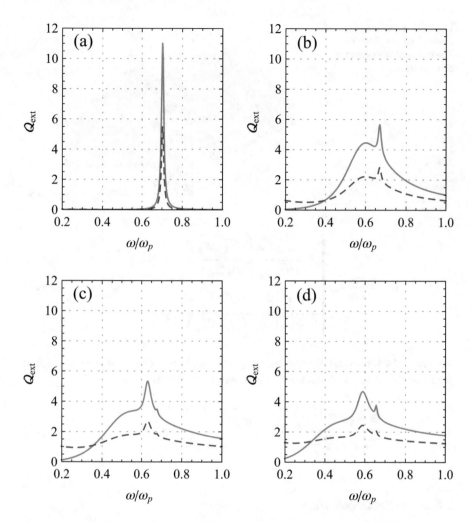

Fig. 3.23 Calculated extinction width (in units of the geometric width) of an EG cylinder in vacuum with respect to the dimensionless frequency ω/ω_p for both TE^z and TM^z polarizations when $\zeta = \pi/4$ and $\gamma = 0.01\omega_p$. The red solid curves are for $\mathbf{E} \perp \mathbf{e}_z$ and the blue dashed ones for $\mathbf{E} \parallel \mathbf{e}_z$. The different panels refer to (**a**) $a = 0.1c/\omega_p$, (**b**) $a = c/\omega_p$, (**c**) $a = 1.5c/\omega_p$, and (**d**) $a = 2c/\omega_p$

Also, using Maxwell's equations, we find [see Sect. 9.1.1]

$$E_\phi = \frac{-i}{\omega\varepsilon_0\varepsilon} \frac{\partial H_z}{\partial \rho} \, , \tag{3.126}$$

thus for tangential electric field component inside and outside the EG, we have

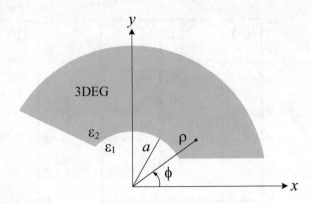

Fig. 3.24 Schematic view of a curved semi-infinite EG. The radius of the system is denoted by a

$$
E_\phi = \frac{-i}{c\varepsilon_0} H_0
\begin{cases}
\dfrac{1}{\sqrt{\varepsilon_1}} \dfrac{J'_m\left(\sqrt{\varepsilon_1}\dfrac{\omega}{c}\rho\right)}{J_m\left(\sqrt{\varepsilon_1}\dfrac{\omega}{c}a\right)}, & \rho \le a, \\[4ex]
-\dfrac{1}{\sqrt{|\varepsilon_2|}} \dfrac{K'_m\left(\sqrt{|\varepsilon_2|}\dfrac{\omega}{c}\rho\right)}{K_m\left(\sqrt{|\varepsilon_2|}\dfrac{\omega}{c}a\right)}, & \rho \ge a,
\end{cases}
\tag{3.127}
$$

Matching the tangential electric and magnetic field components at $\rho = a$, we find the following dispersion relation for p-polarized modes as

$$
\frac{K'_m\left(\sqrt{|\varepsilon_2|}\dfrac{\omega}{c}a\right)}{K_m\left(\sqrt{|\varepsilon_2|}\dfrac{\omega}{c}a\right)} = -\sqrt{\frac{|\varepsilon_2|}{\varepsilon_1}} \frac{J'_m\left(\sqrt{\varepsilon_1}\dfrac{\omega}{c}a\right)}{J_m\left(\sqrt{\varepsilon_1}\dfrac{\omega}{c}\rho\right)}.
\tag{3.128}
$$

3.8.2 s-Polarized Wave

For a s-polarized wave, the electric field **E** is parallel to the z-axis and from Maxwell' equations, we obtain the Helmholtz equation

$$
\frac{d^2 E_z}{d\rho^2} + \frac{1}{\rho}\frac{dE_z}{d\rho} + \left(\varepsilon_1 \frac{\omega^2}{c^2} - \frac{m^2}{\rho^2}\right) E_z = 0,
\tag{3.129}
$$

outside the curved semi-infinite EG, and

$$
\frac{d^2 E_z}{d\rho^2} + \frac{1}{\rho}\frac{dE_z}{d\rho} - \left(|\varepsilon_2| \frac{\omega^2}{c^2} + \frac{m^2}{\rho^2}\right) E_z = 0,
\tag{3.130}
$$

inside the EG. Therefore, the solution of (3.129) and (3.130) may be written as

$$E_z = E_0 \begin{cases} \dfrac{J_m\left(\sqrt{\varepsilon_1}\dfrac{\omega}{c}\rho\right)}{J_m\left(\sqrt{\varepsilon_1}\dfrac{\omega}{c}a\right)} \,, & \rho \le a \,, \\[4ex] \dfrac{K_m\left(\sqrt{|\varepsilon_2|}\dfrac{\omega}{c}\rho\right)}{K_m\left(\sqrt{|\varepsilon_2|}\dfrac{\omega}{c}a\right)} \,, & \rho \ge a \,, \end{cases} \tag{3.131}$$

Also, using

$$H_\phi = \frac{i}{\omega\mu_0}\frac{\partial E_z}{\partial \rho} \,, \tag{3.132}$$

the tangential magnetic field component inside and outside the EG can be given by

$$H_\phi = \frac{i}{c\mu_0} E_0 \begin{cases} \sqrt{\varepsilon_1}\,\dfrac{{J'}_m\left(\sqrt{\varepsilon_1}\dfrac{\omega}{c}\rho\right)}{J_m\left(\sqrt{\varepsilon_1}\dfrac{\omega}{c}a\right)} \,, & \rho \le a \,, \\[4ex] \sqrt{|\varepsilon_2|}\,\dfrac{{K'}_m\left(\sqrt{|\varepsilon_2|}\dfrac{\omega}{c}\rho\right)}{K_m\left(\sqrt{|\varepsilon_2|}\dfrac{\omega}{c}a\right)} \,, & \rho \ge a \,, \end{cases} \tag{3.133}$$

and after imposing the BCs, the dispersion relation for s-polarized modes can be obtained as

$$\frac{{K'}_m\left(\sqrt{|\varepsilon_2|}\dfrac{\omega}{c}a\right)}{K_m\left(\sqrt{|\varepsilon_2|}\dfrac{\omega}{c}a\right)} = \sqrt{\frac{\varepsilon_1}{|\varepsilon_2|}}\,\frac{{J'}_m\left(\sqrt{\varepsilon_1}\dfrac{\omega}{c}a\right)}{J_m\left(\sqrt{\varepsilon_1}\dfrac{\omega}{c}\rho\right)} \,. \tag{3.134}$$

Figure 3.25 is a plot of ω/ω_p versus m when $a = 30c/\omega_p$ and $\varepsilon_1 = 2$ for (a) p-polarized and (b) s-polarized modes. One can see that the dispersion curves of these surface waves possess many branches.

3.9 Plasmonic Properties of Electron Gas Spheres

3.9.1 Surface Plasmon Polariton Modes

To obtain the retarded surface modes and extinction properties of a spherical EG of arbitrary size we start from Helmholtz equation, i.e., (3.93), where k is given by $k_1 = \sqrt{\varepsilon_1}k_0$ inside the EG sphere and by $k_2 = \sqrt{\varepsilon_2}k_0$ outside it. Again, (3.93) can be generated from the spherical solutions of the corresponding scalar wave equation $\nabla^2\psi + k^2\psi = 0$, which in spherical polar coordinates r, θ and φ (see Fig. 3.26) is

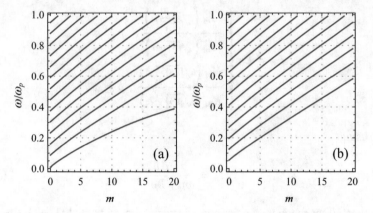

Fig. 3.25 Dispersion curve of SPP modes of a curved EG-insulator system when $a = 30c/\omega_p$ and $\varepsilon_1 = 2$. The different panels refer to (**a**) p-polarized modes and (**b**) s-polarized modes

Fig. 3.26 A spherical polar coordinate system centered on a spherical EG of radius a. Also the figure shows an uniform plane wave incident on the system

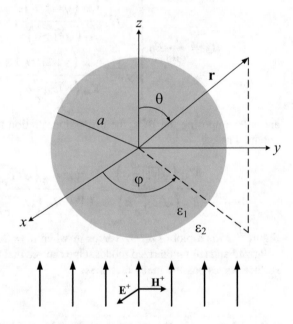

$$\frac{1}{r^2}\frac{\partial}{\partial r}\left(r^2\frac{\partial \psi}{\partial r}\right) + \frac{1}{r^2\sin\theta}\frac{\partial}{\partial\theta}\left(\sin\theta\frac{\partial \psi}{\partial\theta}\right) + \frac{1}{r^2\sin^2\theta}\frac{\partial^2\psi}{\partial\varphi^2} + k^2\psi = 0 \,. \quad (3.135)$$

We seek particular solutions to (3.135) of the form

$$\psi_{e\atop o}\ell m = z_\ell(kr)P_\ell^m(\cos\theta)^{\sin}_{\cos}m\varphi \,, \quad (3.136)$$

where $\ell = 1, 2, \ldots, m = 0, \pm 1, \pm 2, \ldots, \pm \ell$ and subscripts e and o denote *even* and *odd*. Also, $P_\ell^m(\cos\theta)$ is the Legendre function of order ℓ and degree m and $z_\ell(kr)$

is a solution to the *spherical Bessel equation*

$$\frac{1}{r}\frac{d}{dr}\left(r\frac{d}{dr}z_\ell\right) + \left[r^2 - \left(\ell + \frac{1}{2}\right)^2\right]z_\ell = 0\,. \tag{3.137}$$

Here z_ℓ is a spherical Bessel or spherical Hankel function of the first kind,[11] i.e., j_ℓ and h_ℓ. Again, the two types of corresponding solutions of the vector Helmholtz equation, i.e., (3.93) are given by Bohren and Huffman [31]

$$\mathbf{M}_{\substack{e\\o}\ell m} = \nabla \times \left(\mathbf{r}\psi_{\substack{e\\o}\ell m}\right)\,, \tag{3.138a}$$

$$\mathbf{N}_{\substack{e\\o}\ell m} = \frac{\nabla \times \mathbf{M}_{\substack{e\\o}\ell m}}{k}\,, \tag{3.138b}$$

which, in component form, may be written as

$$\mathbf{M}_{\substack{e\\o}\ell m} = \pm m_{\substack{\cos\\\sin}}^{} m\varphi\, \frac{P_\ell^m(\cos\theta)}{\sin\theta} z_\ell(kr)\mathbf{e}_\theta - {}_{\substack{\sin\\\cos}}^{} m\varphi\, \frac{dP_\ell^m(\cos\theta)}{d\theta} z_\ell(kr)\mathbf{e}_\varphi\,, \tag{3.139a}$$

$$\mathbf{N}_{\substack{e\\o}\ell m} = \ell(\ell+1)_{\substack{\sin\\\cos}}^{} m\varphi\, P_\ell^m(\cos\theta)\frac{z_\ell(kr)}{kr}\mathbf{e}_r + {}_{\substack{\sin\\\cos}}^{} m\varphi\, \frac{dP_\ell^m(\cos\theta)}{d\theta}\frac{[krz_\ell(kr)]'}{kr}\mathbf{e}_\theta$$

$$\pm m_{\substack{\cos\\\sin}}^{} m\varphi\, \frac{P_\ell^m(\cos\theta)}{\sin\theta}\frac{[krz_\ell(kr)]'}{kr}\mathbf{e}_\varphi\,, \tag{3.139b}$$

where the prime indicates differentiation with respect to the argument in parentheses. Let us note that unlike the cylindrical case, pure TE^r and TM$^{r\,12}$ modes can in general exist independently. If we choose $\mathbf{E}$ to be given by $M_{o\ell m}$, then $\mathbf{H}$ is proportional to $N_{o\ell m}$ and the solution is called a TE^r mode. The other independent solution, i.e., $\mathbf{E} \propto \mathbf{N}_{e\ell m}$ and $\mathbf{H} \propto \mathbf{M}_{e\ell m}$, is called a TM^r mode [39]. We now derive the equations for the frequencies of the various retarded SPP modes.

3.9.1.1 TE^r Modes

For TE^r polarization $E_r = 0$ and the internal and external fields of the surface mode of type ℓ, m can be expanded as

---

[11]The spherical Hankel function of the first kind h_ℓ can be written in terms of spherical Bessel function of the first and second kind j_ℓ and y_ℓ (the symbol n_ℓ is often used instead of y_ℓ).
[12]Here, TE^r (TM^r) means that the electric (magnetic) field of the wave does not have a r-component, so the field configuration is referred to as transverse electric (magnetic) to r.

$$\mathbf{E} = A_{\ell m} \mathbf{M}_{o\ell m} \ , \tag{3.140a}$$

$$\mathbf{H} = -\frac{ik}{\mu_0 \omega} A_{\ell m} \mathbf{N}_{o\ell m} \ , \tag{3.140b}$$

where the internal and external radial functions are $j_\ell(k_1 r)$ and $h_\ell(k_2 r)$. The tangential field components inside and outside the EG sphere are given by

$$E_\theta = \cos m\varphi \, P_\ell^m(\cos\theta) \frac{m}{\sin\theta} \begin{cases} A_{\ell m 1} j_\ell(k_1 r) \ , \ r \le a \ , \\ A_{\ell m 2} h_\ell(k_2 r) \ , \ r \ge a \ , \end{cases} \tag{3.141}$$

$$E_\varphi = -\sin m\varphi \, \frac{\mathrm{d}P_\ell^m(\cos\theta)}{\mathrm{d}\theta} \begin{cases} A_{\ell m 1} j_\ell(k_1 r) \ , \ r \le a \ , \\ A_{\ell m 2} h_\ell(k_2 r) \ , \ r \ge a \ , \end{cases} \tag{3.142}$$

$$H_\theta = -\frac{i \sin m\varphi}{r \mu_0 \omega} \frac{\mathrm{d}P_\ell^m(\cos\theta)}{\mathrm{d}\theta} \begin{cases} A_{\ell m 1} [k_1 r j_\ell(k_1 r)]' \ , \ r \le a \ , \\ A_{\ell m 2} [k_2 r h_\ell(k_2 r)]' \ , \ r \ge a \ , \end{cases} \tag{3.143}$$

$$H_\phi = -\frac{im \cos m\varphi}{r \mu_0 \omega} \frac{P_\ell^m(\cos\theta)}{\sin\theta} \begin{cases} A_{\ell m 1} [k_1 r j_\ell(k_1 r)]' \ , \ r \le a \ , \\ A_{\ell m 2} [k_2 r h_\ell(k_2 r)]' \ , \ r \ge a \ . \end{cases} \tag{3.144}$$

The continuity conditions at $r = a$ yield the following homogeneous linear system of equations:

$$\begin{pmatrix} j_\ell(k_1 a) & -h_\ell(k_2 a) \\ [k_1 a j_\ell(k_1 a)]' & -[k_2 a h_\ell(k_2 a)]' \end{pmatrix} \begin{pmatrix} A_{\ell m 1} \\ A_{\ell m 2} \end{pmatrix} = 0 \ . \tag{3.145}$$

The equation for the frequencies of the TE^r modes is then obtained by solving the condition that the determinant of the matrix in (3.145) must be zero in order to exist a non-trivial solution, yielding

$$j_\ell(k_1 a) \, [k_2 a h_\ell(k_2 a)]' - h_\ell(k_2 a) \, [k_1 a j_\ell(k_1 a)]' = 0 \ . \tag{3.146}$$

3.9.1.2 TM^r Modes

For TM^r modes $H_r = 0$ and the internal and external fields of the surface mode of type ℓ, m can be expanded as

$$\mathbf{E} = B_{\ell m} \mathbf{N}_{e\ell m} \ , \tag{3.147a}$$

$$\mathbf{H} = -\frac{ik}{\mu_0 \omega} B_{\ell m} \mathbf{M}_{e\ell m} \ . \tag{3.147b}$$

The tangential field components inside and outside the EG sphere are given by

$$E_\theta = \cos m\varphi \frac{\mathrm{d}P_\ell^m(\cos\theta)}{\mathrm{d}\theta} \begin{cases} B_{\ell m1}\dfrac{[k_1 r j_\ell(k_1 r)]'}{k_1 r} , & r \le a , \\[3mm] B_{\ell m2}\dfrac{[k_2 r h_\ell(k_2 r)]'}{k_2 r} , & r \ge a , \end{cases} \tag{3.148}$$

$$E_\varphi = -m \sin m\varphi \frac{P_\ell^m(\cos\theta)}{\sin\theta} \begin{cases} B_{\ell m1}\dfrac{[k_1 r j_\ell(k_1 r)]'}{k_1 r} , & r \le a , \\[3mm] B_{\ell m2}\dfrac{[k_2 r h_\ell(k_2 r)]'}{k_2 r} , & r \ge a , \end{cases} \tag{3.149}$$

$$H_\theta = \frac{i m \sin m\varphi}{r \mu_0 \omega} \frac{P_\ell^m(\cos\theta)}{\sin\theta} \begin{cases} B_{\ell m1} k_1 r j_\ell(k_1 r) , & r \le a , \\[2mm] B_{\ell m2} k_2 r h_\ell(k_2 r) , & r \ge a , \end{cases} \tag{3.150}$$

$$H_\phi = \frac{i \cos m\varphi}{r \mu_0 \omega} \frac{\mathrm{d}P_\ell^m(\cos\theta)}{\mathrm{d}\theta} \begin{cases} B_{\ell m1} k_1 r j_\ell(k_1 r) , & r \le a , \\[2mm] B_{\ell m2} k_2 r h_\ell(k_2 r) , & r \ge a . \end{cases} \tag{3.151}$$

The continuity conditions at $r = a$ yield the following homogeneous linear system of equations:

$$\begin{pmatrix} \dfrac{[k_1 a j_\ell(k_1 a)]'}{k_1 a} & -\dfrac{[k_2 a h_\ell(k_2 a)]'}{k_2 a} \\[3mm] k_1 a j_\ell(k_1 a) & -k_2 a h_\ell(k_2 a) \end{pmatrix} \begin{pmatrix} B_{\ell m1} \\[2mm] B_{\ell m2} \end{pmatrix} = 0 , \tag{3.152}$$

and the equation for the frequencies of the TM^r modes is

$$\varepsilon_1 j_\ell(k_1 a) [k_2 a h_\ell(k_2 a)]' - \varepsilon_2 h_\ell(k_2 a) [k_1 a j_\ell(k_1 a)]' = 0 , \tag{3.153}$$

which is the well-known result [47].

3.9.2   Extinction Property

The theory of scattering of electromagnetic waves from a homogeneous sphere was first given in explicit form by Lorenz [48] in 1890 and some time later by Mie [49] in 1908. A very good summary of the light scattering and absorption solutions by a sphere has been given by Bohren and Huffman [31]. Here, we consider an isolated EG sphere of radius a with a relative dielectric function ε_1 that embedded in an insulator medium with relative dielectric constant ε_2. As shown in Fig. 3.26, the system irradiated by a z-directed, x-polarized plane wave. We consider the incident electric field as

$$\mathbf{E}_i = E_0 e^{ik_2 z}\mathbf{e}_x , \tag{3.154}$$

where $k_2 = \sqrt{\varepsilon_2}k_0$ and $\mathbf{e}_x = \sin\theta\cos\varphi\mathbf{e}_r + \cos\theta\cos\varphi\mathbf{e}_\theta + \sin\varphi\mathbf{e}_\varphi$. For simplicity, it can be expanded in an infinite series of vector spherical harmonics [31]

$$\mathbf{E}_i = \sum_{\ell=1}^{+\infty} E_\ell\left(\mathbf{M}_{o\ell 1} - i\mathbf{N}_{e\ell 1}\right) , \qquad (3.155)$$

where

$$E_\ell = i^\ell \frac{2\ell+1}{\ell(\ell+1)} E_0 .$$

The corresponding incident magnetic field is

$$\mathbf{H}_i = -\frac{k_2}{\omega\mu_0}\sum_{\ell=1}^{+\infty} E_\ell\left(\mathbf{M}_{e\ell 1} + i\mathbf{N}_{o\ell 1}\right) . \qquad (3.156)$$

The transmitted and scattered fields can be represented as

$$\mathbf{E}_t = \sum_{\ell=1}^{+\infty} E_\ell\left(c_\ell\mathbf{M}_{o\ell 1} - id_\ell\mathbf{N}_{e\ell 1}\right) , \qquad (3.157)$$

$$\mathbf{H}_t = -\frac{k_1}{\omega\mu_0}\sum_{\ell=1}^{+\infty} E_\ell\left(d_\ell\mathbf{M}_{e\ell 1} + ic_\ell\mathbf{N}_{o\ell 1}\right) , \qquad (3.158)$$

$$\mathbf{E}_s = -\sum_{\ell=1}^{+\infty} E_\ell\left(b_\ell\mathbf{M}_{o\ell 1} - ia_\ell\mathbf{N}_{e\ell 1}\right) , \qquad (3.159)$$

$$\mathbf{H}_s = \frac{k_2}{\omega\mu_0}\sum_{\ell=1}^{+\infty} E_\ell\left(a_\ell\mathbf{M}_{e\ell 1} + ib_\ell\mathbf{N}_{o\ell 1}\right) . \qquad (3.160)$$

In the above equations, $\mathbf{M}_{o\ell 1}$ and $\mathbf{N}_{o\ell 1}$ are the vector spherical harmonics and have the forms of (3.139a) and (3.139b) where we have $P_\ell^1(\cos\theta)$ that is the Legendre function of order ℓ and degree one. Also, $z_\ell(kr)$ is chosen as follows. Inside the sphere $j_\ell(k_1r)$ is used and outside the system $j_\ell(k_2r)$ and $h_\ell(k_2r)$ are used for the incident and scattered waves, respectively.

To determine the unknown coefficients a_ℓ, b_ℓ, c_ℓ, and d_ℓ, the appropriate BCs must be enforced. We note that the tangential components E_θ, E_φ, H_θ, and H_φ have to be continuous at $r = a$. This requirement yields four simultaneous linear equations in the expansion coefficients and are easily solved for the coefficients of the field inside the EG sphere, as

$$c_\ell = \frac{j_\ell(k_2 a)\,[k_2 a h_\ell(k_2 a)]' - h_\ell(k_2 a)\,[k_2 a j_\ell(k_2 a)]'}{j_\ell(k_1 a)\,[k_2 a h_\ell(k_2 a)]' - h_\ell(k_2 a)\,[k_1 a j_\ell(k_1 a)]'}\,, \tag{3.161}$$

$$d_\ell = \frac{\sqrt{\varepsilon_1}\,j_\ell(k_2 a)\,[k_2 a h_\ell(k_2 a)]' - \sqrt{\varepsilon_1}\,h_\ell(k_2 a)\,[k_2 a j_\ell(k_2 a)]'}{j_\ell(k_1 a)\,[k_2 a h_\ell(k_2 a)]' - h_\ell(k_2 a)\,[k_1 a j_\ell(k_1 a)]'}\,, \tag{3.162}$$

and the scattering coefficients

$$a_\ell = \frac{\varepsilon_1\,j_\ell(k_1 a)\,[k_2 a j_\ell(k_2 a)]' - \varepsilon_2\,j_\ell(k_2 a)\,[k_1 a j_\ell(k_1 a)]'}{\varepsilon_1\,j_\ell(k_1 a)\,[k_2 a h_\ell(k_2 a)]' - \varepsilon_2 h_\ell(k_2 a)\,[k_1 a j_\ell(k_1 a)]'}\,, \tag{3.163}$$

$$b_\ell = \frac{j_\ell(k_1 a)\,[k_2 a j_\ell(k_2 a)]' - j_\ell(k_2 a)\,[k_1 a j_\ell(k_1 a)]'}{j_\ell(k_1 a)\,[k_2 a h_\ell(k_2 a)]' - h_\ell(k_2 a)\,[k_1 a j_\ell(k_1 a)]'}\,. \tag{3.164}$$

The scattering coefficient can be simplified somewhat by introducing the *Ricatti–Bessel functions* $\psi_\ell(x) = x j_\ell(x)$ and $\xi_\ell(x) = x h_\ell(x)$. We have

$$a_\ell = \frac{\sqrt{\varepsilon_1}\,\psi_\ell(k_1 a)\psi'_\ell(k_2 a) - \sqrt{\varepsilon_2}\,\psi_\ell(k_2 a)\psi'_\ell(k_1 a)}{\sqrt{\varepsilon_1}\,\psi_\ell(k_1 a)\xi'_\ell(k_2 a) - \sqrt{\varepsilon_2}\,\xi_\ell(k_2 a)\psi'_\ell(k_1 a)}\,, \tag{3.165}$$

$$b_\ell = \frac{\sqrt{\varepsilon_2}\,\psi_\ell(k_1 a)\psi'_\ell(k_2 a) - \sqrt{\varepsilon_1}\,\psi_\ell(k_2 a)\psi'_\ell(k_1 a)}{\sqrt{\varepsilon_2}\,\psi_\ell(k_1 a)\xi'_\ell(k_2 a) - \sqrt{\varepsilon_1}\,\xi_\ell(k_2 a)\psi'_\ell(k_1 a)}\,. \tag{3.166}$$

The cross sections for extinction, scattering, and absorption are given by Bohren and Huffman [31]

$$Q_{ext} = \frac{2}{(k_2 a)^2} \sum_{\ell=1}^{+\infty} (2\ell + 1)\mathrm{Re}\,(a_\ell + b_\ell)\,, \tag{3.167}$$

$$Q_{sca} = \frac{2}{(k_2 a)^2} \sum_{\ell=1}^{+\infty} (2\ell + 1)\left(|a_\ell|^2 + |b_\ell|^2\right)\,, \tag{3.168}$$

$$Q_{abs} = Q_{ext} - Q_{sca}\,. \tag{3.169}$$

Here the cross sections are in units of the geometric cross section πa^2. In Fig. 3.27 the calculated extinction cross section of an EG sphere is shown for different values of a when $\gamma = 0.01\omega_p$ and $\varepsilon_2 = 1$. When the radius of the sphere increases, additional surface modes appear in the spectrum, and the absorption bands shift to lower frequencies.

The dependence of the extinction on the dielectric constant of the surrounding medium is illustrated in Fig. 3.28. The apparent shifts of the surface mode absorption bands demonstrate the general rule that the surface mode frequencies decrease with increasing ε_2.

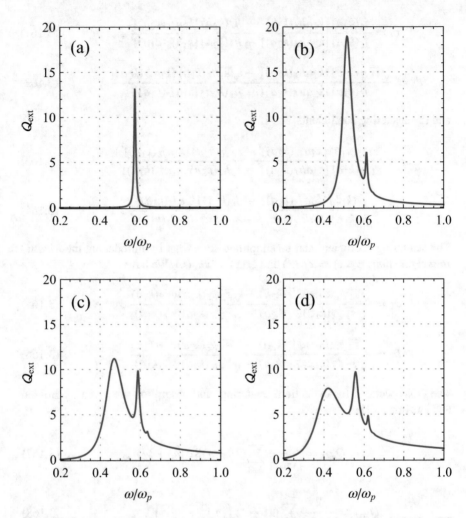

Fig. 3.27 Calculated extinction cross section (in units of the geometric cross section) of an EG sphere in vacuum with respect to the dimensionless frequency ω/ω_p when $\gamma = 0.01\omega_p$. The different panels refer to (**a**) $a = 0.1c/\omega_p$, (**b**) $a = c/\omega_p$, (**c**) $a = 1.5c/\omega_p$, and (**d**) $a = 2c/\omega_p$

Furthermore, in Fig. 3.29 we use both the full electromagnetic formula, i.e., (3.167) and the formula of electrostatic approximation, i.e., (2.130c) to calculate the extinction cross section of a small EG sphere in vacuum when $a = 0.1c/\omega_p$ (about 3.3 nm for a sodium cylinder) and $\gamma = 0.01\omega_p$. The comparison of electrostatic result with the exact calculation using the full electromagnetic formula demonstrates that the electrostatic theory is a very good approximation for a small EG sphere of radius about a few nanometers [50].

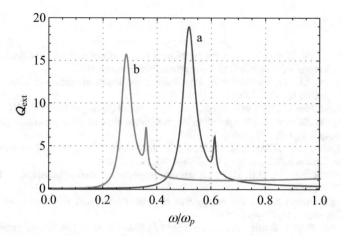

Fig. 3.28 Calculated extinction cross section (in units of the geometric cross section) of an EG sphere of radius $a = c/\omega_p$ for two different surrounding media when $\gamma = 0.01\omega_p$. The different curves refer to (**a**) air, $\varepsilon_2 = 1$ and (**b**) silicon dioxide (SiO$_2$), $\varepsilon_2 = 3.9$

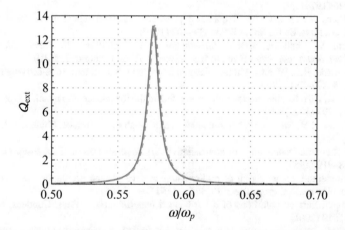

Fig. 3.29 Comparison of extinction cross section (in units of the geometric cross section) of a small EG sphere in vacuum from (3.167) when $a = 0.1c/\omega_p$ (about 3.3 nm for a sodium cylinder), and $\gamma = 0.01\omega_p$ with the result of electrostatic approximation, i.e., (2.130c). The red solid curve is the result of full electromagnetic formula and the green dashed one shows the result of electrostatic approximation

References

1. P. Markos, C.M. Soukoulis, *Wave Propagation: From Electrons to Photonic Crystals and Left-Handed Materials* (Princeton University Press, Princeton, 2008)
2. A.V. Zayats, I.I. Smolyaninov, A.A. Maradudin, Nano-optics of surface plasmon polaritons. Phys. Rep. **408**, 131–314 (2005)
3. S.A. Maier, *Plasmonics: Fundamentals and Applications* (Springer, Berlin, 2007)
4. W.L. Barnes, Surface plasmon-polariton length scales: a route to sub-wavelength optics. J. Opt. A: Pure Appl. Opt. **8**, S87 (2006)
5. J.M. Pitarke, V.M. Silkin, E.V. Chulkov, P.M. Echenique, Theory of surface plasmons and surface-plasmon polaritons. Rep. Prog. Phys. **70**, 1–87 (2007)
6. J. Nkoma, R. Loudon, D.R. Tilley, Elementary properties of surface polaritons. J. Phys. C **7**, 3547–3559 (1974)
7. A. Moradi, Energy density and energy flow of magnetoplasmonic waves on graphene. Solid State Commun. **253**, 63–66 (2017)
8. A. Moradi, Energy density and energy flow of plasmonic waves in bilayer graphene. Opt. Commun. **394**, 135–138 (2017)
9. Y. Bliokh, A.Y. Bekshaev, F. Nori, Optical momentum and angular momentum in complex media: from the Abraham-Minkowski debate to unusual properties of surface plasmon polaritons. New J. Phys. **19**, 123014 (2017)
10. D.L. Mills, E. Burstein, Polaritons: the electromagnetic modes of media. Rep. Prog. Phys. **37**, 817–926 (1974)
11. J.S. Nkoma, S.K. Rwaboona, Frequency dependence of some functions associated with surface polaritons. Phys. Status Solidi B **93**, 397–402 (1979)
12. D. Sarid, W. Challener, *Modern Introduction to Surface Plasmons: Theory, Mathematica Modeling, and Applications* (Cambridge University Press, Cambridge, 2010)
13. R.H. Ritchie, R.E. Wilems, Photon-plasmon interaction in a nonuniform electron gas. I. Phys. Rev. **178**, 372–381 (1969)
14. J. Crowell, R.H. Ritchie, Surface-plasmon effect in the reflectance of a metal. J. Opt. Soc. Am. **60**, 794–799 (1970)
15. J.M. Elson, R.H. Ritchie, Photon interactions at a rough metal surface. Phys. Rev. B **4**, 4129–4138 (1971)
16. Y.O. Nakamura, Quantization of non-radiative surface plasma oscillations. Prog. Theor. Phys. **70**, 908–919 (1983)
17. Y.O. Nakamura, Quantization of non-radiative surface plasma oscillations as a constrained Hamiltonian system. Prog. Theor. Phys. **74**, 11911205 (1985)
18. R. Ruppin, Surface polaritons of a left-handed material slab. J. Phys. Condens. Matter **13**, 1811–1819 (2001)
19. A. Moradi, Plasmonic waves of a semi-infinite random nanocomposite. Phys. Plasmas **20**, 104507 (2013)
20. A. Moradi, Plasmonic waves of random metal-dielectric nanocomposite films. Photon. Nanostruct. Fundam. Appl. **15**, 41–45 (2015)
21. T. Turbadar, Complete absorption of light by thin metal films. Proc. Phys. Soc. Lond. **73**, 40–44 (1959)
22. A. Otto, Excitation of nonradiative surface plasma waves in silver by the method of frustrated total reflection. Z. Phys. **216**, 398–410 (1968)
23. A. Moradi, Optical properties of random metal-dielectric nanocomposite films: nanoparticle size effects. Phys. Scr. **90**, 095803 (2015)
24. E. Kretschmann, The determination of the optical constants of metals by excitation of surface plasmons. Z. Phys. **241**, 31324 (1971)
25. J.J. Brion, R.F. Wallis, A. Hartstein, E. Burstein, Theory of surface magnetoplasmons in semiconductors. Phys. Rev. Lett. **28**, 1455–1458 (1972)

26. B. Hu, Y. Zhang, Q. Wang, Surface magneto plasmons and their applications in the infrared frequencies. Nanophotonics **4**, 383 (2015)
27. A. Moradi, Comment on: propagation of surface waves on a semi-bounded quantum magnetized collisional plasma. Phys. Plasmas **23**, 044701 (2016)
28. A. Moradi, Comment on: surface electromagnetic wave equations in a warm magnetized quantum plasma. Phys. Plasmas **23**, 074701 (2016)
29. P. Yeh, *Optical Waves in Layered Media* (Wiley, New Jersey, 1998)
30. H. Hojo, A. Mase, Dispersion relation of electromagnetic waves in one-dimensional plasma photonic crystals. J. Plasma Fusion Res. **80**, 89–90 (2004)
31. C.F. Bohren, D.R. Huffman, *Absorption and Scattering of Light by Small Particles* (Wiley, New York, 1983)
32. C.A. Pfeiffer, E.N. Economou, K.L. Ngai, Surface polaritons in a circularly cylindrical interface: surface plasmons. Phys. Rev. B **10**, 3038–3051 (1974)
33. W. von Ignatowski, Reflexion elektromagnetischer Wellen an einem dünnen drähte. Ann. Phys. **18**, 495–522 (1905)
34. W. Seitz, Die beugung des lichtes an einem dünnen, zylindrischen drähte. Ann. Phys. **21**, 1013–1029 (1906)
35. L. Rayleigh, The dispersal of light by a dielectric cylinder. Philos. Mag. **36**, 365–376 (1918)
36. J.R. Wait, Scattering of a plane wave from a circular dielectric cylinder at oblique incidence. Can. J. Phys. **33**, 189–195 (1955)
37. H. Khosravi, D.R. Tilley, R. Loudon, Surface polaritons in cylindrical optical fibers. J. Opt. Soc. Am. A **8**, 112–122 (1991)
38. A. Moradi, Comment on: a theoretical model to explain the mechanism of light wave propagation through non-metallic nanowires. Opt. Commun. **357**, 193–194 (2015)
39. A.D. Boardman, *Electromagnetic Surface Modes* (Wiley, New York, 1982)
40. D. Wu, X. Liu, B. Li, Localized surface plasmon resonance properties of two-layered gold nanowire: effects of geometry, incidence angle, and polarization. J. Appl. Phys. **109**, 083540 (2011)
41. A. Moradi, Plasmonic modes and extinction properties of a random nanocomposite cylinder. Phys. Plasmas **21**, 042112 (2014)
42. A. Moradi, Oblique incidence scattering from single-walled carbon nanotubes. Phys. Plasmas **17**, 033504 (2010)
43. A. Moradi, Extinction properties of single-walled carbon nanotubes: two-fluid model. Phys. Plasmas **21**, 032106 (2014)
44. J. Polanco, R.M. Fitzgerald, A.A. Maradudin, Propagation of s-polarized surface polaritons circumferentially around a locally cylindrical surface. Phys. Lett. A **376**, 1573–1575 (2012)
45. A. Moradi, Surface polaritons of a metal-insulator-metal curved slab. Superlattices Microstruct. **97**, 335–340 (2016)
46. J. Polanco, R.M. Fitzgerald, A.A. Maradudin, Propagation of p-polarized surface plasmon polaritons circumferentially around a locally cylindrical surface. Opt. Commun. **316**, 120–126 (2014)
47. I. Prigogine, S.A. Rice, *Advances in Chemical Physics: Aspects of the Study of Surfaces* (Wiley, New York, 1974)
48. L.V. Lorenz, On the light reflected and refracted by a transparent sphere. K. Dan. Vidensk. Selsk. Skr. **6**, 1–62 (1890)
49. G. Mie, Beiträge zur optik trüber medien, speziell kolloidaler metallösungen. Ann. Phys **25**, 377–445 (1908)
50. A. Moradi, Extinction properties of an isolated C_{60} molecule. Solid State Commun. **192**, 24–26 (2014)

Chapter 4
Problems in Electrostatic Approximation: Spatial Nonlocal Effects

Abstract In this chapter, by considering the spatial nonlocal effects, some electrostatic boundary-value problems involving bounded electron gases are studied. Using the nonlocal hydrodynamic theory discussed in Chap. 1, Poisson's equation in planar, cylindrical, and spherical geometries is solved. Comparisons are made with the results of Chap. 2. For brevity, in many sections of this chapter the $\exp(-i\omega t)$ time factor is suppressed. Furthermore, all media under consideration are nonmagnetic and attention is only confined to the linear phenomena.

4.1 Plasmonic Properties of Semi-Infinite Electron Gases: Standard Hydrodynamic Model

4.1.1 Surface Plasmon Frequency

We consider a semi-infinite homogeneous and lossless electron plasma occupying the half-space $z > 0$ in a Cartesian coordinate system with the position vector $\mathbf{r} = (x, y, z)$ [1]. The plane $z = 0$ is the EG-insulator interface (the relative dielectric constant of insulator is ε_1).

Now a SP is supposed to propagate parallel to the interface $z = 0$ along the x-direction. Thus, the homogeneous electron plasma with the density n_0 (per unit volume) will be perturbed by the surface wave and may be regarded as a charged fluid with 3D scalar density field $n_0 + n(x, z, t)$ and the electrons velocity $\mathbf{v}(x, z, t)$, where n and $\mathbf{v}$ represent the first-order perturbed values of the electron plasma density and velocity, respectively. Based on the SHD theory (see Sect. 1.2), in the linear approximation, we have

$$\frac{\partial \mathbf{v}}{\partial t} = \frac{e}{m_e}\nabla\Phi - \frac{\alpha^2}{n_0}\nabla n \,, \tag{4.1}$$

The original version of this chapter was revised. The correction to this chapter is available at https://doi.org/10.1007/978-3-030-43836-4_11

151

$$\frac{\partial n}{\partial t} + n_0 \nabla \cdot \mathbf{v} = 0 \,, \tag{4.2}$$

$$\nabla^2 \Phi(x, z) = \begin{cases} 0 \,, & z \leq 0 \,, \\ en/\varepsilon_0\varepsilon_b \,, & z \geq 0 \,, \end{cases} \tag{4.3}$$

where ε_b in general is frequency-dependent as we remarked in Chap. 1 and describing the remaining dielectric response of the system that is not due to the free electrons, such as *interband transitions* [2, 3]. We note again that in the right-hand side of (4.1), the first term is the force on electrons due to the electric field, i.e., $\mathbf{E} = -\nabla\Phi$, where Φ is the self-consistent[1] electrostatic potential and the second term represents interactions in the electron fluid (see Chap. 1). By eliminating the velocity field $\mathbf{v}(x, z, t)$ from (4.1) and (4.2), one can obtain the following equation

$$\left(\alpha^2 \nabla^2 - \frac{\partial^2}{\partial t^2} - \frac{\omega_p^2}{\varepsilon_b} \right) n = 0 \,, \tag{4.4}$$

where $\omega_p = (e^2 n_0/\varepsilon_0 m_e)^{1/2}$ is the classical plasma frequency in the homogeneous electron plasma, as mentioned before. Due to the planar symmetry of the present system, one can replace the quantities n and Φ in (4.3) and (4.4) by expressions of the form

$$\Phi(x, z) = \tilde{\Phi}(z) \exp(ik_x x) \,, \tag{4.5}$$

$$n(x, z) = \tilde{n}(z) \exp(ik_x x) \,, \tag{4.6}$$

where k_x is the longitudinal wavenumber in the x-direction. After substitution, one finds

$$\left(\frac{d^2}{dz^2} - k_x^2 \right) \tilde{\Phi}(z) = \begin{cases} 0 \,, & z \leq 0 \,, \\ e\tilde{n}/\varepsilon_0\varepsilon_b \,, & z \geq 0 \,, \end{cases} \tag{4.7}$$

$$\left(\frac{d^2}{dz^2} - \kappa_L^2 \right) \tilde{n}(z) = 0 \,, \tag{4.8}$$

where $\kappa_L^2 = k_x^2 - k_L^2$ and

$$k_L^2 = \frac{\omega^2 - \omega_p^2/\varepsilon_b}{\alpha^2} \,. \tag{4.9}$$

For the present system, we look for a solution of (4.7) and (4.8) of the form

---

[1] The problem is considered as a BVP by use of a self-consistent solution of the coupled linearized SHD and Poisson equations.

$$\tilde{\Phi}(z) = \begin{cases} A_1 e^{k_x z} , & z \le 0 , \\ A_2 e^{-k_x z} + B e^{-\kappa_L z} , & z \ge 0 , \end{cases} \tag{4.10}$$

$$\tilde{n}(z) = \begin{cases} 0 , & z \le 0 , \\ C e^{-\kappa_L z} , & z \ge 0 , \end{cases} \tag{4.11}$$

Combining (4.7) with (4.10) and (4.11), we can easily obtain

$$C = \varepsilon_0 \varepsilon_b e^{-1} \left(\kappa_L^2 - k_x^2 \right) B . \tag{4.12}$$

The relations between the coefficients A_1, A_2, and B in the above equations can be determined from the matching appropriate BCs at $z = 0$. Two of them are the usual electrostatic BCs at an EG-insulator boundary, i.e., the continuity of the electrostatic potential and the continuity of the normal component of the displacement vector field, as

$$\Phi_1|_{z=0} = \Phi_2|_{z=0} , \tag{4.13a}$$

$$\varepsilon_1 \frac{\partial \Phi_1}{\partial z}\bigg|_{z=0} = \varepsilon_b \frac{\partial \Phi_2}{\partial z}\bigg|_{z=0} , \tag{4.13b}$$

where subscripts 1 and 2 refer to outside and inside the EG, respectively. Also, we need an ABC which may be motivated by a physical requirement that the normal component of the electron velocity field **v** should vanish at $z = 0$, as

$$\mathbf{e}_z \cdot \mathbf{v}|_{z=0} = 0 , \tag{4.14}$$

that is in agreement with the specular reflection model of a metal surface [4]. Using (4.1) the condition (4.14) can be written in a more convenient form, as

$$\varepsilon_0 \omega_p^2 \frac{\partial \Phi_2}{\partial z}\bigg|_{z=0} = e\alpha^2 \frac{\partial n}{\partial z}\bigg|_{z=0} . \tag{4.15}$$

Now, applying the mentioned BCs we will be able to find the relationship between the constants A_1, A_2, B, and C, as

$$A_2 = \frac{\kappa_L}{k_x - \kappa_L} \left(1 + \frac{\varepsilon_1}{\varepsilon_b} \right) \left[\varepsilon_b \frac{\alpha^2}{\omega_p^2} \left(\kappa_L^2 - k_x^2 \right) - 1 \right] A_1 , \tag{4.16a}$$

$$B = \frac{k_x}{k_x - \kappa_L} \left(1 + \frac{\varepsilon_1}{\varepsilon_b} \right) A_1 , \tag{4.16b}$$

$$C = \frac{\varepsilon_0 \varepsilon_b}{e} \frac{k_x}{k_x - \kappa_L} \left(\kappa_L^2 - k_x^2 \right) \left(1 + \frac{\varepsilon_1}{\varepsilon_b} \right) A_1 . \tag{4.16c}$$

Also, we obtain a relation between ω and k_x, as

$$\varepsilon_b \left(\varepsilon_b + \varepsilon_1 \right) \alpha^2 \kappa_L \left(\kappa_L + k_x \right) = \varepsilon_1 \omega_p^2 , \tag{4.17}$$

which in the special case $\varepsilon_b = 1 = \varepsilon_1$ can be written in a more convenient form, as

$$\omega = \frac{\omega_p}{\sqrt{2}} \left\{ 1 + \frac{\alpha^2 k_x^2}{\omega_p^2} + \sqrt{2} \frac{\alpha k_x}{\omega_p} \left[1 + \frac{\alpha^2 k_x^2}{2\omega_p^2} \right]^{1/2} \right\}^{1/2} . \tag{4.18}$$

At long wavelengths, where $\alpha k_x / \omega_p \ll 1$ (but still in the nonretarded regime where $k_x > \omega_p / c$), (4.18) yields the well-known Ritchie formula [5]

$$\omega \approx \frac{\omega_p}{\sqrt{2}} \left(1 + \frac{\alpha k_x}{\sqrt{2}\omega_p} \right) . \tag{4.19}$$

This shows that dispersion relation of SPs of a semi-infinite EG in the SHD model is, initially, linear. According to (4.18), the SPs dispersion curve of a flat EG-vacuum interface is plotted in Fig. 4.1, where an appropriate choice of units (ω is measured in ω_p units and k_x in k_s units, where $k_s \equiv \omega_p / \alpha$ is the inverse TF screening length of the EG, as mentioned in Chap. 1) yields universal curve independent of any choice of specific parameters such as α. Note that SPs can propagate only for frequencies $\omega > \omega_s$.

This represents a cutoff (when $k_x \rightarrow 0$ or the phase velocity goes to infinity) for SPs, that means for low frequencies the phase velocity of a SP is greater than c,

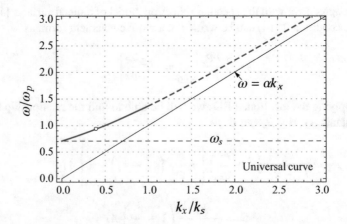

Fig. 4.1 Dispersion curve of SPs of a semi-infinite EG as given by (4.18), when $\varepsilon_b = 1 = \varepsilon_1$. The dashed horizontal line indicates the asymptotic SP frequency of the system in the local model, i.e., $\omega_s = \omega_p / \sqrt{2}$. Because of the use of appropriate units shown along the axes, these curves do not depend on any physical parameters

Table 4.1 Typical plasma parameters in some metals at room temperature [6, 7]

Metal	Symbol	Nature	v_F(m/s)	v_p(1/s)
Sodium	Na	Alkali	1.07×10^6	1.428×10^{15}
Potassium	P	Alkali	8.52×10^5	1.038×10^{15}
Gold	Au	Noble	1.39×10^6	0.219×10^{16}
Silver	Ag	Noble	1.39×10^6	0.223×10^{16}
Copper	Cu	Noble	1.57×10^6	0.213×10^{16}

Note that $\omega_p = 2\pi v_p$

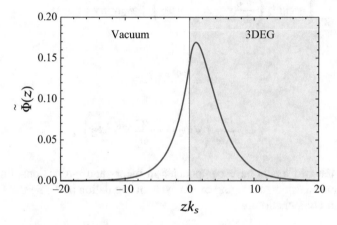

Fig. 4.2 Normalized profile $\tilde{\Phi}(z)$ of a SP on a flat EG-vacuum interface, when $k_x/k_s = 0.4$ and $\varepsilon_b = 1$ corresponding to the labeled point in Fig. 4.1

the speed of light in vacuum, but we note that the retardation effects have not been considered here. Furthermore, as the wave frequency increases to higher and higher values, the phase velocity decreases toward the speed α of the electrons and a wave-particle interaction between the SPs and the electrons occurs that causes a strong damping known as *Landau damping*, while we have assumed no damping in our model. Therefore the present description is no longer valid at such low and/or high frequencies and the range of validity of the present approach may be characterized by

$$\frac{\alpha}{c} \ll \frac{k_x}{k_s} \ll 1 , \qquad (4.20)$$

as indicated by the dashed part of the curve in Fig. 4.1, when $k_x > k_s$ and $k_x < 0.001 k_s$. Note that for typical metals the values of v_F are of the order of Bohr velocity, as can be seen in Table 4.1. Therefore $\alpha/c \sim 0.001$ is an expected outcome. Also, by using (4.10) and (4.16a)–(4.16c), the normalized profile of electrostatic potential of a SP is calculated in Fig. 4.2 for $k_x/k_s = 0.4$ corresponding to the labeled point in Fig. 4.1.

4.1.2   Power Flow

For the power flow density associated with a SP on a flat EG-insulator interface, using (4.1) and (4.2) and also (1.80), we have, in the two media,

$$
\mathbf{S} = \begin{cases}
-\varepsilon_0\varepsilon_1\Phi_1\dfrac{\partial}{\partial t}\nabla\Phi_1 , & z \leq 0 , \\[2mm]
-\varepsilon_0\Phi_2\left(\dfrac{en_0}{\varepsilon_0}\mathbf{v} + \varepsilon_b\dfrac{\partial}{\partial t}\nabla\Phi_2\right) + m_e\alpha^2 n\mathbf{v} , & z \geq 0 ,
\end{cases}
\tag{4.21}
$$

where

$$
n = \frac{\varepsilon_0\varepsilon_b}{e}\nabla^2\Phi_2 ,
\tag{4.22}
$$

$$
\mathbf{v} = \frac{ie}{m_e\omega}\left(\nabla\Phi_2 - \varepsilon_b\frac{\alpha^2}{\omega_p^2}\nabla\nabla^2\Phi_2\right) ,
\tag{4.23}
$$

and $\mathbf{S}$ in the two media have components in the x- and z-directions, but their z-components vanish on averaging over a cycle of oscillation of the fields. The cycle-averaged x-components are

$$
S_x = -\frac{\varepsilon_0\omega k_x}{2}
\begin{cases}
\varepsilon_1 A_1^2 e^{2k_x z} , & z \leq 0 , \\[2mm]
\left(\varepsilon_b - \dfrac{\omega_p^2}{\omega^2}\right)\left[A_2^2 e^{-2k_x z} - \varepsilon_b\dfrac{\omega^2}{\omega_p^2}B^2 e^{-2\kappa_L z}\right] , & z \geq 0 .
\end{cases}
\tag{4.24}
$$

In Fig. 4.3 we calculate the normalized profiles S_x of a SP on a flat EG-vacuum interface for $k_x/k_s = 0.4$ and $\varepsilon_b = 1 = \varepsilon_1$, by using (4.16a)–(4.16c) and (4.24). It can be seen that power flow densities are largest at the boundaries, and their amplitudes decay exponentially with increasing distance into each medium from the interfaces. Also, it is evident that the power flow density in the EG region occurs in the $+x$-direction, while in the vacuum region, the power flow density occurs in the $-x$-direction, i.e., opposite to the direction of phase propagation. The total power flow density (per unit width) associated with a SP is determined by an integration over z. We obtain

$$
\langle S_x \rangle = -\frac{\varepsilon_0\omega}{4}\left[\varepsilon_1 A_1^2 + \left(\varepsilon_b - \frac{\omega_p^2}{\omega^2}\right)\left(A_2^2 - \varepsilon_b\frac{k_x}{\kappa_L}\frac{\omega^2}{\omega_p^2}B^2\right)\right] ,
\tag{4.25}
$$

where $\langle \cdots \rangle \equiv \int_{-\infty}^{+\infty} \cdots \, dz$. This total power flow density (per unit width) is positive, when k_x is positive. From Fig. 4.3 it is clear that the power flow density in the EG region is greater than that in vacuum, given a net power flow density to the x-direction, when k_x is positive. This means that SPs group velocity should be positive.

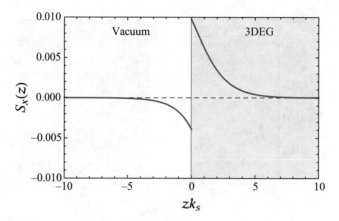

Fig. 4.3 Normalized profile $S_x(z)$ of a SP on a flat EG-vacuum interface, when $k_x/k_s = 0.4$ and $\varepsilon_b = 1$

4.1.3 Energy Distribution

Now let us consider the energy density distribution in the transverse direction. As described in Sect. 1.4.6, the energy density of the electrostatic field in a lossless medium can be expressed by (1.79). In the presence of the spatial dispersion, for the cycle-averaged of energy density distribution associated with a SP on a flat EG-insulator interface, we have, in the two media,

$$
U = \frac{\varepsilon_0}{4}
\begin{cases}
\varepsilon_1 |\nabla \Phi_1|^2 , & z \leq 0 , \\
\varepsilon_b |\nabla \Phi_2|^2 + \dfrac{m_e n_0}{\varepsilon_0} |\mathbf{v}|^2 + \dfrac{m_e}{\varepsilon_0 n_0} \alpha^2 |n|^2 , & z \geq 0 ,
\end{cases}
\tag{4.26}
$$

After substitution (4.10), (4.22), and (4.23) into (4.26), we obtain

$$
U = \frac{\varepsilon_0 k_x^2}{2}
\begin{cases}
\varepsilon_1 A_1^2 e^{2k_x z} , & z \leq 0 , \\
\left(\varepsilon_b + \dfrac{\omega_p^2}{\omega^2} \right) A_2^2 e^{-2k_x z} + \varXi B^2 e^{-2\kappa_L x} + \varUpsilon A_2 B e^{-(k_x + \kappa_L)z} , & z \geq 0 ,
\end{cases}
\tag{4.27}
$$

where

$$
\varXi = \varepsilon_b \frac{\kappa_L^2}{k_x^2} \left(1 + \varepsilon_b \frac{k_x^2}{\kappa_L^2} \frac{\omega^2}{\omega_p^2} \right) ,
\tag{4.28}
$$

$$
\varUpsilon = 2\varepsilon_b \left(1 + \frac{\kappa_L}{k_x} \right) .
\tag{4.29}
$$

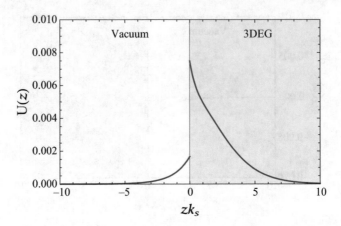

Fig. 4.4 Normalized profile $U(z)$ of a SP on a flat EG-vacuum interface, when $k_x/k_s = 0.4$ and $\varepsilon_b = 1$

All contributions to the energy density are positive, as can be seen in Fig. 4.4. The total energy density associated with a SP is again determined by integration over z, the energy per unit surface area being

$$\langle U \rangle = \frac{\varepsilon_0 k_x}{4} \left\{ \varepsilon_1 A_1^2 + \left(\varepsilon_b + \frac{\omega_p^2}{\omega^2} \right) A_2^2 + \varepsilon_b \frac{\kappa_L}{k_x} \left(1 + \varepsilon_b \frac{k_x^2}{\kappa_L^2} \frac{\omega^2}{\omega_p^2} \right) B^2 + 4\varepsilon_b A_2 B \right\}.$$

$$(4.30)$$

4.1.4   Energy Velocity

The energy velocity of a SP is given as the ratio of the total power flow density (per unit width) and total energy density (per unit area), such as

$$v_e = -\frac{\omega}{k_x} \frac{\varepsilon_1 A_1^2 + \left(\varepsilon_b - \frac{\omega_p^2}{\omega^2} \right) \left(A_2^2 - \varepsilon_b \frac{k_x}{\kappa_L} \frac{\omega^2}{\omega_p^2} B^2 \right)}{\varepsilon_1 A_1^2 + \left(\varepsilon_b + \frac{\omega_p^2}{\omega^2} \right) A_2^2 + \varepsilon_b \frac{\kappa_L}{k_x} \left(1 + \varepsilon_b \frac{k_x^2}{\kappa_L^2} \frac{\omega^2}{\omega_p^2} \right) B^2 + 4\varepsilon_b A_2 B}.$$

$$(4.31)$$

The expression on the right-hand side is precisely that obtained from the usual definition of the SP group velocity, i.e., (2.51). According to the above equation, the group velocity curve of SPs on a semi-infinite EG is plotted in Fig. 4.5, when $\varepsilon_b = 1 = \varepsilon_1$. We note that an appropriate choice of units (v is measured in α units and k_x in k_s units) yields universal curves independent of any choice of specific parameters such as α.

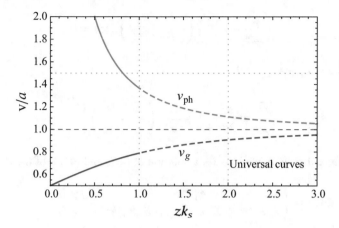

Fig. 4.5 Group velocity curve of SPs of a semi-infinite EG as given by (4.31) for $\varepsilon_b = 1 = \varepsilon_1$. The red curve shows the phase velocity of SPs. Because of the use of appropriate units shown along the axes, these curves do not depend on any physical parameters

4.2 Plasmonic Properties of Semi-Infinite Electron Gases: Quantum Hydrodynamic Model

4.2.1 Surface Plasmon Frequency

If we want to fully retain quantum effects which give rise to a QHD model, we have to keep both α and β finite [1, 8], as remarked in Chap. 1. Therefore, (4.2), (4.3), and (4.5)–(4.7) remain unchanged, but (4.1) and (4.4) should be read as

$$\frac{\partial \mathbf{v}}{\partial t} = \frac{e}{m_e} \nabla \Phi - \frac{\alpha^2}{n_0} \nabla n + \xi \frac{\beta^2}{n_0} \nabla \nabla^2 n \, , \tag{4.32}$$

$$\left(\xi \beta^2 \nabla^4 - \alpha^2 \nabla^2 + \frac{\partial^2}{\partial t^2} + \frac{\omega_p^2}{\varepsilon_b} \right) n = 0 \, , \tag{4.33}$$

where (4.33) is a fourth-order partial differential equation (PDE) in the spatial variables and for the present system we have $\xi = 1$ [9, 10]. Therefore, in the presence of the Bohm potential correction, we obtain

$$\left[\left(\frac{d^2}{dz^2} - k_x^2 \right)^2 - \frac{\alpha^2}{\beta^2} \left(\frac{d^2}{dz^2} - k_x^2 \right) - \frac{\omega^2 - \omega_p^2/\varepsilon_b}{\beta^2} \right] \tilde{n}(z) = 0 \, . \tag{4.34}$$

To make progress, it is convenient to rewrite (4.34) in a factored form [11],

$$\left(\frac{d^2}{dz^2} - \kappa_{L-}^2\right)\left(\frac{d^2}{dz^2} - \kappa_{L+}^2\right)\tilde{n}(z) = 0 , \tag{4.35}$$

where $\kappa_{L\pm}^2 = k_x^2 - k_{L\mp}^2$, and

$$k_{L\mp}^2 = -\frac{\alpha^2}{2\beta^2} \mp \frac{\alpha^2}{2\beta^2}\left[1 + \frac{4\beta^2}{\alpha^4}\left(\omega^2 - \frac{\omega_p^2}{\varepsilon_b}\right)\right]^{1/2} , \tag{4.36}$$

For the present system, we look for a solution of (4.7) and (4.35) of the form

$$\tilde{\Phi}(z) = \begin{cases} A_1 e^{k_x z} , & z \leq 0 , \\ A_2 e^{-k_x z} + B_+ e^{-\kappa_{L+} z} + B_- e^{-\kappa_{L-} z} , & z \geq 0 , \end{cases} \tag{4.37}$$

$$\tilde{n}(z) = \begin{cases} 0 , & z \leq 0 , \\ C_+ e^{-\kappa_{L+} z} + C_- e^{-\kappa_{L-} z} , & z \geq 0 , \end{cases} \tag{4.38}$$

Combining (4.7) with (4.37) and (4.38), we can easily obtain

$$C_\pm = \varepsilon_0 \varepsilon_b e^{-1}\left(\kappa_{L\pm}^2 - k_x^2\right) B_\pm . \tag{4.39}$$

Again, the relations between the coefficients A_1, A_2, and $B_\pm$ in the above equations can be determined from the matching appropriate BCs at $z = 0$. In addition to (4.13a) and (4.13b), we also need two ABC to be imposed on the induced charge density at $z = 0$, because the equation for the induced charge density, i.e., (4.35) is a fourth-order PDE in the spatial variables. In addition to the physical requirement that the normal component of the electron velocity should vanish at $z = 0$, as

$$\varepsilon_0 \omega_p^2 \frac{\partial \Phi_2}{\partial z}\bigg|_{z=0} = e\left[\alpha^2 - \beta^2\left(\frac{\partial^2}{\partial z^2} - k_x^2\right)\right]\frac{\partial n}{\partial z}\bigg|_{z=0} , \tag{4.40}$$

also the Bohm potential, which is proportional to $\nabla^2 n$ according to (4.32), should be vanished at $z = 0$ [11]. Therefore, we have

$$\left(\frac{\partial^2}{\partial z^2} - k_x^2\right)n|_{z=0} = 0 . \tag{4.41}$$

Applying these four BCs and doing some algebra, for $\varepsilon_b = 1 = \varepsilon_1$, we obtain a relation between ω and k_x, as

$$\kappa_{L-}\left[1 - \frac{\beta^2}{\alpha^2}\left(\kappa_{L-}^2 - k_x^2\right)\right] - \frac{\kappa_{L-}^2 - k_x^2}{\kappa_{L+}^2 - k_x^2}\kappa_{L+}\left[1 - \frac{\beta^2}{\alpha^2}\left(\kappa_{L+}^2 - k_x^2\right)\right]$$

$$= \frac{\omega_p^2}{2\alpha^2} \frac{1}{\kappa_{L-} + k_x} \left(1 - \left[\frac{\kappa_{L-}^2 - k_x^2}{\kappa_{L+}^2 - k_x^2} \right]^2 \frac{\kappa_{L-} + k_x}{\kappa_{L+} + k_x} \right), \qquad (4.42)$$

where in the absence of the Bohm potential or gradient correction, i.e., $\beta = 0$, (4.42) reduces to (4.17). To see clearly behavior of the dispersion curve of the quantum surface electrostatic wave of the system, we plot dimensionless frequency ω/ω_p, versus the dimensionless variable k_x/k_s in Fig. 4.6, for different values of dimensionless plasmonic coupling parameter $H = l_c k_s$, where $l_c = \beta/\alpha$ as defined in Chap. 1. One can see that the dispersion curves of SPs obtained with both SHD and QHD models increase and continue to behave in highly similar manner when $\omega < \omega_p$. This means that the Bohm potential does not exert much influence when the value of eigenfrequency ω is below the bulk plasmon frequency ω_p. However, for $\omega > \omega_p$, by considering the Bohm potential effects, SPs with blue-shifted frequencies emerge.

4.2.2 Surface Magneto Plasmon Frequency: Voigt Configuration

Now, the quantum EG half-space of the previous section is considered to be embedded in an external strong magnetic field $\mathbf{B}_0 = B_0 \mathbf{e}_y$, where B_0 is the strength of the background magnetic field [12]. Also, a SMP with wavenumber k_x, and the form $f(z) \exp(\pm i k_x x - i \omega t)$ is assumed to propagate parallel to the interface $z = 0$, along the x-direction. We note that the upper and lower signs refer to waves propagating in the positive and negative x-directions, respectively.

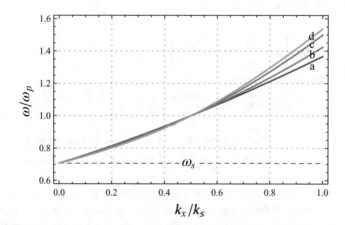

Fig. 4.6 Dispersion curves of SPs of a semi-infinite EG as given by (4.42), when $\varepsilon_b = 1 = \varepsilon_1$. The different curves refer to (a) $H = 0$, (b) $H = 0.5$, (c) $H = 1$, and (d) $H = 1.5$

In this case, the term $-(e/m_e)\mathbf{v} \times \mathbf{B}_0$ must be added to the right-hand side of (4.32). Thus, using the new form of (4.32), the velocity components are

$$v_x = \frac{-\omega}{\omega^2 - \omega_c^2} \left(\frac{\omega_c}{\omega} \frac{\partial}{\partial z} \pm k_x \right) \left\{ \frac{e}{m_e} \Phi - \left[\alpha^2 - \beta^2 \left(\frac{\partial^2}{\partial z^2} - k_x^2 \right) \right] \frac{n}{n_0} \right\} ,$$
(4.43)

$$v_z = \frac{i\omega}{\omega^2 - \omega_c^2} \left(\frac{\partial}{\partial z} \pm \frac{\omega_c}{\omega} k_x \right) \left\{ \frac{e}{m_e} \Phi - \left[\alpha^2 - \beta^2 \left(\frac{\partial^2}{\partial z^2} - k_x^2 \right) \right] \frac{n}{n_0} \right\} ,$$
(4.44)

where $\omega_c = eB_0/m_e$, as defined in Chap. 2. For the present BVP, we use (4.7) and (4.35), but $\kappa_{L\pm}$ must be read as

$$\kappa_{L\pm}^2 = k_x^2 + \frac{\alpha^2}{2\beta^2} \pm \frac{\alpha^2}{2\beta^2} \left[1 + \frac{4\beta^2}{\alpha^4} \left(\omega^2 - \omega_c^2 - \frac{\omega_p^2}{\varepsilon_b} \right) \right]^{1/2} .$$
(4.45)

Now, we have a family of solutions for the electrostatic potential and for the perturbed charge density, as shown by (4.37) and (4.38), respectively. Again, it is necessary to impose four BCs for the present system. Three of them are (4.13a), (4.13b), and (4.41). The fourth one is

$$\varepsilon_0 \omega_p^2 \left(\frac{\partial}{\partial z} \pm \frac{\omega_c}{\omega} k_x \right) \Phi_2|_{z=0} = e \left(\frac{\partial}{\partial z} \pm \frac{\omega_c}{\omega} k_x \right) \left[\alpha^2 - \beta^2 \left(\frac{\partial^2}{\partial z^2} - k_x^2 \right) \right] n|_{z=0} .$$
(4.46)

Applying the BCs and doing some algebra, for $\varepsilon_b = 1 = \varepsilon_1$, we obtain a relation between ω and k_x, as

$$(\omega \kappa_{L-} \mp k_x \omega_c) \left[1 - \frac{\beta^2}{\alpha^2} \left(\kappa_{L-}^2 - k_x^2 \right) \right]$$

$$- \frac{\kappa_{L-}^2 - k_x^2}{\kappa_{L+}^2 - k_x^2} (\omega \kappa_{L+} \mp k_x \omega_c) \left[1 - \frac{\beta^2}{\alpha^2} \left(\kappa_{L+}^2 - k_x^2 \right) \right]$$

$$= \frac{\omega_p^2}{2\alpha^2} \frac{\omega \pm \omega_c}{\kappa_{L-} + k_x} \left(1 - \left[\frac{\kappa_{L-}^2 - k_x^2}{\kappa_{L+}^2 - k_x^2} \right]^2 \frac{\kappa_{L-} + k_x}{\kappa_{L+} + k_x} \right) .$$
(4.47)

One notes immediately that this dispersion relation is nonreciprocal, i.e., positive and negative values of the wavevector k_x are not equivalent. Also, in the case $\omega_c = 0$, we get the result in previous section. On the other hand, if we neglect the Bohm potential effects, i.e., $\beta = 0$ Eq. (4.47) leads to

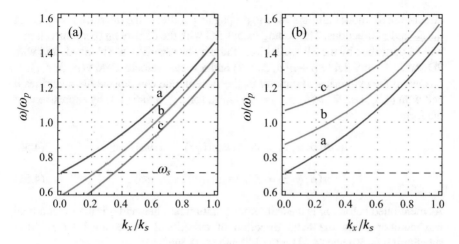

Fig. 4.7 Dispersion curves of SMPs of a semi-infinite EG as given by (4.47), when $H = 0.7$ and $\varepsilon_b = 1 = \varepsilon_1$. The different panels refer to (**a**) forward-going SMPs and (**b**) backward-going SMPs. Also in each panel, the different curves refer to (**a**) $\omega_c/\omega_p = 0$, (**b**) $\omega_c/\omega_p = 0.3$, and (**c**) $\omega_c/\omega_p = 0.6$

$$\omega_p^2 (\omega \pm \omega_c) = 2\alpha^2 (\kappa_L + k_x)(\omega \kappa_L \mp k_x \omega_c) , \qquad (4.48)$$

where

$$\kappa_L^2 = k_x^2 - \frac{\omega^2 - \omega_c^2 - \omega_p^2/\varepsilon_b}{\alpha^2} . \qquad (4.49)$$

which was derived by Nakamura and Paranjape [13] and also Aers et al. [14]. Furthermore, if we neglect the parameter α, from (4.48), we find (2.40) that is quite well-known. In Fig. 4.7, we plot dimensionless frequency ω/ω_p, versus dimensionless variable k_x/k_s for forward- and backward-going SMPs with several values of ω_c/ω_p, when $H = 0.7$. It can be seen that for forward-going SMPs, the increasing of ω_c/ω_p, red-shifts the SMP frequency of the system, while for backward-going SMPs the frequency of surface oscillations is blue-shifted.

4.3 Plasmonic Properties of Circular Electron Gas Cylinders: Quantum Hydrodynamic Model

4.3.1 Surface Plasmon Modes

Let us consider a circular EG cylinder of radius a embedded in a homogeneous insulator environment with relative permittivity ε_2 and take cylindrical coordinates for an arbitrary point in space. In the presence of the quantum spatial dispersion effects, the SP modes are supposed to propagate parallel to the interface $\rho = a$ along the ϕ- and z-directions. The homogeneous EG with the density n_0 (per unit volume) will be perturbed by the SP oscillations and may be regarded as a charged fluid with 3D scalar density field $n_0 + n(\rho, \phi, z, t)$ and a vector velocity field $\mathbf{v}(\rho, \phi, z, t)$.

Due to the cylindrical symmetry of the system, one can replace the quantities n and Φ in (4.3), i.e., $\nabla^2 \Phi(\rho, \phi, z) = en/\varepsilon_0 \varepsilon_b$, for $\rho \leq a$ and (4.4) by expressions of the form

$$\Phi(\rho, \phi, z) = \tilde{\Phi}(\rho) \exp(iqz) \exp(im\phi) , \tag{4.50}$$

$$n(\rho, \phi, z) = \tilde{n}(\rho) \exp(iqz) \exp(im\phi) . \tag{4.51}$$

As mentioned before, m is the azimuthal quantum number, and q is the longitudinal wavenumber along the axial direction of cylindrical EG denoted by z. After substituting (4.50) and (4.51) into (4.3) and (4.4), we find

$$\left[\frac{d^2}{d\rho^2} + \frac{1}{\rho} \frac{d}{d\rho} - \left(q^2 + \frac{m^2}{\rho^2} \right) \right] \tilde{\Phi}(\rho) = \begin{cases} e\tilde{n}/\varepsilon_0 \varepsilon_b , & \rho \leq a , \\ 0 , & \rho \geq a , \end{cases} \tag{4.52}$$

$$\left[\frac{d^2}{d\rho^2} + \frac{1}{\rho} \frac{d}{d\rho} - \left(\kappa_{L+}^2 + \frac{m^2}{\rho^2} \right) \right] \left[\frac{d^2}{d\rho^2} + \frac{1}{\rho} \frac{d}{d\rho} - \left(\kappa_{L-}^2 + \frac{m^2}{\rho^2} \right) \right] \tilde{n}(\rho) = 0 , \tag{4.53}$$

where $\kappa_{L\pm}^2 = q^2 - k_{L\mp}^2$ and $k_{L\mp}$ is given by (4.36). For the present system, we seek for a solution of (4.52) and (4.53) of the type

$$\tilde{\Phi}(\rho) = \begin{cases} A_1 I_m(q\rho) + B_+ I_m(\kappa_{L+}\rho) + B_- I_m(\kappa_{L-}\rho) , & \rho \leq a , \\ A_2 K_m(q\rho) , & \rho \geq a , \end{cases} \tag{4.54}$$

$$\tilde{n}(\rho) = \begin{cases} C_+ I_m(\kappa_{L+}\rho) + C_- I_m(\kappa_{L-}\rho) , & \rho \leq a , \\ 0 , & \rho \geq a , \end{cases} \tag{4.55}$$

Combining (4.52) with (4.54) and (4.55), we can easily obtain

$$C_\pm = \varepsilon_0 \varepsilon_b e^{-1} \left(\kappa_{L\pm}^2 - q^2 \right) B_\pm . \tag{4.56}$$

The relations between the coefficients A_1, A_2, and $B_\pm$ in the above equations can be determined from the matching BCs at the cylinder surface $\rho = a$. We have [15]

$$\Phi_1|_{\rho=a} = \Phi_2|_{\rho=a} , \tag{4.57a}$$

$$\varepsilon_b \frac{\partial \Phi_1}{\partial \rho} \bigg|_{\rho=a} = \varepsilon_2 \frac{\partial \Phi_2}{\partial \rho} \bigg|_{\rho=a} , \tag{4.57b}$$

$$\varepsilon_0 \omega_p^2 \frac{\partial \Phi_1}{\partial \rho}\bigg|_{\rho=a} = e\left[\alpha^2 - \beta^2 \nabla^2\right]\frac{\partial n}{\partial \rho}\bigg|_{\rho=a} , \qquad (4.57c)$$

$$\nabla^2 n|_{\rho=a} = 0 , \qquad (4.57d)$$

where subscript 1 denotes the region inside the EG cylinder and subscript 2 denotes the region outside the system and

$$\nabla^2 = \frac{\partial^2}{\partial \rho^2} + \frac{1}{\rho}\frac{\partial}{\partial \rho} + \frac{1}{\rho^2}\frac{\partial^2}{\partial \phi^2} + \frac{\partial^2}{\partial z^2} . \qquad (4.58)$$

The application of the mentioned BCs for $\varepsilon_b = 1 = \varepsilon_2$, gives

$$A_2 = \frac{1}{\Upsilon - \Pi}(qa)I_m'(qa)\left[(qa)\Xi I_m'(qa) - \Pi I_m(qa)\right]A_1 , \qquad (4.59a)$$

$$B_- = \frac{1}{\Upsilon - \Pi}\frac{(qa)I_m'(qa)}{(\kappa_{L-}a)^2 - (qa)^2}A_1 , \qquad (4.59b)$$

$$B_+ = -\frac{\Gamma}{\Upsilon - \Pi}\frac{(qa)I_m'(qa)}{(\kappa_{L+}a)^2 - (qa)^2}A_1 , \qquad (4.59c)$$

where

$$\Upsilon = \frac{\alpha^2}{a^2\omega_p^2}\left\{\left(1 - \frac{\beta^2}{a^2\alpha^2}\left[(\kappa_{L-}a)^2 - (qa)^2\right]\right)(\kappa_{L-}a)I_m'(\kappa_{L-}a)\right.$$
$$\left. -\Gamma\left(1 - \frac{\beta^2}{a^2\alpha^2}\left[(\kappa_{L+}a)^2 - (qa)^2\right]\right)(\kappa_{L+}a)I_m'(\kappa_{L+}a)\right\} ,$$

$$\Pi = \frac{(\kappa_{L-}a)I_m'(\kappa_{L-}a)}{(\kappa_{L-}a)^2 - (qa)^2} - \Gamma\frac{(\kappa_{L+}a)I_m'(\kappa_{L+}a)}{(\kappa_{L+}a)^2 - (qa)^2} ,$$

$$\Xi = \frac{I_m(\kappa_{L-}a)}{(\kappa_{L-}a)^2 - (qa)^2} - \Gamma\frac{I_m(\kappa_{L+}a)}{(\kappa_{L+}a)^2 - (qa)^2} ,$$

$$\Gamma = \frac{(\kappa_{L-}a)^2 - (qa)^2}{(\kappa_{L+}a)^2 - (qa)^2}\frac{I_m(\kappa_{L-}a)}{I_m(\kappa_{L+}a)} ,$$

and the prime denotes differentiation with respect to the argument. Furthermore, we obtain a relation between ω and q, as

$$(qa)K_m'(qa)\left[\Pi I_m(qa) - (qa)\Xi I_m'(qa)\right] + \Upsilon = 0 . \qquad (4.60)$$

The roots of (4.60) provide the dispersion relation $\omega_{mn}(q)$ of the modes, where for a given m, each value of n defines a particular dispersion relation, including both SPs (for real values of $\kappa_{L\pm}$ and $n = 0$) and bulk plasmons (for imaginary $\kappa_{L\pm}$ and $n \neq 0$). We note the solutions corresponding to imaginary values of $\kappa_{L\pm}$ yield electron densities characterized by an oscillatory behavior inside the system, while those corresponding to real values of $\kappa_{L\pm}$ yield electron densities that are localized in the region close to the surface, as discussed in detail in [16]. Therefore, we use the notation $n = 0$ for SP modes, and $n = 1, 2, ...$, for bulk modes. Also, we note that in the absence of the Bohm potential or gradient correction, i.e., $\Gamma = 0$ and $\beta = 0$, (4.60) becomes identical to the standard nonlocal dispersion relation result [16, 17], as

$$\omega^2 = \omega_p^2(qa)I_m'(qa)K_m(qa)\left[1 - \frac{I_m(\kappa_L a)}{(\kappa_L a)I_m'(\kappa_L a)}\frac{(ka)K_m'(qa)}{K_m(qa)}\right], \tag{4.61}$$

where

$$\kappa_L^2 = q^2 - \frac{\omega^2 - \omega_p^2/\varepsilon_b}{\alpha^2} . \tag{4.62}$$

We present in Fig. 4.8, the dependence of the dimensionless frequency ω/ω_p on the dimensionless variable qa, when $m = 0$ and $a = 15/k_s$. For the other parameters the sodium values in Table 4.1 have been employed. In Fig. 4.9 we show the radial dependence of the electrostatic potential of the surface and bulk plasmon modes, when $m = 0$ and $a = 15/k_s$. In this example we have taken $qa = 2$, corresponding to the labeled points in panel (a) and (b) of Fig. 4.8. The dashed line in panel (a) of this figure corresponds to the potential of the non-dispersive model $\alpha \to 0$. For the SP case, in panel (a) we see that by increasing values of ak_s the potential produces a cusp behavior at $\rho = a$ as in previous model. This is so because when ak_s increases the induced electronic charge tends to accumulate at the surface as it is shown in Fig. 4.10 of this same figure (in the limit $ak_s \to \infty$ we have a surface charge density described by a Dirac delta function).

4.3.2 Surface Plasmon Resonance

In Chap. 2 we described the SP resonance of an infinitely long and thin EG cylinder in a purely classical manner, which gives rise to a non-dispersive, or local description of the EG. In this way, we considered a thin cylindrical EG with radius a centered at the origin and extending along the z-axis to infinity, embedded in a homogeneous insulator environment with relative permittivity ε_2. We assumed that the cylinder was illuminated by a uniform, quasi-static electric field $\mathbf{E} = \mathbf{e}_x E_0$ (see Fig. 2.14).

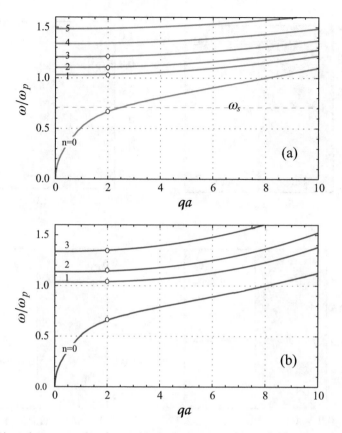

Fig. 4.8 Dispersion curves of surface and bulk plasmon modes of an EG cylinder free-standing in vacuum, when $m = 0$ and $a = 15/k_s$. Panel (**a**): results of (4.61). Panel (**b**): results of (4.60). For the other parameters the sodium values in Table 4.1 have been employed. For each panel, the lowest curve $n = 0$ corresponds to the SP mode. Also, the dashed horizontal green line in panel (**a**) indicates the asymptotic SP frequency of the system in the local (non-dispersive) model, i.e., $\omega_s = \omega_p/\sqrt{2}$

We here improve the theory by returning to the full spatial quantum effects which give rise to a QHD model by using (4.2) and (4.32) after adding the damping term $-\gamma \mathbf{v}$ to the right-hand side of (4.32) [18]. Due to the cylindrical symmetry of the present system, one can replace the quantities n and Φ in (4.3), i.e., $\nabla^2 \Phi(\rho, \phi) = en/\varepsilon_0 \varepsilon_b$, for $\rho \leq a$ and (4.33) by expressions of the form

$$\Phi(\rho, \phi) = \tilde{\Phi}(\rho) \cos m\phi , \tag{4.63}$$

$$n(\rho, \phi) = \tilde{n}(\rho) \cos m\phi . \tag{4.64}$$

Therefore, the normal mode solutions of the plasmon frequency of the present problem satisfy

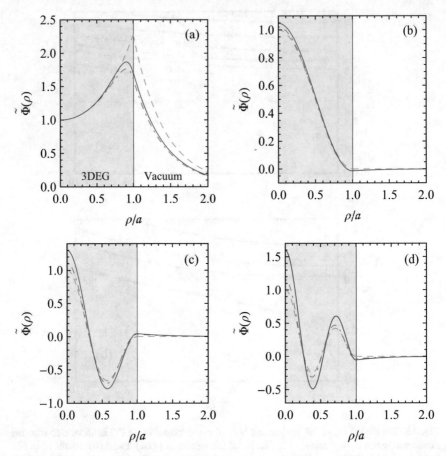

Fig. 4.9 Normalized profile $\tilde{\Phi}(\rho)$ of surface and bulk plasmon modes of an EG cylinder, when $m = 0$ and $a = 15/k_s$. For the other parameters the sodium values in Table 4.1 have been employed. For each panel, green dashed, red dash dotted, and blue solid lines represent the results of local (non-dispersive), SHD and QHD models, respectively. Panel (**a**) corresponds to the SP modes with $n = 0$, while panels (**b**), (**c**), and (**d**) relate to the bulk modes, for $n = 1, 2$, and 3, respectively. Here we have taken $qa = 2$, corresponding to the labeled points in panel (a) and (b) of Fig. 4.8

$$\left[\frac{d^2}{d\rho^2} + \frac{1}{\rho} \frac{d}{d\rho} - \frac{1}{\rho^2} \right] \tilde{\Phi}(\rho) = \begin{cases} e\tilde{n}/\varepsilon_0 \varepsilon_b \,, & \rho \leq a \,, \\ 0 \,, & \rho \geq a \,, \end{cases} \tag{4.65}$$

$$\left[\frac{d^2}{d\rho^2} + \frac{1}{\rho} \frac{d}{d\rho} + \left(k_{L+}^2 - \frac{1}{\rho^2} \right) \right] \left[\frac{d^2}{d\rho^2} + \frac{1}{\rho} \frac{d}{d\rho} + \left(k_{L-}^2 - \frac{1}{\rho^2} \right) \right] \tilde{n}(\rho) = 0 \,, \tag{4.66}$$

where the quantum nonlocal longitudinal wavenumber is given as

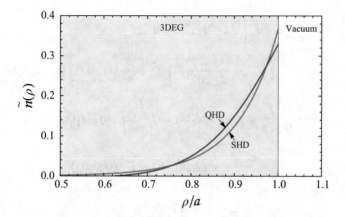

Fig. 4.10 Normalized profile $\tilde{n}(\rho)$ of a SP mode of an EG cylinder, when $m = 0 = n$ and $a = 15/k_s$. For the other parameters the sodium values in Table 4.1 have been employed. Red and blue solid lines represent the results of SHD and QHD models, respectively. Here, we have taken $qa = 2$, corresponding to the lower labeled points in panel (a) and (b) of Fig. 4.8

$$k_{L\mp}^2 = -\frac{\alpha^2}{2\beta^2} \mp \frac{\alpha^2}{2\beta^2} \left[1 + \frac{4\beta^2}{\alpha^4} \left[\omega(\omega + i\gamma) - \frac{\omega_p^2}{\varepsilon_b} \right] \right]^{1/2}. \tag{4.67}$$

Now, the solution of (4.65) and (4.66) is considered as

$$\Phi(\rho, \phi) = \cos\phi \begin{cases} A_1\rho + B_+ J_1(k_{L+}\rho) + B_- J_1(k_{L-}\rho) , & \rho \leq a , \\ E_0\rho + B_1\rho^{-1} , & \rho \geq a , \end{cases} \tag{4.68}$$

$$n(\rho, \phi) = \cos\phi \begin{cases} C_+ J_1(k_{L+}\rho) + C_- J_1(k_{L-}\rho) , & \rho \leq a , \\ 0 , & \rho \geq a , \end{cases} \tag{4.69}$$

where J_1 is the Bessel function of the first kind and order 1. Combining (4.65) with (4.68) and (4.69), we can easily obtain $C_\pm = -\varepsilon_0 e^{-1} \varepsilon_b k_{L\pm}^2 B_\pm$. The coefficients A_1, $B_\pm$, and B_1 in the above equations can be determined from the matching BCs of the fields, i.e., (4.57a)–(4.57d) at the surface $\rho = a$ of the system, we find

$$\Phi_1 = -\frac{2\varepsilon_2 \left(\varXi - a\Theta/\rho + \varepsilon_b \Pi/a^2\omega_p^2 \right)}{\varepsilon_2(\varXi - \Theta) + \varepsilon_b(\varepsilon_2 + \varepsilon_b)\Pi/a^2\omega_p^2} E_0\rho \cos\phi , \tag{4.70}$$

$$\Phi_2 = -\left[1 + \frac{a^2}{\rho^2} \frac{\varepsilon_2(\varXi - \Theta) + \varepsilon_b(\varepsilon_2 - \varepsilon_b)\Pi/a^2\omega_p^2}{\varepsilon_2(\varXi - \Theta) + \varepsilon_b(\varepsilon_2 + \varepsilon_b)\Pi/a^2\omega_p^2} \right] E_0\rho \cos\phi , \tag{4.71}$$

where subscript 1 denotes the region inside the EG cylinder and subscript 2 denotes the region outside the system and

$$\Pi = \left[\alpha^2 + \frac{\beta^2}{a^2}(k_{L-}a)^2\right](k_{L-}a)J_1'(k_{L-}a)$$

$$- \frac{(k_{L+}a)^2}{(k_{L-}a)^2}\Gamma\left[\alpha^2 + \frac{\beta^2}{a^2}(k_{L+}a)^2\right](k_{L+}a)J_1'(k_{L+}a) ,$$

$$\varXi = \frac{1}{(k_{L-}a)^2}\left[(k_{L-}a)J_1'(k_{L-}a) - (k_{L+}a)\Gamma J_1'(k_{L+}a)\right] ,$$

$$\Theta = \frac{1}{(k_{L-}a)^2}\left[J_1(k_{L-}a) - \Gamma J_1(k_{L+}a)\right] ,$$

$$\Gamma = \frac{(k_{L-}a)^4}{(k_{L+}a)^4}\frac{J_1(k_{L-}a)}{J_1(k_{L+}a)} ,$$

and the prime denotes differentiation with respect to the argument. Thus, the solution for the electric field $E = -\nabla\Phi$ turns out to be

$$\mathbf{E}_1 = E_0 \frac{2\varepsilon_2\left(\varXi - a\Theta/\rho + \varepsilon_b\Pi/a^2\omega_p^2\right)}{\varepsilon_2\left(\varXi - \Theta\right) + \varepsilon_b(\varepsilon_2 + \varepsilon_b)\Pi/a^2\omega_p^2}\mathbf{e}_x , \qquad (4.72)$$

$$\mathbf{E}_2 = E_0\mathbf{e}_x - E_0\frac{\varepsilon_2\left(\varXi - \Theta\right) + \varepsilon_b(\varepsilon_2 - \varepsilon_b)\Pi/a^2\omega_p^2}{\varepsilon_2\left(\varXi - \Theta\right) + \varepsilon_b(\varepsilon_2 + \varepsilon_b)\Pi/a^2\omega_p^2}\frac{a^2}{\rho^2}\left(1 - 2\sin^2\phi\right)\mathbf{e}_x$$

$$- 2E_0\frac{\varepsilon_2\left(\varXi - \Theta\right) + \varepsilon_b(\varepsilon_2 - \varepsilon_b)\Pi/a^2\omega_p^2}{\varepsilon_2\left(\varXi - \Theta\right) + \varepsilon_b(\varepsilon_2 + \varepsilon_b)\Pi/a^2\omega_p^2}\frac{a^2}{\rho^2}\sin\phi\cos\phi\mathbf{e}_y . \qquad (4.73)$$

Now, from (2.78) the corresponding normalized transversal polarizability of the system can be written as

$$\alpha_{pol} = 2\frac{\varepsilon_1 - \varepsilon_2\left(1 + \delta_{nl}\right)}{\varepsilon_1 + \varepsilon_2\left(1 + \delta_{nl}\right)} , \qquad (4.74)$$

where

$$\delta_{nl} = \frac{\varepsilon_1 - \varepsilon_b}{\varepsilon_b}\frac{J_1\left(k_{L-}a\right) - \Gamma J_1\left(k_{L+}a\right)}{(k_{L-}a)J_1'\left(k_{L-}a\right) - \Gamma\left(k_{L+}a\right)J_1'\left(k_{L+}a\right)} , \qquad (4.75)$$

and $\varepsilon_1 = \varepsilon_b - \omega_p^2/\left(\omega^2 + i\gamma\omega\right)$ is the classical Drude permittivity. In the absence of the Bohm potential effects, i.e., $\Gamma = 0$ and $\beta \to 0$, Eq. (4.75) reduces to

$$\delta_{nl} = \frac{\varepsilon_1 - \varepsilon_b}{\varepsilon_b}\frac{J_1(k_La)}{(k_La)J_1'(k_La)} , \qquad (4.76)$$

where

$$k_L^2 = \frac{\omega(\omega + i\gamma) - \omega_p^2/\varepsilon_b}{\alpha^2}. \tag{4.77}$$

If we consider the case $\delta_{nl} = 0$, Eq. (4.74) becomes identical to the local polarizability of the system, i.e., (2.79). One can see that nonlocal effects enter the local polarizability as an elegant and simple rescaling of either the EG permittivity from ε_1 to $\tilde{\varepsilon}_1 = \varepsilon_1 (1 + \delta_{nl})^{-1}$ or of the background permittivity from ε_2 to $\tilde{\varepsilon}_2 = \varepsilon_2 (1 + \delta_{nl})$ [19]. It is apparent that the normalized transversal polarizability experiences a resonant enhancement under the condition that $|\tilde{\varepsilon}_1 + \varepsilon_2|$ is a minimum, which for the case of small or slowly varying $\mathrm{Im}\left[\tilde{\varepsilon}_1\right]$ around the resonance simplifies to

$$\mathrm{Re}\left[\tilde{\varepsilon}_1\right] = -\varepsilon_2. \tag{4.78}$$

To illustrate the gradient correction effects on the normalized transversal polarizability of the system, we present the variation of the $\mathrm{Im}\left[\alpha_{pol}\right]$ with respect to the dimensionless variable ω/ω_p in Fig. 4.11, when $\varepsilon_b = 1 = \varepsilon_2$, $\gamma = 0.01\omega_p$, and $a = 15/k_s$. For the other parameters the sodium values in Table 4.1 have been employed. The dashed green and dash dotted red curves show the result of the local and SHD theories, respectively. It is clear that by inclusion of gradient correction effects with respect to the local theory, the main peak shifts toward the high frequencies side, and subsidiary peaks appear above the bulk plasmon frequency. Note that by increasing the radius of the system, the number of subsidiary peaks increases in the considered energy interval, but their relative amplitude decreases.

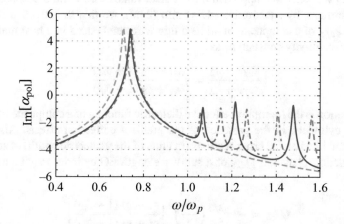

Fig. 4.11 Variation of the $\mathrm{Im}\left[\alpha_{pol}\right]$ with respect to the dimensionless variable ω/ω_p, when $\varepsilon_b = 1 = \varepsilon_2$, $\gamma = 0.01\omega_p$, and $a = 15/k_s$. For the other parameters the sodium values in Table 4.1 have been employed. The dashed green and dash dotted red curves show the result of the local and SHD theories, respectively

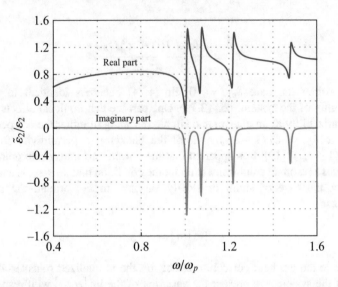

Fig. 4.12 Real and imaginary parts of the normalized rescaled background permittivity $\tilde{\varepsilon}_2/\varepsilon_2$ with respect to the dimensionless variable ω/ω_p, when $\varepsilon_b = 1 = \varepsilon_2$, $\gamma = 0.01\omega_p$, and $a = 15/k_s$. For the other parameters the sodium values in Table 4.1 have been employed

In Fig. 4.12, we examine the frequency dependency of the real and imaginary parts of the rescaled background permittivity $\tilde{\varepsilon}_2$ for different value of a. Above the plasma frequency both $\text{Re}[\tilde{\varepsilon}_2]/\varepsilon_2$ and $\text{Im}[\tilde{\varepsilon}_2]/\varepsilon_2$ show periodic variations, which give rise to the pressure resonances in the normalized transversal polarizability curve. As the radius increases the frequency dependence of both $\text{Re}[\tilde{\varepsilon}_2]/\varepsilon_2$ and $\text{Im}[\tilde{\varepsilon}_2]/\varepsilon_2$ weakens, and approach the classical values, i.e., 1 and 0, respectively.

Furthermore, by substituting (4.74) into (2.82), we find the effective dielectric function ε_{eff} of a composite of aligned thin wires embedded in a host matrix with relative permittivity constant ε_2 as

$$\frac{\varepsilon_{eff} - \varepsilon_2}{\varepsilon_{eff} + \varepsilon_2} = f \frac{\varepsilon_1 - \varepsilon_2 (1 + \delta_{nl})}{\varepsilon_1 + \varepsilon_2 (1 + \delta_{nl})} . \qquad (4.79)$$

This equation is the (complex) effective dielectric function of a composite of aligned thin EG cylinders, in the presence of the gradient correction effects. Also, using (2.81a) and (2.81b), the cross sections (in units of the geometric width) of scattering and extinction Q_{sca} and Q_{ext} of a sub-wavelength EG cylinder may be expressed as [20]

$$Q_{sca} = \frac{\pi^2}{4} \varepsilon_2^{3/2} \left(\frac{a\omega}{c} \right)^3 \left| \frac{\varepsilon_1 - \varepsilon_2 (1 + \delta_{nl})}{\varepsilon_1 + \varepsilon_2 (1 + \delta_{nl})} \right|^2 , \qquad (4.80)$$

$$Q_{ext} = -\pi \varepsilon_2^{1/2} \left(\frac{a\omega}{c} \right) \text{Re} \left[i \frac{\varepsilon_1 - \varepsilon_2 (1 + \delta_{nl})}{\varepsilon_1 + \varepsilon_2 (1 + \delta_{nl})} \right] . \qquad (4.81)$$

4.4 Plasmonic Properties of Electron Gas Spheres: Quantum Hydrodynamic Model

4.4.1 Surface Plasmon Modes

In order to carry out the calculations of the spatial nonlocal effects on the surface and bulk plasmon modes of an EG sphere of radius a, embedded in a homogeneous insulator environment with relative permittivity ε_2, we start with the linearized QHD theory of an EG.

Due to the spherical symmetry of the present system, one can replace the quantities n and Φ in (4.3), i.e., $\nabla^2 \Phi(r, \theta, \varphi) = en/\varepsilon_0\varepsilon_b$, for $r \leq a$ and (4.4) by expressions of the form

$$\Phi(r, \theta, \varphi) = \tilde{\Phi}(r) Y_{\ell m}(\theta, \varphi) , \tag{4.82}$$

$$n(r, \theta, \varphi) = \tilde{n}(r) Y_{\ell m}(\theta, \varphi) . \tag{4.83}$$

Therefore, we have

$$\left[\frac{d^2}{dr^2} + \frac{2}{r}\frac{d}{dr} - \frac{\ell(\ell+1)}{r^2} \right] \tilde{\Phi}(r) = \begin{cases} e\tilde{n}/\varepsilon_0\varepsilon_b , & r \leq a , \\ 0 , & r \geq a , \end{cases} \tag{4.84}$$

$$\left[\frac{d^2}{dr^2} + \frac{2}{r}\frac{d}{dr} - \left(\kappa_{L+}^2 + \frac{\ell(\ell+1)}{r^2} \right) \right] \left[\frac{d^2}{dr^2} + \frac{2}{r}\frac{d}{dr} - \left(\kappa_{L-}^2 + \frac{\ell(\ell+1)}{r^2} \right) \right] \tilde{n}(r) = 0, \tag{4.85}$$

where $\kappa_{L\pm} = -k_{L\mp}$ and $k_{L\mp}$ is given by (4.36). Now, we consider the solution of (4.84) and (4.85) as

$$\tilde{\Phi}(r) = \begin{cases} A_1 r^\ell + B_+ i_\ell(\kappa_{L+}r) + B_- i_\ell(\kappa_{L-}r) , & r \leq a , \\ A_2 r^{-(\ell+1)} , & r \geq a , \end{cases} \tag{4.86}$$

$$\tilde{n}(r) = \begin{cases} C_+ i_\ell(\kappa_{L+}r) + C_- i_\ell(\kappa_{L-}r) , & r \leq a , \\ 0 , & r \geq a , \end{cases} \tag{4.87}$$

where $i_\ell(x)$ is the spherical modified Bessel function of the first kind and order ℓ and $C_\pm = \varepsilon_0\varepsilon_b e^{-1}\kappa_{L\pm}^2 B_\pm$. The relations between the coefficients A_1, A_2, and $B_\pm$ in the above equations can be determined from the matching appropriate BCs at the surface of the system $r = a$. We have [21]

$$\Phi_1|_{r=a} = \Phi_2|_{r=a} , \tag{4.88a}$$

$$\varepsilon_b \frac{\partial \Phi_1}{\partial r}\bigg|_{r=a} = \varepsilon_2 \frac{\partial \Phi_2}{\partial r}\bigg|_{r=a} , \tag{4.88b}$$

$$\varepsilon_0 \omega_p^2 \frac{\partial \Phi_1}{\partial r}\bigg|_{r=a} = e \left[\alpha^2 - \beta^2 \nabla^2\right] \frac{\partial n}{\partial r}\bigg|_{r=a} , \tag{4.88c}$$

$$\nabla^2 n|_{r=a} = 0 , \tag{4.88d}$$

where subscript 1 denotes the region inside the EG sphere and subscript 2 denotes the region outside the system and

$$\nabla^2 = \frac{\partial^2}{\partial r^2} + \frac{2}{r} \frac{\partial}{\partial r} + \frac{1}{r^2 \sin\theta} \frac{\partial}{\partial \theta} \left(\sin\theta \frac{\partial}{\partial \theta}\right) + \frac{1}{r^2 \sin^2\theta} \frac{\partial^2}{\partial \varphi^2} . \tag{4.89}$$

The application of the mentioned BCs for $\varepsilon_b = 1 = \varepsilon_2$ gives

$$A_2 = \frac{\ell a^{2\ell+1}}{2\ell+1} \frac{\ell \Xi - \Pi}{\Upsilon - \Pi} A_1 , \tag{4.90a}$$

$$B_- = \frac{\ell a^\ell}{(\kappa_{L-} a)^2 (\Upsilon - \Pi)} A_1 , \tag{4.90b}$$

$$B_+ = -\Gamma \frac{\ell a^\ell}{(\kappa_{L+} a)^2 (\Upsilon - \Pi)} A_1 , \tag{4.90c}$$

where

$$\Upsilon = \frac{\alpha^2}{a^2 \omega_p^2} \left\{ \left(1 - \frac{\beta^2}{\alpha^2} \kappa_{L-}^2\right) (\kappa_{L-} a) i_\ell'(\kappa_{L-} a) - \Gamma \left(1 - \frac{\beta^2}{\alpha^2} \kappa_{L+}^2\right) (\kappa_{L+} a) i_\ell'(\kappa_{L+} a) \right\} ,$$

$$\Pi = \frac{i_\ell'(\kappa_{L-} a)}{\kappa_{L-} a} - \Gamma \frac{i_\ell'(\kappa_{L+} a)}{\kappa_{L+} a} ,$$

$$\Xi = \frac{i_\ell(\kappa_{L-} a)}{(\kappa_{L-} a)^2} - \Gamma \frac{i_\ell(\kappa_{L+} a)}{(\kappa_{L+} a)^2} ,$$

$$\Gamma = \frac{(\kappa_{L-} a)^2 i_\ell(\kappa_{L-} a)}{(\kappa_{L+} a)^2 i_\ell(\kappa_{L+} a)} ,$$

and the prime denotes differentiation with respect to the argument. Also, we obtain a relation between ω and q, as

$$(\ell + 1)[\Pi - \ell \Xi] - (2\ell + 1)\Upsilon = 0 . \tag{4.91}$$

The roots of this transcendental equation provide a relationship between the angular frequencies ω of the various multipoles (corresponding to different ℓ values) and the physical parameters that characterize the sphere: a, α, β, and ω_p.

The plasmon mode frequencies may be specified by a total angular component ℓ and a radial component n. For a given ℓ, each value of n defines a particular dispersion relation, including both SP modes (for real values of $\kappa_{L\pm}$ and $n = 0$) and bulk plasmons (for imaginary $\kappa_{L\pm}$ and $n \neq 0$). Thus, again we use the notation $n = 0$ for SP modes, and $n = 1, 2, ...,$ for bulk modes. We should note here that if $\ell = 0$ the SP mode is not possible [22], while for bulk modes the situation is different where for $\ell = 0$ there is an infinite number of bulk modes ($n = 1, 2, 3, ...$) corresponding to imaginary values of $\kappa_{L\pm}$. In the absence of the Bohm potential effects, i.e., $\Gamma_\ell = 0$ and $\beta \to 0$, Eq. (4.91) becomes identical to the standard nonlocal dispersion relation result in [22], as

$$\omega^2 = \omega_p^2 \frac{\ell}{2\ell + 1} \left[1 + (\ell + 1) \frac{i_\ell(\kappa_L a)}{(\kappa_L a)i_\ell'(\kappa_L a)} \right], \qquad (4.92)$$

where

$$\kappa_L^2 = -\frac{\omega^2 - \omega_p^2/\varepsilon_b}{\alpha^2}. \qquad (4.93)$$

To see the influence of the Bohm potential on the surface and bulk plasmon modes of a spherical EG, the numerical results of dispersion relation are illustrated in Figs. 4.13 and 4.14. For the other parameters the sodium values in Table 4.1 have been employed. Also, in Fig. 4.15 the radial dependence of the normalized electrostatic potential of the SP modes is shown, when $a = 20/k_s$ corresponding to the labeled points in Fig. 4.13.

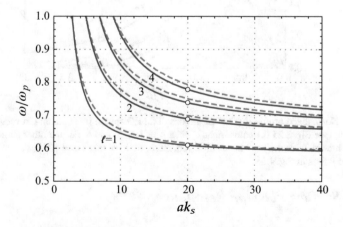

Fig. 4.13 Dispersion curves of SP ($n = 0$) modes of an EG sphere free-standing in vacuum, when $\ell = 1, 2, 3,$ and 4. Blue solid curves: results of (4.91). Red dashed curves: results of (4.92). For the other parameters the sodium values in Table 4.1 have been employed. For $\ell = 0$ the SP mode is not possible

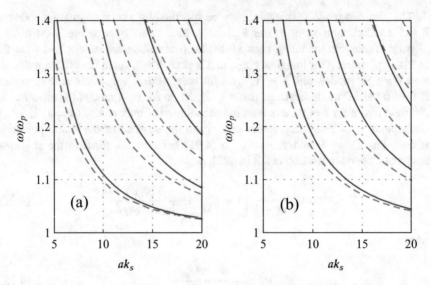

Fig. 4.14 Same as Fig. 4.13, but for bulk modes ($n \neq 0$). The different panels refer to (**a**) $\ell = 0$ and (**b**) $\ell = 1$

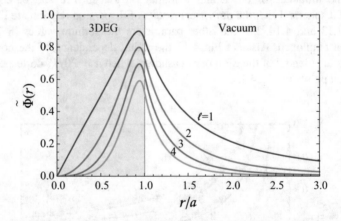

Fig. 4.15 Normalized profile $\tilde{\Phi}(r)$ of SP modes of an EG sphere free-standing in vacuum, using QHD mode (see Fig. 2.25 for comparison), when $\ell = 1, 2, 3$, and 4. For the other parameters the sodium values in Table 4.1 have been employed. Here, we have taken $a = 20/k_s$ corresponding to the labeled points in Fig. 4.13

4.4.2   Surface Plasmon Resonance

Here, by returning to the full spatial quantum effects which give rise to a QHD model, we improve the SP resonance theory of a small spherical EG that was discussed in Chap. 2. We use (4.2) and (4.32) after adding the damping term $-\gamma \mathbf{v}$ to the right-hand side of (4.32) [23]. Due to the spherical symmetry of the present system, one can replace the quantities n and Φ in (4.3), i.e., $\nabla^2 \Phi(r, \theta, \varphi) =$

$en/\varepsilon_0\varepsilon_b$, for $r \leq a$ and (4.4) by expressions of the form

$$\Phi(r, \theta) = \tilde{\Phi}(r)\cos\theta , \qquad (4.94)$$

$$n(r, \theta) = \tilde{n}(r)\cos\theta . \qquad (4.95)$$

Therefore, the normal mode solutions of the plasmon frequency of the present problem satisfy

$$\left[\frac{d^2}{dr^2} + \frac{2}{r}\frac{d}{dr} - \frac{\ell(\ell+1)}{r^2}\right]\tilde{\Phi}(r) = \begin{cases} e\tilde{n}/\varepsilon_0\varepsilon_b , & r \leq a , \\ 0 , & r \geq a , \end{cases} \qquad (4.96)$$

$$\left[\frac{d^2}{dr^2} + \frac{2}{r}\frac{d}{dr} + \left(k_{L+}^2 - \frac{2}{r^2}\right)\right]\left[\frac{d^2}{dr^2} + \frac{2}{r}\frac{d}{dr} + \left(k_{L-}^2 - \frac{2}{r^2}\right)\right]\tilde{n}(r) = 0 , \qquad (4.97)$$

where the quantum nonlocal longitudinal wavenumber $k_{L\pm}$ is given by (4.67). Now, we consider the solution of (4.96) and (4.97) as

$$\Phi(r, \theta) = \cos\theta \begin{cases} A_1 r + B_+ j_1(k_{L+}r) + B_- j_1(k_{L-}r) , & r \leq a , \\ E_0 r + B_1 r^{-2} , & r \geq a , \end{cases} \qquad (4.98)$$

$$n(r, \theta) = \cos\theta \begin{cases} C_+ j_1(k_{L+}r) + C_- j_1(k_{L-}r) , & r \leq a , \\ 0 , & r \geq a , \end{cases} \qquad (4.99)$$

where $C_\pm = -\varepsilon_0 e^{-1}\varepsilon_b k_{L\pm}^2 B_\pm$. The coefficients A_1, B_1, and $B_\pm$ in the above equations can be determined from the matching BCs at the surface of sphere $r = a$. After doing some algebra, we find

$$\Phi_2 = -\left[1 - \frac{a^3}{r^3}\frac{\varepsilon_1 - \varepsilon_2(1+\delta_{nl1})}{\varepsilon_1 + 2\varepsilon_2(1+\delta_{nl1})}\right]E_0 r \cos\varphi , \qquad (4.100)$$

where

$$\delta_{nl1} = \frac{\varepsilon_1 - \varepsilon_b}{\varepsilon_b}\frac{j_1(k_{L+}a) - \Gamma_1 j_1(k_{L-}a)}{(k_{L+}a) j_1'(k_{L+}a) - \Gamma_1(k_{L-}a) j_1'(k_{L-}a)} , \qquad (4.101)$$

$$\Gamma_1 = \frac{(k_{L+}a)^4}{(k_{L-}a)^4}\frac{j_1(k_{L+}a)}{j_1(k_{L-}a)} . \qquad (4.102)$$

Following the usual approach to introducing the polarizability [24, 25], we determine the normalized polarizability α_{pol} to be

$$\alpha_{pol} = 3\frac{\varepsilon_1 - \varepsilon_2(1+\delta_{nl1})}{\varepsilon_1 + 2\varepsilon_2(1+\delta_{nl1})} . \qquad (4.103)$$

We note in the absence of the quantum nonlocal effects, i.e., $\Gamma_1 = 0$ and $\beta \to 0$, Eq. (4.103) becomes identical to the dipole standard nonlocal polarizability result[2] [19]. On the other hand, we note excitation sources, such as an *electron beam* or a dipole emitter can excite higher-order multipoles, even in spheres with diameters below 20 nm [26–30]. Thus, it is relevant to consider the multipolar response of a small spherical EG. The result for the normalized polarizability of order ℓ may be found as

$$\alpha_{pol} = 3\ell a^{2(\ell-1)} \frac{\varepsilon_1 - \varepsilon_2 (1 + \delta_{n l \ell})}{\ell \varepsilon_1 + (\ell + 1)\varepsilon_2 (1 + \delta_{n l \ell})} , \qquad (4.104)$$

where ℓ is the multipole order with $\ell = 1$ denoting the dipole mode, $\ell = 2$ the quadrupole mode, and so on. Also,

$$\delta_{n l \ell} = \ell \frac{\varepsilon_1 - \varepsilon_b}{\varepsilon_b} \frac{j_\ell (k_{L+}a) - \Gamma_\ell j_\ell (k_{L-}a)}{(\chi_+ a) j_\ell' (k_{L+}a) - \Gamma_\ell (k_{L-}a) j_\ell' (k_{L-}a)} , \qquad (4.105)$$

$$\Gamma_\ell = \frac{(k_{L+}a)^4}{(k_{L-}a)^4} \frac{j_\ell (k_{L+}a)}{j_\ell (k_{L-}a)} . \qquad (4.106)$$

Comparing (4.105) with corresponding local result, i.e., (2.144) shows that quantum nonlocal effects enter the local polarizability as an elegant and simple rescaling of either the EG permittivity from ε_1 to $\tilde{\varepsilon}_1 = \varepsilon_1(1 + \delta_{n l \ell})^{-1}$ or of the background permittivity from ε_2 to $\tilde{\varepsilon}_2 = \varepsilon_2(1 + \delta_{n l \ell})$. Figure 4.16 shows the imaginary part of the multipole polarizabilities of a small spherical EG of radius $a = 20\omega_p/\alpha$ versus the variable ℓ, when $\varepsilon_b = 1 = \varepsilon_2$, $\gamma = 0.01\omega_p$, and $\omega = 1.2\omega_p$.

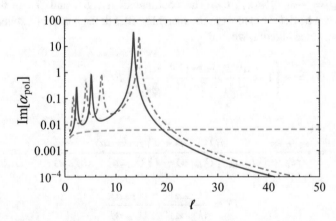

Fig. 4.16 Variation of Im[α_{pol}] with respect to the variable ℓ, when $\varepsilon_b = 1 = \varepsilon_2$, $\gamma = 0.01\omega_p$, $\omega = 1.2\omega_p$, and $a = 20/k_s$. For the other parameters the sodium values in Table 4.1 have been employed. The dashed green and dash dotted red curves show the result of the local and SHD theories, respectively

---

[2]See Eq. (2.129).

Also, using (2.130a) and (2.130b), the cross sections (in units of the geometric cross section) of scattering and extinction Q_{sca} and Q_{abs} of a sub-wavelength EG sphere may be expressed as

$$Q_{sca} = \frac{8}{3}\varepsilon_2^2 \left(\frac{a\omega}{c}\right)^4 \left|\frac{\varepsilon_1 - \varepsilon_2 (1 + \delta_{nl1})}{\varepsilon_1 + 2\varepsilon_2 [1 + \delta_{nl1}]}\right|^2 , \tag{4.107a}$$

$$Q_{abs} = 4\varepsilon_2^{1/2} \left(\frac{a\omega}{c}\right) \text{Im}\left[\frac{\varepsilon_1 - \varepsilon_2 (1 + \delta_{nl1})}{\varepsilon_1 + 2\varepsilon_2 (1 + \delta_{nl1})}\right] . \tag{4.107b}$$

Furthermore, the effective dielectric function ε_{eff} of a composite of small EG spheres embedded in a host matrix with relative dielectric constant ε_2 may be written, as

$$\frac{\varepsilon_{eff} - \varepsilon_2}{\varepsilon_{eff} + \varepsilon_2} = f \frac{\varepsilon_1 - \varepsilon_2 (1 + \delta_{nl1})}{\varepsilon_1 + 2\varepsilon_2 (1 + \delta_{nl1})} , \tag{4.108}$$

where f is the volume fraction of the embedded spheres.

References

1. A. Moradi, Surface plasmon oscillations on a quantum plasma half-space. Phys. Plasmas **22**, 014501 (2015)
2. W. Yan, M. Wubs, N.A. Mortensen, Hyperbolic metamaterials: nonlocal response regularizes broadband supersingularity. Phys. Rev. B **86**, 205429 (2012)
3. A. Moradi, Quantum ion-acoustic wave oscillations on a quantum plasma half-space. Phys. Scr. **90**, 085601 (2015)
4. R.H. Ritchie, A.L. Marusak, The surface plasmon dispersion relation for an electron gas. Surf. Sci. **4**, 234–240 (1966)
5. R.H. Ritchie, On surface plasma oscillations in metal foils. Prog. Theor. Phys. **29**, 607–609 (1963)
6. A.R. Melnyk, M.J. Harrison, Theory of optical excitation of plasmons in metals. Phys. Rev. B **2**, 835–850 (1970)
7. S. Kawata, *Near-Field Optics and Surface Plasmon Polariton* (Springer, New York, 2001)
8. M. Lazar, P.K. Shukla, A. Smolyakov, Surface waves on a quantum plasma half-space. Phys. Plasmas **14**, 124501 (2007)
9. Z.A. Moldabekov, M. Bonitz, T.S. Ramazanov, Theoretical foundations of quantum hydrodynamics for plasmas. Phys. Plasmas **25**, 031903 (2018)
10. A. Moradi, Propagation of electrostatic energy through a quantum plasma. Contrib. Plasma Phys. **59**, 173–180 (2019)
11. Y.-Y. Zhang, S.-B. An, Y.-H. Song, N. Kang, Z.L. Mišković, Y.-N. Wang, Plasmon excitation in metal slab by fast point charge: the role of additional boundary conditions in quantum hydrodynamic model. Phys. Plasmas **21**, 102114 (2014)
12. A. Moradi, Electrostatic surface waves on a magnetized quantum plasma half-space. Phys. Plasmas **23**, 034501 (2016)
13. Y.O. Nakamura, B.V. Paranjape, Surface plasmon in a parallel magnetic field. Solid State Commun. **16**, 467–470 (1975)

14. G.C. Aers, A.D. Boardman, E.D. Issac, High-frequency electrostatic surface waves on a warm gaseous magnetoplasma half-space. IEEE Trans. Plasma Sci. **PS-5**, 123–130 (1977)
15. A. Moradi, Plasmon modes of metallic nanowires including quantum nonlocal effects. Phys. Plasmas **22**, 032112 (2015)
16. I. Villo-Perez, N.R. Arista, Hydrodynamical model for bulk and surface plasmons in cylindrical wires. Surf. Sci. **603**, 1–13 (2009)
17. A. Moradi, E. Ebrahimi, Plasmon spectra of cylindrical nanostructures including nonlocal effects. Plasmonics **9**, 209–218 (2014)
18. A. Moradi, Quantum nonlocal polarizability of metallic nanowires. Plasmonics **10**, 1225–1230 (2015)
19. S. Raza, W. Yan, N. Stenger, M. Wubs, N.A. Mortensen, Blueshift of the surface plasmon resonance in silver nanoparticles: substrate effects. Opt. Express **21**, 27344–27355 (2013)
20. A. Moradi, Extinction properties of metallic nanowires: quantum diffraction and retardation effects. Phys. Lett. A **379**, 2379–2383 (2015)
21. A. Moradi, Plasmon modes of spherical nanoparticles: the effects of quantum nonlocality. Surf. Sci. **637**, 53–57 (2015)
22. I. Villo-Perez, Z.L. Mišković, N.R. Arista, Plasmon spectra of nano-structures: a hydrodynamic model, in *Trends in Nanophysics*, ed. by A. Barsan, V. Aldea (Springer, Berlin, 2010)
23. A. Moradi, Quantum nonlocal polarizability of spherical metal nanoparticles. Int. J. Mod. Phys. B **30**, 1650048 (2016)
24. L. Novotny, B. Hecht, *Principles of Nano-Optics* (Cambridge University Press, New York, 2006)
25. S.A. Maier, *Plasmonics: Fundamentals and Applications* (Springer, New York, 2007)
26. R. Fuchs, F. Claro, Multipolar response of small metallic spheres: nonlocal theory. Phys. Rev. B **35**, 3722–3727 (1987)
27. F.J. Garcia de Abajo, Nonlocal effects in the plasmons of strongly interacting nanoparticles, dimers, and waveguides. Phys. Chem. C **112**, 17983–17987 (2008)
28. C. David, F.J. Garcia de Abajo, Spatial nonlocality in the optical response of metal nanoparticles. J. Phys. Chem. C **115**, 19470–19475 (2011)
29. F.J. Garcia de Abajo, Relativistic energy loss and induced photon emission in the interaction of a dielectric sphere with an external electron beam. Phys. Rev. B **59**, 3095–3107 (1999)
30. T. Christensen, W. Yan, S. Raza, A.-P. Jauho, N.A. Mortensen, M. Wubs, Nonlocal response of metallic nanospheres probed by light, electrons, and atoms. ACS Nano **8**, 1745–1758 (2014)

Chapter 5
Problems in Electromagnetic Theory: Spatial Nonlocal Effects

Abstract In this chapter, by considering the spatial nonlocal effects, some electromagnetic boundary-value problems involving the bounded electron gases with planar, cylindrical, and spherical geometries are studied. We use the nonlocal hydrodynamic theory discussed in Chap. 1 and investigate planar geometry in details. For brevity, in many sections of this chapter the $\exp(-i\omega t)$ time factor is suppressed. Furthermore, all media under consideration are nonmagnetic and attention is only confined to the linear phenomena.

5.1 Plasmonic Properties of Semi-Infinite Electron Gases: Standard Hydrodynamic Model

5.1.1 Total Reflection of a Plane Wave

Let us consider an incident p-polarized plane electromagnetic wave coming from below $z < 0$ and propagating in an insulator medium with a relative real permittivity ε_1, reflected by an EG interface located at $z = 0$, as shown in Fig. 3.1. The EG being characterized by a longitudinal dielectric function ε_{2L} and a transverse dielectric function ε_{2T}. The spatial effects can be accounted by employing the SHD model, and this is the approach adopted here. In this way, we employ the usual, Drude type, transverse dielectric function

$$\varepsilon_{2T} = \varepsilon_b - \frac{\omega_p^2}{\omega^2}. \tag{5.1}$$

The original version of this chapter was revised. The correction to this chapter is available at https://doi.org/10.1007/978-3-030-43836-4_11

A. Moradi, *Canonical Problems in the Theory of Plasmonics*, Springer Series in Optical Sciences 230, https://doi.org/10.1007/978-3-030-43836-4_5

Also, as mentioned in Chap. 1, the longitudinal plasma dielectric function of an EG is

$$\varepsilon_{2L} = \varepsilon_b - \frac{\omega_p^2}{\omega^2 - \alpha^2 k_L^2} , \tag{5.2}$$

where ω is the angular frequency, k_L is the nonlocal longitudinal wavenumber. The longitudinal plasma wave obeys the dispersion relation

$$\varepsilon_{2L}(k_L, \omega) = 0 . \tag{5.3}$$

As mentioned before, a longitudinal wave is curl free, which means it satisfies $\nabla \times \mathbf{E}_L = 0$ and there is no accompanying magnetic field. Therefore, we may define $\mathbf{E}_L = -\nabla\Phi$, and electric potential Φ may be given by

$$\Phi(x, z) = -A_{2L} e^{-\kappa_{2L} z} e^{ik_x x} , \tag{5.4}$$

where $\kappa_{2L}^2 = k_x^2 - k_L^2$, A_{2L} is the amplitude of the electric potential Φ and by assumption, the x and z dependences of the electric potential are $\exp(ik_x x)$ and $\exp(-\kappa_{2L} z)$, respectively, and $\partial\Phi/\partial y = 0$. Also, the transverse electromagnetic waves satisfy the dispersion relation

$$k_{2T}^2 = \varepsilon_{2T} \frac{\omega^2}{c^2} . \tag{5.5}$$

The wavevectors of the longitudinal and transverse waves that are excited in the EG can be obtained from (5.3) and (5.5), respectively. From (5.3), the expression of k_L can be written as shown by (4.9). We note that for a s-polarized wave there can be no interaction between transverse and longitudinal waves in isotropic media [1]. For p-polarized case, the total magnetic field H_y, in the insulator space (i.e., $z \leq 0$), is taken to be in the form of a superposition of incident and reflected waves. Thus

$$H_y = -A_1 \left[e^{ik_z z} + r_p e^{-ik_z z} \right] e^{ik_x x} , \tag{5.6}$$

where r_p is, by definition, the Fresnel reflection coefficient for p-polarized waves, A_1 is the amplitude of the incident magnetic field H_y and $k_z^2 = \varepsilon_1 k_0^2 - k_x^2$. Also, using (3.1b) and (3.1c) the total electric fields along the x- and z-directions in the insulator space can be written in the form

$$E_x = \frac{-k_z}{\omega\varepsilon_0\varepsilon_1} A_1 \left[e^{ik_z z} - r_p e^{-ik_z z} \right] e^{ik_x x} , \tag{5.7}$$

$$E_z = \frac{k_x}{\omega\varepsilon_0\varepsilon_1} A_1 \left[e^{ik_z z} + r_p e^{-ik_z z} \right] e^{ik_x x} . \tag{5.8}$$

In the EG, the magnetic field can be written, as

$$H_{yT} = -A_{2T} e^{-\kappa_{2T} z} e^{ik_x x} , \qquad (5.9)$$

where $\kappa_{2T}^2 = k_x^2 - \varepsilon_T k_0^2$, $A_{2T} = t_p A_1$ and t_p is, by definition, the Fresnel transmission coefficient for p-polarized waves. Again, using (3.1b) and (3.1c) the electric fields along the x- and z-directions in the EG space can be written in the form

$$E_{xT} = \frac{-i\kappa_{2T}}{\omega\varepsilon_0\varepsilon_2} A_{2T} e^{-\kappa_{2T} z} e^{ik_x x} , \qquad (5.10)$$

$$E_{zT} = \frac{k_x}{\omega\varepsilon_0\varepsilon_2} A_{2T} e^{-\kappa_{2T} z} e^{ik_x x} . \qquad (5.11)$$

Furthermore, the electric fields, associated with the longitudinal wave, along the x- and z-directions in the EG space can be written as

$$E_{xL} = ik_x A_{2L} e^{-\kappa_{2L} z} e^{ik_x x} , \qquad (5.12)$$

$$E_{zL} = -\kappa_{2L} A_{2L} e^{-\kappa_{2L} z} e^{ik_x x} . \qquad (5.13)$$

The unknown relations between the coefficients A_1, A_{2L}, and A_{2T} can be determined by using the appropriate BCs at the surface of the system. The usual two classical BCs at the EG-insulator interface require the continuity of the tangential components of the electric and magnetic fields across the interface. We note that in the EG both the transverse and longitudinal (usually neglected) waves give a contribution to the value of electric field. We have

$$1 + r_p = t_p , \qquad (5.14)$$

$$A_1 \frac{1}{\omega\varepsilon_0} \left[\frac{ik_z}{\varepsilon_1} \left(1 - r_p \right) + \frac{\kappa_{2T}}{\varepsilon_{2T}} t_p \right] = k_x A_{2L} . \qquad (5.15)$$

Also, since we have allowed for the excitation of the longitudinal waves inside the EG, the two mentioned BCs are not sufficient to determine the unknown coefficients and an ABC must be introduced at the system interface. According to the results obtained by Yan et al. [2], this ABC at the EG-insulator interface requires the continuity of the normal component of the displacement field. We have

$$A_1 \frac{k_x}{\omega\varepsilon_0} \left[1 + r_p - \frac{\varepsilon_b}{\varepsilon_{2T}} t_p \right] = -\varepsilon_b A_{2L} \kappa_{2L} . \qquad (5.16)$$

Now, we can solve the system of (5.14)–(5.16) to obtain

$$r_p = \frac{\varepsilon_{2T} k_z - i\varepsilon_1 \kappa_{2T} \left(1 - \delta_{nl} \right)}{\varepsilon_{2T} k_z + i\varepsilon_1 \kappa_{2T} \left(1 - \delta_{nl} \right)} , \qquad (5.17)$$

where

$$\delta_{nl} = \frac{k_x^2}{\kappa_{2L}\kappa_{2T}}\left(1 - \frac{\varepsilon_{2T}}{\varepsilon_b}\right). \tag{5.18}$$

The Fresnel reflection coefficient indicates that the bulk plasmon is not excited at normal incidence for $k_x = 0$ because in that case only one component of the electric field is present in the incident and reflected fields [3]. When the angle of incidence increases the excitation of the bulk plasmon is more and more important because of the increasing E_z component. In the present problem again the reflectance coefficient is

$$R_p = |r_p|^2 = 1. \tag{5.19}$$

5.1.2   An Alternative Derivation of Fresnel Reflection Coefficient

For the BVP of the previous section, the normal wave impedance η_n, at the interface, is defined by [4]

$$\eta_n = \lim_{z \to 0} \frac{E_x}{H_y}. \tag{5.20}$$

Combining (5.6) and (5.7), it is seen that

$$\eta_n = \frac{k_z}{\omega\varepsilon_0\varepsilon_1}\frac{1 - r_p}{1 + r_p}. \tag{5.21}$$

The Fresnel reflection coefficient may then be expressed conveniently in terms of the normal surface impedance via the relation

$$r_p = \frac{\dfrac{k_z}{\omega\varepsilon_0\varepsilon_1} - \eta_n}{\dfrac{k_z}{\omega\varepsilon_0\varepsilon_1} + \eta_n}. \tag{5.22}$$

Also, combining (5.9), (5.10), and (5.12), it is now seen that the normal wave impedance may be expressed by

$$\eta_n = \frac{i\kappa_{2T}}{\omega\varepsilon_0\varepsilon_2} - ik_x\frac{A_{2L}}{A_{2T}}. \tag{5.23}$$

The ratio A_{2L}/A_{2T} is obtained in a very straightforward manner by imposing the ABC at the EG-insulator interface, i.e., the continuity of the normal component of the displacement field. Thus by making use of (5.8), (5.11), (5.13), and (5.14), it is found that

$$\frac{A_{2L}}{A_{2T}} = -\frac{k_x}{\omega \varepsilon_0 \varepsilon_b \kappa_{2L}} \left(1 - \frac{\varepsilon_b}{\varepsilon_{2T}} \right) . \tag{5.24}$$

Then substituting (5.23) into (5.22) and using (5.24) gives (5.17).

5.1.3 Dispersion Relation

We assume that a p-polarized SPP propagates along a semi-infinite EG surface (in the half-space $z > 0$) in the x-direction. For the present system, the solution of (3.17) has the form

$$E_x(z) = \begin{cases} A_1 e^{+\kappa_1 z} , & z \le 0 , \\ A_{2T} e^{-\kappa_{2T} z} , & z \ge 0 , \end{cases} \tag{5.25}$$

where $\kappa_1^2 = k_x^2 - \varepsilon_1 k_0^2$ and $\kappa_{2T}^2 = k_x^2 - \varepsilon_{2T} k_0^2$. Therefore, from (3.16) we obtain

$$H_y(z) = i\omega\varepsilon_0 \begin{cases} A_1 \dfrac{\varepsilon_1}{\kappa_1} e^{+\kappa_1 z} , & z \le 0 , \\ -A_{2T} \dfrac{\varepsilon_{2T}}{\kappa_{2T}} e^{-\kappa_{2T} z} , & z \ge 0 , \end{cases} \tag{5.26}$$

$$E_z(z) = i k_x \begin{cases} -A_1 \dfrac{1}{\kappa_1} e^{+\kappa_1 z} , & z \le 0 , \\ A_{2T} \dfrac{1}{\kappa_{2T}} e^{-\kappa_{2T} z} . & z \ge 0 , \end{cases} \tag{5.27}$$

Furthermore, the electric fields associated with the longitudinal wave, along the x- and z-directions in the EG space can be represented by (5.12) and (5.13), respectively. Now, applying the appropriate BCs (as mentioned in previous section) we will be able to find the relationship between the constants A_1, A_{2T}, and A_{2L}, as

$$A_{2T} = -\frac{\varepsilon_1}{\kappa_1} \frac{\kappa_{2T}}{\varepsilon_{2T}} A_1 , \tag{5.28}$$

$$A_{2L} = -i k_x \frac{\dfrac{\varepsilon_b}{\varepsilon_{2T}} + \dfrac{\varepsilon_1}{\kappa_1} \dfrac{\kappa_{2T}}{\varepsilon_{2T}}}{k_x^2 + \kappa_1 \kappa_{2L} \dfrac{\varepsilon_b}{\varepsilon_1}} A_1 . \tag{5.29}$$

Also, we obtain a relation between ω and k_x, as

$$\frac{\varepsilon_1}{\kappa_1} + \frac{\varepsilon_{2T}}{\kappa_{2T}} = \frac{\varepsilon_1}{\kappa_1} \delta_{nl} , \tag{5.30}$$

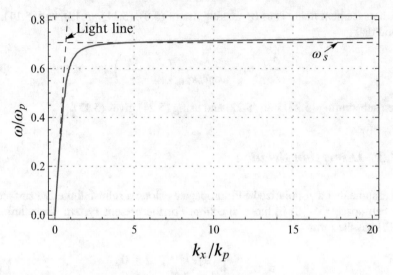

Fig. 5.1 Dispersion curve of SPPs at a flat EG-vacuum interface, as obtained from (5.30), when $\varepsilon_b = 1$. For the other parameters the sodium values in Table 4.1 have been employed. Here $k_p = \omega_p/c$ and the dashed horizontal line indicates the asymptotic SP frequency of the system in the local model, i.e., $\omega_s = \omega_p/\sqrt{2}$

where δ_{nl} is defined by (5.18). If the right-hand side of (5.30) is equated to zero, the usual dispersion equation for SPPs is obtained, i.e., (3.19). Figure 5.1 displays the dispersion curve for propagation of SPPs at an EG-vacuum interface as obtained from the numerical solution of (5.30). Again, we see that for small values of k_x, the dispersion curve approaches the light line. Furthermore for large valves of k_x, SPPs can propagate with frequencies larger than ω_s.

5.1.4 Power Flow

In [5], Matsuo and Tsuji derived the expressions for energy density and power flow density of SPPs on an EG-vacuum interface by using the SHD approximation, when $\varepsilon_b = 1$. Here, we extend the work by E. Matsuo and M. Tsuji [5], when $\varepsilon_b \neq 1$ and $\varepsilon_1 \neq 1$. For the power flow density associated with SPPs on a semi-infinite EG, we have

$$S_x = \begin{cases} [\mathbf{E}_1 \times \mathbf{H}_1]_x \, , & z \leq 0, \\ [(\mathbf{E}_{2T} + \mathbf{E}_{2L}) \times \mathbf{H}_2]_x + m_e \alpha^2 n \, (v_{xT} + v_{xL}) \, , & z \geq 0, \end{cases} \qquad (5.31)$$

in the real number representation. In (5.31), subscript 1 denotes the region outside the EG, while subscript 2 denotes the region inside the EG. In this chapter and also in the above equation, the vector quantities can be separated into transverse and longitudinal parts. For example, the electrons velocity field $\mathbf{v}$ is written as the sum of two terms $\mathbf{v} = \mathbf{v}_L + \mathbf{v}_T$, where $\mathbf{v}_L$ is called the longitudinal or irrotational velocity and has $\nabla \times \mathbf{v}_L = 0$, while $\mathbf{v}_T$ is called the transverse or solenoidal velocity and has $\nabla \cdot \mathbf{v}_T = 0$. In this way, from (4.1) and (4.2) we obtain

$$\frac{\partial \mathbf{v}_L}{\partial t} = -\frac{e}{m_e} \mathbf{E}_L - \frac{\alpha^2}{n_0} \nabla n \ , \tag{5.32}$$

$$\frac{\partial n}{\partial t} + n_0 \nabla \cdot \mathbf{v}_L = 0 \ . \tag{5.33}$$

Also, we have

$$\frac{\partial \mathbf{v}_T}{\partial t} = -\frac{e}{m_e} \mathbf{E}_T \ . \tag{5.34}$$

Now, by eliminating v_{xT}, v_{xL}, and n in (5.31), using (5.32)–(5.34), we obtain

$$S_x = -\frac{1}{2} \mathrm{Re} \begin{cases} E_{1z} H_{1y}^* \ , & z \leq 0 \ , \\ (E_{2Tz} + E_{2Lz}) H_{2y}^* - \varepsilon_0 \varepsilon_{2T} \dfrac{\omega^3}{k_x \omega_p^2} \left(\varepsilon_b E_{2Lx} + \dfrac{\omega_p^2}{\omega^2} E_{2Tx} \right) E_{2Lx}^* \ , & z \geq 0 \ , \end{cases} \tag{5.35}$$

that shows the cycle-averaged x-components of power flow density in the complex number representation. At this stage, by eliminating the components of $\mathbf{E}$ and $\mathbf{H}$ in (5.35), we find

$$S_x = \frac{\varepsilon_0 \omega k_x}{2}$$

$$\times \begin{cases} \dfrac{\varepsilon_1}{\kappa_1^2} |A_1|^2 e^{2\kappa_1 z} \ , & z \leq 0 \ , \\ \varepsilon_{2T} \left[\dfrac{1}{\kappa_{2T}^2} |A_{2T}|^2 e^{-2\kappa_{2T} z} + \varepsilon_b \dfrac{\omega^2}{\omega_p^2} |A_{2L}|^2 e^{-2\kappa_{2L} z} - \dfrac{i}{k_x} \dfrac{A_{2T} A_{2L}^*}{e^{(\kappa_{2L} + \kappa_{2T})z}} \right] \ , & z \geq 0 \ , \end{cases} \tag{5.36}$$

where the distribution in (5.36) is discontinuous at the interface $z = 0$, as mentioned before. Since SPPs are localized along the z coordinate, the total power flow density (per unit width) associated with SPPs are determined by an integration over z. We find

$$\langle S_x \rangle = \frac{\varepsilon_0 \omega k_x}{4} \left[\frac{\varepsilon_1}{\kappa_1^3} |A_1|^2 + \varepsilon_{2T} \left(\frac{1}{\kappa_{2T}^3} |A_{2T}|^2 + \varepsilon_b \frac{\omega^2}{\kappa_{2L} \omega_p^2} |A_{2L}|^2 - \frac{2i}{k_x \kappa_{2T}} A_{2T} A_{2L}^* \right) \right] . \tag{5.37}$$

The first term on the right-hand side of (5.37) corresponds to the power flow density in the insulator. The second and third terms are the power flow densities due to the transverse and longitudinal fields in the EG, respectively. Finally, the forth is the interference term between the both fields.

5.1.5   Energy Distribution

In the presence of the spatial dispersion, we have the energy density U associated with SPs of a semi-infinite EG in the real number representation, as

$$U = \frac{1}{2} \begin{cases} \varepsilon_0 \varepsilon_1 |\mathbf{E}_1|^2 + \mu_0 |\mathbf{H}_1|^2, & z \leq 0, \\ \varepsilon_0 \varepsilon |\mathbf{E}_{2T} + \mathbf{E}_{2L}|^2 + \mu_0 |\mathbf{H}_2|^2 + m_e n_0 (\mathbf{v}_T + \mathbf{v}_L)^2 + \dfrac{m_e \alpha^2}{n_0} n^2, & z \geq 0, \end{cases}$$
(5.38)

where $\varepsilon = \varepsilon_b + \omega(\partial \varepsilon_b / \partial \omega)$. By eliminating $\mathbf{v}_T$, $\mathbf{v}_L$, and n in (5.38), using (5.32)–(5.34), we find

$$U = \frac{1}{4} \begin{cases} \varepsilon_0 \varepsilon_1 |\mathbf{E}_1|^2 + \mu_0 |\mathbf{H}_1|^2, & z \leq 0, \\[2mm] \varepsilon_0 \left(\varepsilon + \dfrac{\omega_p^2}{\omega^2} \right) |\mathbf{E}_{2T}|^2 + \mu_0 |\mathbf{H}_2|^2 \\[2mm] + \varepsilon_0 \left(\varepsilon + \varepsilon_b^2 \dfrac{\omega^2}{\omega_p^2} \right) |\mathbf{E}_{2L}|^2 + 2\varepsilon_0 (\varepsilon_b + \varepsilon) \, \mathbf{E}_{2L} \cdot \mathbf{E}_{2L}^* \\[2mm] + \varepsilon_0 \varepsilon_b \varepsilon_{2T} \dfrac{\omega^2}{\omega_p^2} \dfrac{k_x^2 - \kappa_L^2}{k_x^2} |E_{2Lx}|^2, & z \geq 0. \end{cases}$$
(5.39)

that shows the cycle-averaged of energy density in the complex number representation. Then, by eliminating the components of $\mathbf{E}$ and $\mathbf{H}$ in (5.39), we get

$$U = \frac{\varepsilon_0}{4} \begin{cases} 2\varepsilon_1 \dfrac{k_x^2}{\kappa_1^2} |A_1|^2 e^{2\kappa_1 z}, & z \leq 0, \\[2mm] \left[\left(\varepsilon + \dfrac{\omega_p^2}{\omega^2} \right) \left(1 + \dfrac{k_x^2}{\kappa_{2T}^2} \right) - \varepsilon_{2T} \left(1 - \dfrac{k_x^2}{\kappa_{2T}^2} \right) \right] |A_{2T}|^2 e^{-2\kappa_{2T} z} \\[2mm] + \left[\left(\varepsilon + \varepsilon_b^2 \dfrac{\omega^2}{\omega_p^2} \right) (k_x^2 + \kappa_{2L}^2) + \varepsilon_b \varepsilon_{2T} (k_x^2 - \kappa_{2L}^2) \dfrac{\omega^2}{\omega_p^2} \right] |A_{2L}|^2 e^{-2\kappa_{2L} z} \\[2mm] - 2i k_x (\varepsilon_b + \varepsilon) \left(1 + \dfrac{\kappa_{2L}}{\kappa_{2T}} \right) A_{2T} A_{2L}^* e^{-(\kappa_{2L} + \kappa_{2T}) z}, & z \geq 0. \end{cases}$$
(5.40)

The total energy density associated with the SPPs is again determined by integration over z, the energy per unit surface area being

$$
\langle U \rangle = \frac{\varepsilon_0}{4} \left\{ \varepsilon_1 \frac{k_x^2}{\kappa_1^3} |A_1|^2 + \frac{1}{2\kappa_{2T}} \left[\left(\varepsilon + \frac{\omega_p^2}{\omega^2} \right) \left(1 + \frac{k_x^2}{\kappa_{2T}^2} \right) - \varepsilon_{2T} \left(1 - \frac{k_x^2}{\kappa_{2T}^2} \right) \right] |A_{2T}|^2 \right.
$$

$$
+ \frac{1}{2\kappa_{2L}} \left[\left(\varepsilon + \varepsilon_b^2 \frac{\omega^2}{\omega_p^2} \right) \left(k_x^2 + \kappa_{2L}^2 \right) + \varepsilon_b \varepsilon_{2T} \left(k_x^2 - \kappa_{2L}^2 \right) \frac{\omega^2}{\omega_p^2} \right] |A_{2L}|^2
$$

$$
\left. - 2i \left(\varepsilon_b + \varepsilon \right) \frac{k_x}{\kappa_{2T}} A_{2T} A_{2L}^* \right\} , \qquad (5.41)
$$

where the terms on the right-hand side are, respectively, energy density in the insulator, energy densities due to the transverse and longitudinal fields in the EG, respectively, and the interference energy density term between both the fields. Now, the energy velocity of SPPs can be obtained as $v_e = \langle S_x \rangle / \langle U \rangle$ that is equal to the group velocity of SPPs. However, this equality of energy propagation velocity and group velocity of SPPs is broken in the presence of the Bohm potential in the short-wavelength limit, as shown in [6]. Furthermore, if we take a limit $\alpha \to 0$, then $\kappa_{2L} \to \infty$ and all longitudinal and interference terms in (5.37) and (5.41) vanish and expressions reduce to those for the usual local model.

Note that in the retarded region, i.e., $k_x \ll \omega_p/c$, the dispersion curve in Fig. 5.1 passes near the light line, i.e., $\omega = ck_x$. In this region it is easy to find that the energy density and power flow density are almost contained in the insulator and the role of the longitudinal field is negligibly small compared with that of the transverse field because $\alpha \ll c$. This is consistent with the fact that the nonlocal effect is negligible in the region near the light line. In this limit (5.36) shows that the power flow density in the EG is negative.

In the opposite limit $k_x \gg \omega_p/c$ (nonretarded region), the dispersion relation (5.30) reduces to (4.18) which is also rewritten in the usual form as shown by (4.19). In this nonretarded region, when ω is near to $\omega_p/\sqrt{2}$, the longitudinal and interference terms are negligibly small compared with the insulator and transverse terms. As ω increases from $\omega_p/\sqrt{2}$, however, all terms tend to have the same order of magnitude as discussed in Sect. 4.1. We should note that the sign of the power flow density in the EG in the present study (in the nonretarded region with the local approximation) is negative that is *vice versa* of the corresponding result obtained in Sect. 4.1.

5.2   Surface Plasmon Polaritons of Semi-Infinite Electron Gases: An Alternative Derivation of Dispersion Relation

We assume that a p-polarized SPP propagates along a semi-infinite EG surface (in the half-space $z > 0$) in the x-direction. In the insulator region, i.e., $z < 0$, our equations are the full set of Maxwell's equations, as

$$\nabla \cdot \mathbf{H} = 0 , \tag{5.42}$$

$$\nabla \cdot \mathbf{E} = 0 , \tag{5.43}$$

$$\nabla \times \mathbf{E} = -\mu_0 \frac{\partial \mathbf{H}}{\partial t} , \tag{5.44}$$

$$\nabla \times \mathbf{H} = \varepsilon_0 \varepsilon_1 \frac{\partial \mathbf{E}}{\partial t} , \tag{5.45}$$

whose solution is, using $\nabla \cdot \mathbf{E} = 0$,

$$E_x = A_1 e^{-\kappa_1 z} , \tag{5.46a}$$

$$E_z = \frac{-i k_x}{\kappa_1} A_1 e^{-\kappa_1 z} , \tag{5.46b}$$

$$H_y = \frac{i \omega \varepsilon_0 \varepsilon_1}{\kappa_1} A_1 e^{-\kappa_1 z} , \tag{5.46c}$$

where $\kappa_1^2 = k_x^2 - \varepsilon_1 k_0^2$. In the EG region, we have[1]

$$\nabla \cdot \mathbf{H} = 0 , \tag{5.47}$$

$$\nabla \cdot \mathbf{E} = -\frac{e}{\varepsilon_0 \varepsilon_b} n , \tag{5.48}$$

$$\nabla \times \mathbf{E} = -\mu_0 \frac{\partial \mathbf{H}}{\partial t} , \tag{5.49}$$

$$\nabla \times \mathbf{H} = -e n_0 \mathbf{v} + \varepsilon_0 \varepsilon_b \frac{\partial \mathbf{E}}{\partial t} . \tag{5.50}$$

Equations (5.47)–(5.50) describe the influence of the EG on the $\mathbf{E}$ and $\mathbf{H}$ fields. The back action is described by the SHD equations for the EG in the linearized form, as

---

[1] Here, the vector quantities are not separated into transverse and longitudinal parts.

$$\frac{\partial \mathbf{v}}{\partial t} = -\frac{e}{m_e}\mathbf{E} - \frac{\alpha^2}{n_0}\nabla n \ , \tag{5.51}$$

$$\frac{\partial n}{\partial t} + n_0 \nabla \cdot \mathbf{v} = 0 \ . \tag{5.52}$$

In the EG region $z > 0$, the governing Maxwell and SHD equations can be arranged for two quantities $\mathbf{e}_y \cdot \nabla \times \mathbf{E}$ (or $\mathbf{H}$) and $\nabla \cdot \mathbf{E}$ as done by Boardman in [7]. Taking the divergence of (5.50) and using (5.51), it is straightforward to obtain

$$\left(\nabla^2 + k_{2L}^2\right)\nabla \cdot \mathbf{E} = 0 \ , \tag{5.53}$$

where k_{2L} can be obtained from (5.3) and is given by (4.9). Similarly, taking curl of (5.50), we can derive

$$\left(\nabla^2 + \varepsilon_{2T}\frac{\omega^2}{c^2}\right)\mathbf{e}_y \cdot \nabla \times \mathbf{E} = 0 \ . \tag{5.54}$$

Now, we look for a solution of (5.53) and (5.54) of the form

$$\nabla \cdot \mathbf{E} = A_{2L}e^{-\kappa_{2L}z} \ , \tag{5.55}$$

which is an irrotational solution, and

$$\mathbf{e}_y \cdot \nabla \times \mathbf{E} = A_{2T}e^{-\kappa_{2T}z} \ , \tag{5.56}$$

which is a divergence-free solution. Equations (5.55) and (5.56) constitute simultaneous equations for the electric field components E_x and E_z. Solving these equations, we find

$$E_x = \frac{-k_x}{\kappa_{2L}^2 - k_x^2}A_{2L}e^{-\kappa_{2L}z} + \frac{i\kappa_{2T}}{\kappa_{2T}^2 - k_x^2}A_{2T}e^{-\kappa_{2T}z} \ , \tag{5.57}$$

$$E_z = \frac{-i\kappa_{2L}}{\kappa_{2L}^2 - k_x^2}A_{2L}e^{-\kappa_{2L}z} - \frac{k_x}{\kappa_{2T}^2 - k_x^2}A_{2T}e^{-\kappa_{2T}z} \ . \tag{5.58}$$

Also, using (5.49) the magnetic field in the EG is

$$H_y = \frac{-1}{\omega\mu_0}A_{2T}e^{-\kappa_{2T}z} \ . \tag{5.59}$$

At this stage, using the appropriate BCs (i.e., the continuity of E_x, H_y, and εE_z) we find (5.30).

5.2.1   *Dispersion Relation of Surface Magneto Plasmon Polariton: Voigt Configuration*

Here, we extend the BVP of the previous section and derive the dispersion relation of a SMPP on the interface of an insulator and a magnetized EG. We consider the Voigt configuration. For other configurations the field and fluid variables are coupled to render analytic treatment difficult. Now, a SMPP with wavenumber k_x, and the form $f(z) \exp(\pm i k_x x - i\omega t)$ is assumed to propagate parallel to the interface $z = 0$, along the x-direction.[2] Therefore, (5.51), (5.53), and (5.54) are modified to

$$\mathbf{v} = \frac{\omega}{\omega^2 - \omega_c^2} \left(-\frac{ie}{m_e}\mathbf{E} + \frac{e}{m_e}\frac{\omega_c}{\omega}\mathbf{e}_y \times \mathbf{E} + \frac{\alpha^2}{n_0}\left[\frac{\omega_c}{\omega}\mathbf{e}_y \times \nabla - i\nabla\right]n \right), \qquad (5.60)$$

$$\left(\alpha^2\nabla^2 + \omega^2 - \omega_c^2 - \frac{\omega_p^2}{\varepsilon_b}\right)\nabla \cdot \mathbf{E} = -i\frac{\omega_p^2}{\varepsilon_b}\frac{\omega_c}{\omega}\mathbf{e}_y \cdot \nabla \times \mathbf{E}, \qquad (5.61)$$

$$\left(\frac{c^2}{\omega^2}\nabla^2 + \varepsilon_b - \frac{\omega_p^2}{\omega^2 - \omega_c^2}\right)\mathbf{e}_y \cdot \nabla \times \mathbf{E} = \frac{i\omega_c}{\omega(\omega^2 - \omega_c^2)}\left(\omega_p^2 - \varepsilon_b\alpha^2\nabla^2\right)\nabla \cdot \mathbf{E}. \qquad (5.62)$$

Equations (5.61) and (5.62) are two coupled wave equations for the present magnetized EG. Eliminating $\alpha^2\nabla^2\nabla \cdot \mathbf{E}$ from these two equations, we obtain

$$\left(c^2\nabla^2 + \varepsilon_b\omega^2 - \omega_p^2\right)\mathbf{e}_y \cdot \nabla \times \mathbf{E} = i\varepsilon_b\omega\omega_c\nabla \cdot \mathbf{E}. \qquad (5.63)$$

Then, (5.61) and (5.63) give the wave equation, as

$$\left[\left(\alpha^2\nabla^2 + \omega^2 - \omega_c^2 - \frac{\omega_p^2}{\varepsilon_b}\right)\left(c^2\nabla^2 + \varepsilon_b\omega^2 - \omega_p^2\right) - \omega_p^2\omega_c^2\right]\left(\begin{array}{c}\nabla \cdot \mathbf{E} \\ \mathbf{e}_y \cdot \nabla \times \mathbf{E}\end{array}\right) = 0. \qquad (5.64)$$

To make progress, it is convenient to rewrite (5.64) in a factored form, as

$$\left(\frac{\mathrm{d}^2}{\mathrm{d}z^2} - \kappa_-^2\right)\left(\frac{\mathrm{d}^2}{\mathrm{d}z^2} - \kappa_+^2\right)\left(\begin{array}{c}\nabla \cdot \mathbf{E} \\ \mathbf{e}_y \cdot \nabla \times \mathbf{E}\end{array}\right) = 0, \qquad (5.65)$$

where

---

[2] As mentioned before, the upper and lower signs refer to waves propagating in the positive and negative x-directions, respectively.

$$\kappa_\pm^2 = k_x^2 - \frac{1}{2}\left\{k_T^2 + k_L^2 \pm \left[\left(k_T^2 + k_L^2\right)^2 - 4\frac{\omega^2}{\varepsilon_b}\frac{\varepsilon_{2T}^2\omega^2 - \varepsilon_b^2\omega_c^2}{c^2\alpha^2}\right]^{1/2}\right\}, \qquad (5.66)$$

with

$$k_L^2 = \frac{\omega^2 - \omega_c^2 - \omega_p^2/\varepsilon_b}{\alpha^2} . \qquad (5.67)$$

The asymptotic (electrostatic limit, i.e., $c \to \infty$ and local limit, i.e., $\alpha \to 0$) values of $\kappa_\pm$ are useful in the forthcoming analysis and we evaluate $\kappa_\pm$ by expanding in the powers of α^2/c^2 as done by Lee and Cho in [8]. Keeping only the lowest order of α^2/c^2, we obtain from (5.66)

$$\kappa_+^2 = k_x^2 - \frac{\omega^2}{\varepsilon_b}\frac{\varepsilon_{2T}^2\omega^2 - \varepsilon_b^2\omega_c^2}{c^2\alpha^2 k_L^2} , \qquad (5.68a)$$

$$\kappa_-^2 = k_x^2 - k_L^2 - \varepsilon_{2T}\frac{\omega^2}{c^2} + \frac{\omega^2}{\varepsilon_b}\frac{\varepsilon_{2T}^2\omega^2 - \varepsilon_b^2\omega_c^2}{c^2\alpha^2 k_L^2} . \qquad (5.68b)$$

Now, we look for a solution of (5.65) of the form

$$\nabla \cdot \mathbf{E} = A_{2+}e^{-\kappa_+ z} + A_{2-}e^{-\kappa_- z} . \qquad (5.69)$$

Using (5.69) in (5.61), we get

$$\mathbf{e}_y \cdot \nabla \times \mathbf{E} = i\xi_+ A_{2+}e^{-\kappa_+ z} + i\xi_- A_{2-}e^{-\kappa_- z} , \qquad (5.70)$$

where

$$\xi_\pm = \varepsilon_b\alpha^2\frac{\omega}{\omega_c\omega_p^2}\left(\kappa_\pm^2 - k_x^2 + k_L^2\right) . \qquad (5.71)$$

Therefore, solving (5.69) and (5.70), we obtain

$$E_x = i\frac{\pm k_x - \xi_+\kappa_+}{\kappa_+^2 - k_x^2}A_{2+}e^{-\kappa_+ z} + i\frac{\pm k_x - \xi_-\kappa_-}{\kappa_-^2 - k_x^2}A_{2-}e^{-\kappa_- z} , \qquad (5.72)$$

$$E_z = -\frac{\kappa_+ \mp k_x\xi_+}{\kappa_+^2 - k_x^2}A_{2+}e^{-\kappa_+ z} - \frac{\kappa_- \mp k_x\xi_-}{\kappa_-^2 - k_x^2}A_{2-}e^{-\kappa_- z} . \qquad (5.73)$$

Also, using (5.49) the magnetic field in the magnetized EG is

$$H_y = \frac{\xi_+}{\omega\mu_0}A_{2+}e^{-\kappa_+ z} + \frac{\xi_-}{\omega\mu_0}A_{2-}e^{-\kappa_- z} . \qquad (5.74)$$

To connect the insulator solution, i.e., (5.46a)–(5.46c) to the EG solution, i.e., (5.72)–(5.74), we need three BCs at $z = 0$ that are

$$E_{1x}|_{z=0} = E_{2x}|_{z=0} ,$$ (5.75a)

$$\varepsilon_1 E_{1z}|_{z=0} = \varepsilon_b E_{2z}|_{z=0} ,$$ (5.75b)

$$\mathbf{e}_z \cdot \mathbf{v}|_{z=0} = 0 ,$$ (5.75c)

where subscript 1 denotes the region outside the EG, while subscript 2 denotes the region inside it. Note that condition denotes by (5.75b) is equivalent to $H_{1y}|_{z=0} = H_{2y}|_{z=0}$. Applying the mentioned BCs, we obtain a relation between ω and k_x, as

$$\begin{vmatrix} -1 & i\left(\mp k_x + \xi_+\kappa_+\right) & i\left(\mp k_x + \xi_-\kappa_-\right) \\ \pm i\varepsilon_1 \dfrac{k_x}{\kappa_1} & \varepsilon_b\left(\kappa_+ \mp k_x\xi_+\right) & \varepsilon_b\left(\kappa_- \mp k_x\xi_-\right) \\ i\left(\pm\dfrac{\varepsilon_1}{\varepsilon_b}\dfrac{k_x}{\kappa_1} + \dfrac{\omega_c}{\omega}\right) & \dfrac{k_x^2 - \kappa_+^2}{\omega_p^2/\varepsilon_b\alpha^2}\left(\pm\dfrac{\omega_c}{\omega}k - \kappa_+\right) & \dfrac{k_x^2 - \kappa_-^2}{\omega_p^2/\varepsilon_b\alpha^2}\left(\pm\dfrac{\omega_c}{\omega}k - \kappa_-\right) \end{vmatrix} = 0 .$$ (5.76)

Equation (5.76) is the dispersion relation of SMPP on the interface of an insulator and a magnetized EG in the presence of the standard nonlocal effects. However, (5.76) is a very complex equation, but its correctness can be checked by evaluating various asymptotic expressions, such as electrostatic nonlocal limit and cold SMPP limit.

5.3 Plasmonic Properties of Multilayer Planar Structures: Standard Hydrodynamic Model

5.3.1 Insulator-Electron Gas-Insulator Structures

Consider now an EG slab of thickness d with longitudinal dielectric function ε_{2L} and transverse dielectric function ε_{2T}, located between two semi-infinite insulator media with relative dielectric constants ε_1 (in the region $z < 0$) and ε_3 (in the region $z > d$), respectively. The p-polarized SPPs are supposed to propagate parallel to the interface $z = 0$ and $z = d$ along the x-direction. Thus, the solution of (3.17) for the tangential component of the electric field has the form

$$E_x(z) = \begin{cases} A_1 e^{+\kappa_1 z} , & z \le 0 , \\ A_{2T} e^{+\kappa_2 z} + A_{3T} e^{-\kappa_2 z} , & 0 \le z \le d , \\ A_4 e^{-\kappa_3 z} , & z \ge d , \end{cases}$$ (5.77)

where the relations between the coefficients A_1, A_{2T}, A_{3T} and A_4 can be determined from the matching BCs at the separation surfaces. Also, from (3.16) we obtain

$$
H_y(z) = i\omega\varepsilon_0 \begin{cases} A_1 \dfrac{\varepsilon_1}{\kappa_1} e^{+\kappa_1 z} , & z \leq 0 , \\[2mm] \dfrac{\varepsilon_{2T}}{\kappa_{2T}} \left(A_{2T} e^{\kappa_{2T} z} - A_{3T} e^{-\kappa_{2T} z} \right) , & 0 \leq z \leq d , \\[2mm] -A_4 \dfrac{\varepsilon_3}{\kappa_3} e^{-\kappa_3 z} , & z \geq d , \end{cases}
\tag{5.78}
$$

$$
E_z(z) = -ik_x \begin{cases} A_1 \dfrac{1}{\kappa_1} e^{+\kappa_1 z} , & z \leq 0 , \\[2mm] \dfrac{1}{\kappa_{2T}} \left(A_{2T} e^{\kappa_{2T} z} - A_{3T} e^{-\kappa_{2T} z} \right) , & 0 \leq z \leq d , \\[2mm] -A_3 \dfrac{1}{\kappa_3} e^{-\kappa_3 z} , & z \geq d , \end{cases}
\tag{5.79}
$$

where $\kappa_{2T}^2 = k_x^2 - \varepsilon_T k_0^2$ and $\kappa_{1,3}^2 = k_x^2 - \varepsilon_{1,3} k_0^2$. Also in the thin EG film, at the same angular frequency ω, there are the longitudinal waves (bulk plasmon waves) described by the following electric fields:

$$
E_{xL}(z) = ik_x \left(A_{2L} e^{+\kappa_{2L} z} + A_{3L} e^{-\kappa_{2L} z} \right) ,
\tag{5.80}
$$

$$
E_{zL}(z) = \kappa_{2L} \left(A_{2L} e^{+\kappa_{2L} z} - A_{3L} e^{-\kappa_{2L} z} \right) ,
\tag{5.81}
$$

where $\kappa_{2L}^2 = k_x^2 - k_L^2$. Using the electromagnetic BCs at $z = 0$, and $z = d$, i.e., the continuity of E_x, H_y, and εE_z, when both superstrate and substrate of thin EG film are the same, i.e., $\varepsilon_3 = \varepsilon_1$, we obtain a relation between ω and k_x, as

$$
\tanh\left(\frac{\kappa_{2T} d}{2} \right) = -\frac{\varepsilon_{2T}}{\varepsilon_1} \frac{\kappa_1}{\kappa_{2T}} + \delta_{nl} \tanh\left(\frac{\kappa_{2L} d}{2} \right) ,
\tag{5.82a}
$$

$$
\coth\left(\frac{\kappa_{2T} d}{2} \right) = -\frac{\varepsilon_{2T}}{\varepsilon_1} \frac{\kappa_1}{\kappa_{2T}} + \delta_{nl} \coth\left(\frac{\kappa_{2L} d}{2} \right) ,
\tag{5.82b}
$$

for odd and even modes, respectively, with respect to the tangential component of the electric field.[3] In the nonretarded limit, (5.82a) and (5.82b) simplify to

$$
\tanh\left(\frac{k_x d}{2} \right) = -\frac{\varepsilon_{2T}}{\varepsilon_1} + \delta_{nl} \tanh\left(\frac{\kappa_{2L} d}{2} \right) ,
\tag{5.83a}
$$

---

[3]If the symmetry considerations apply to the magnetic field, then they can be classified as even and odd modes, respectively, as done by Raza et al. in [9].

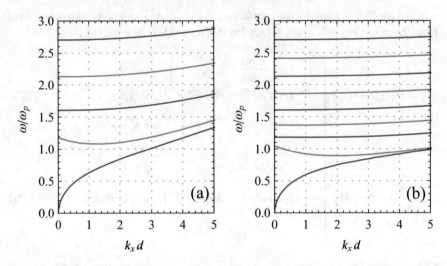

Fig. 5.2 Dispersion curves of surface and bulk plasmon modes of a vacuum-EG-vacuum system, as obtained from (5.85). Red and blue lines represent odd and even modes, respectively. The different panels refer to (**a**) $d = 5/k_s$ and (**b**) $d = 10/k_s$

$$\coth\left(\frac{k_x d}{2}\right) = -\frac{\varepsilon_{2T}}{\varepsilon_1} + \delta_{nl} \coth\left(\frac{\kappa_{2L} d}{2}\right). \tag{5.83b}$$

For completeness, we note when $\varepsilon_b = 1 = \varepsilon_1$ is assumed, (5.83a) and (5.83b) reduce to

$$\omega_+ = \frac{\omega_p}{\sqrt{2}}\sqrt{1 + e^{-k_x d}}\left[1 + \frac{k_x}{\kappa_{2L}} \tanh\left(\frac{\kappa_{2L} d}{2}\right)\right]^{1/2}, \tag{5.84a}$$

$$\omega_- = \frac{\omega_p}{\sqrt{2}}\sqrt{1 - e^{-k_x d}}\left[1 + \frac{k_x}{\kappa_{2L}} \coth\left(\frac{\kappa_{2L} d}{2}\right)\right]^{1/2}. \tag{5.84b}$$

Equations (5.84a) and (5.84b) can be combined into a single equation by introducing a quantity ϑ [10], where $\vartheta = +1$ for odd modes and $\vartheta = -1$ for even modes. Therefore, we have

$$\omega_\vartheta = \frac{\omega_p}{\sqrt{2}}\sqrt{1 + \vartheta e^{-k_x d}}\left[1 + \frac{k_x}{\kappa_{2L}}\left[\tanh\left(\frac{\kappa_{2L} d}{2}\right)\right]^{\vartheta}\right]^{1/2}. \tag{5.85}$$

In the above equation, we note that real value of κ_{2L} corresponds to SP modes, while κ_{2L} imaginary corresponds to bulk plasmon modes. In Fig. 5.2, we show dimensionless frequency ω/ω_p versus dimensionless variable $k_x d$ for different values of d. It is clear that the even mode of the SPs monotonically increases, while the odd SP mode is red-shifted first and then blue-shifted, after the local minimum.

5.3.2 Electron Gas-Insulator-Electron Gas Structures

We now reverse the EG slab geometry and study an insulator gap (with relative dielectric constant ε_1) in an EG with longitudinal dielectric function ε_{2L} and transverse dielectric function ε_{2T}. Then we have a relation between ω and k_x, as

$$\coth\left(\frac{\kappa_1 d}{2}\right) = -\frac{\varepsilon_1}{\varepsilon_{2T}}\frac{\kappa_{2T}}{\kappa_1}(1 - \delta_{nl}) \, , \tag{5.86a}$$

$$\tanh\left(\frac{\kappa_1 d}{2}\right) = -\frac{\varepsilon_1}{\varepsilon_{2T}}\frac{\kappa_{2T}}{\kappa_1}(1 - \delta_{nl}) \, , \tag{5.86b}$$

for odd and even modes, respectively, with respect to the tangential component of the electric field. In the nonretarded limit, (5.86a) and (5.86b) simplify to

$$\coth\left(\frac{k_x d}{2}\right) = -\frac{\varepsilon_1}{\varepsilon_{2T}}(1 - \delta_{nl}) \, , \tag{5.87a}$$

$$\tanh\left(\frac{k_x d}{2}\right) = -\frac{\varepsilon_1}{\varepsilon_{2T}}(1 - \delta_{nl}) \, , \tag{5.87b}$$

and if we consider $\varepsilon_b = 1 = \varepsilon_1$, then (5.87a) and (5.87b) reduce to

$$\omega_\vartheta = \frac{\omega_p}{\sqrt{2}}\sqrt{1 + \vartheta e^{-k_x d}}\left[1 + \frac{k_x}{\kappa_{2L}}\left[\tanh\left(\frac{\kappa_x d}{2}\right)\right]^\vartheta\right]^{1/2} . \tag{5.88}$$

From comparing (5.85) to (5.88) with $\alpha \to 0$ or $\kappa_{2L} \to \infty$, it is clear that dispersion relations for the SP modes of the vacuum-EG-vacuum and EG-vacuum-EG structures are identical [11]. However, it is well-known that when retardation is taken into account, this symmetry is broken, which is also clear from comparing (5.82) to (5.86) with $\delta_{nl} = 0$. In addition, (5.85) and (5.88) show explicitly that in the nonretarded limit the symmetry is also broken by the inclusion of nonlocal effects due to the presence of nonlocal pressure waves [9, 12].

5.4 Plasmonic Properties of Circular Electron Gas Cylinders: Quantum Hydrodynamic Model

Let us consider a cylindrical EG of radius a and infinite length that aligned along the z-axis and surrounded by a homogeneous insulator environment with relative permittivity ε_2. We use cylindrical coordinates (ρ, ϕ, z) for an arbitrary point in space. In the present BVP, the EG supports both the usual transverse and longitudinal waves and above the electron plasma frequency both types of

waves can propagate. Here it is assumed that the quantum nonlocal responses of
the system are dominated by the quantum nonlocality induced by free EG, while
the *bound electrons* only contribute to local responses. Thus in the QHD model,
the dielectric properties of the system are characterized by both the usual Drude
transverse dielectric function, as

$$\varepsilon_{1T} = \varepsilon_b - \frac{\omega_p^2}{\omega\,(\omega + i\gamma)} \,, \tag{5.89}$$

and the longitudinal dielectric function [13]

$$\varepsilon_{1L} = \varepsilon_b - \frac{\omega_p^2}{\omega\,(\omega + i\gamma) - \alpha^2 k_L^2 - \beta^2 k_L^4} \,. \tag{5.90}$$

Now, we assume that the system be exposed by a normally incident beam with TE^z
polarization. This means that the projection of the propagation vector k_2 on the xz
plane makes an angle $\zeta = \pi/2$ with the z-axis and therefore $q = 0$, as can be seen
in Fig. 3.20.

The standard vector cylindrical harmonics functions can be defined according
to (3.97). However, for the present system we also have a new vector cylindrical
harmonics function, as [14, 15]

$$\mathbf{L}_m = \nabla \psi_m \,, \tag{5.91}$$

where $\psi_m = Z_m(\kappa\rho)\exp(im\phi)$. In component form these mentioned vector
harmonics are

$$\mathbf{M}_m = k\left(im\frac{Z_m(k\rho)}{k\rho}\mathbf{e}_\rho - Z_m'(k\rho)\mathbf{e}_\phi\right)\exp(im\phi) \,, \tag{5.92a}$$

$$\mathbf{N}_m = \mathbf{e}_z k Z_m(k\rho)\exp(im\phi) \,, \tag{5.92b}$$

$$\mathbf{L}_m = \left(k Z_m'(k\rho)\mathbf{e}_\rho + \frac{im Z_m(k\rho)}{\rho}\mathbf{e}_\phi\right)\exp(im\phi) \,. \tag{5.92c}$$

Here k is given by $k_{1T} = \sqrt{\varepsilon_{1T}}\omega/c$ for the transverse modes, while it is given
by $k_{1L\pm}$ for the longitudinal modes inside the EG cylinder (where $k_{L\pm}$ is defined
by (4.67)) and by $k_2 = \sqrt{\varepsilon_2}\omega/c$ outside it. Also $Z_m(k\rho)$ represents a cylindrical
Bessel or Hankel function, and is chosen as Sect. 3.7. Inside the cylinder $J_m(k_{1T}\rho)$
is used for the transverse modes and $J_m(k_{1L}^\pm\rho)$ is used for the quantum longitudinal
modes. Outside the cylinder $J_m(k_2\rho)$ and $H_m(k_2\rho)$ are used for the incident and
scattered waves, respectively [16, 17]. In this case, the incident electric field $\mathbf{E}_i$ can
be expanded as [18]

$$\mathbf{E}_i = -i \sum_{m=-\infty}^{+\infty} E_m \mathbf{M}_m(k_2\rho) , \qquad (5.93)$$

where $E_m = E_0(-i)^m / k_2$. The transmitted and scattered electric fields can be represented as

$$\mathbf{E}_t = \sum_{m=-\infty}^{+\infty} g_m E_m \mathbf{M}_m(k_{1T}\rho) , \qquad (5.94)$$

$$\mathbf{E}_s = i \sum_{m=-\infty}^{+\infty} a_{m\perp} E_m \mathbf{M}_m(k_2\rho) . \qquad (5.95)$$

Furthermore, in the EG cylinder, at the same angular frequency ω, there are quantum longitudinal waves that can be described as

$$\mathbf{E}_l^\pm = \sum_{m=-\infty}^{+\infty} h_m^\pm E_m \mathbf{L}_m(k_{1L}^\pm \rho) . \qquad (5.96)$$

The unknown expansion coefficients a_m, g_m, and $h_m^\pm$ can be determined by the four appropriate BCs at the surface of the cylinder. Three of them are

$$(H_{iz} + H_{sz}) \,|_{\rho=a} = H_{tz}|_{\rho=a} , \qquad (5.97)$$

$$\left(E_{i\phi} + E_{s\phi}\right)\big|_{\rho=a} = \left(E_{t\phi} + E_{l\phi}^+ + E_{l\phi}^-\right)\big|_{\rho=a} , \qquad (5.98)$$

$$\varepsilon_2 \left(E_{i\rho} + E_{s\rho}\right)\big|_{\rho=a} = \varepsilon_b \left(E_{t\rho} + E_{l\rho}^+ + E_{l\rho}^-\right)\big|_{\rho=a} . \qquad (5.99)$$

Finally, as mentioned in Chap. 4, the forth BC at the cylinder surface requires $\nabla^2 n|_{\rho=a} = 0$, where n is the first-order perturbed value of the EG density [19] and hence of

$$\nabla^2 \left(\nabla \cdot \mathbf{E}\right)|_{\rho=a} = 0 . \qquad (5.100)$$

Solving the system of (5.97)–(5.100), we find that the coefficients of the scattered wave are given by

$$a_{m\perp} = \frac{\left[c_m + k_2 J_m'(k_{1T}a)\right] J_m(k_2 a) - k_{1T} J_m(k_{1T}a) J_m'(k_2 a)}{\left[c_m + k_2 J_m'(k_{1T}a)\right] H_m(k_2 a) - k_{1T} J_m(k_{1T}a) H_m'(k_2 a)} , \qquad (5.101)$$

where

$$c_m = m^2 J_m \left(k_{1T} a\right) \left(\frac{\varepsilon_2}{\varepsilon_b} \frac{k_{1T}}{k_2 a} - \frac{k_2}{k_{1T} a}\right) \frac{J_m \left(k_{1L}^+ a\right) - \Gamma_m J_m \left(k_{1L}^- a\right)}{\left(k_{1L}^+ a\right) J_m' \left(k_{1L}^+ a\right) - \Gamma_m \left(k_{1L}^- a\right) J_m' \left(k_{1L}^- a\right)},$$

$$(5.102)$$

$$\Gamma_m = \frac{\left(k_{1L}^+ a\right)^4 J_m \left(k_{1L}^+ a\right)}{\left(k_{1L}^- a\right)^4 J_m \left(k_{1L}^- a\right)}. \qquad (5.103)$$

Here the primes denote differentiation with respect to the argument of the radial functions. In the non-dispersive limit, in which longitudinal modes cannot propagate in the EG, $k_{1L}^{\pm}$ tends to infinitely large imaginary values, c_m tends to zero and $a_{m\perp}$ given by (5.101), reduces to the usual local expression [18]. Also, the standard nonlocal response result can be obtained when $\Gamma_m = 0$ and $k_{1L}^+ = k_L$, where k_L is defined by (4.77).

5.4.1   Dispersion Relation

The denominator of $a_{m\perp}$ vanishes at the frequencies of the radiative TE^z SPP modes with $q = 0$ and $\gamma = 0$. The radiative dispersion relation associated with these SPPs is

$$\left[c_m + k_2 J_m' (k_{1T} a)\right] H_m(k_2 a) - k_{1T} J_m (k_{1T} a) H_m'(k_2 a) = 0, \qquad (5.104)$$

where the symbols have the same meaning as in previous section.

5.4.2   Extinction Property

To determine the extinction properties of the system, we calculate the extinction width in units of the geometric width $2a$, i.e., (3.113). Let us note that in the nonretarded approximation (or dipole approximation, where $k_{1T} a$ and $k_2 a$ are sufficiently small), the scattering coefficient $a_{1\perp}$ is approximately

$$a_{1\perp} = -\frac{i\pi}{8} (k_2 a)^2 \alpha_{pol}, \qquad (5.105)$$

where α_{pol} is the (complex) normalized transversal polarizability of a thin EG cylinder in the electrostatic approximation (see Sect. 4.3) and can be written as (4.74). From (3.113), the formula for extinction cross section in the nonretarded approximation is

$$Q_{ext\perp} = \frac{4}{k_2 a} \text{Re}\,[a_{1\perp}]\;. \tag{5.106}$$

In Fig. 5.3, we plot the extinction spectra of a sodium nanowire with respect to the variable ω/ω_p for different values of γ/ω_p, a, and ε_2, respectively, by employing QHD model. In a thin nanowire, the electron collisions with the surface suppress the electron mean free path. This means that the damping parameter γ appearing in the dielectric functions will increase. As a result, it will further suppress the presence of the secondary peaks, while a broadening of the frequency for surface mode can be observed [see panel (a)]. Furthermore, from panel (b) it is clear that by increasing the wire radius, the number of subsidiary peaks in the extinction spectra of the system increases, but their relative amplitude decreases. Also, we observe that the shift of the SP peak from it classical local position decreases, while its amplitude increases. Numerical results presented in panel (c) show by considering the dielectric effect, the SP energy shifts to lower values, while the bulk mode frequencies are stable.

5.5 Plasmonic Properties of Electron Gas Spheres: Quantum Hydrodynamic Model

Let us consider an isolated spherical EG of radius a that surrounded by a homogeneous insulator environment with relative permittivity ε_2. In the QHD model, the dielectric properties of the present system are characterized by both the usual Drude transverse dielectric function, as (5.89) and the quantum longitudinal dielectric function, as (5.90). Now, we assume the system be irradiated by a z-directed, x-polarized plane wave, as shown in Fig. 3.26. We note that the longitudinal electron plasma waves obey the dispersion relation

$$\varepsilon_{1L}(k_{1L}^{\pm}, \omega) = 0\;, \tag{5.107}$$

and the transverse electromagnetic waves satisfy the relation

$$k_{1T} = \sqrt{\varepsilon_{1T}}\,\frac{\omega}{c}\;. \tag{5.108}$$

From (5.107), the expression of $k_{1L}^{\pm}$ can be written as (4.67). We consider the incident electric field as (3.154). For simplicity, it can be expanded in an infinite series of vector spherical harmonics, i.e., (3.155)–(3.160). Also, in the EG sphere, at the same angular frequency ω, there are longitudinal quantum waves that can be described by the following electric fields:

$$\mathbf{E}_l^{\pm} = \sum_{\ell=-\infty}^{+\infty} i f_{\ell}^{\pm} E_{\ell} \mathbf{L}_{e\ell 1}(k_{1L}^{\pm} r)\;, \tag{5.109}$$

Fig. 5.3 Calculated extinction width (in units of the geometric width) of a sodium nanowire with respect to the dimensionless variable ω/ω_p. Panel (**a**): different values of γ/ω_p, when $\varepsilon_b = 1 = \varepsilon_2$ and $a = 2\,\mathrm{nm}$. Panel (**b**): different values of a, when $\varepsilon_b = 1 = \varepsilon_2$ and $\gamma/\omega_p = 0.01$. Panel (**c**): different values of ε_2, when $\varepsilon_b = 1$, $a = 2\,\mathrm{nm}$, and $\gamma/\omega_p = 0.01$. For the other parameters the sodium values in Table 4.1 have been employed

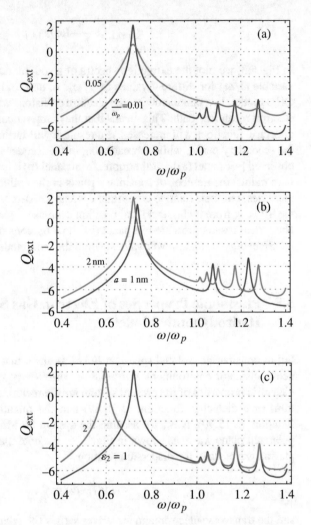

where $\mathbf{L}_{e\ell 1}$ have the form

$$
\mathbf{L}_{e\ell 1} = k \cos\varphi\, P_\ell^1(\cos\theta)\frac{\mathrm{d}}{\mathrm{d}(kr)} Z_\ell(kr)\mathbf{e}_r
$$

$$
+ \cos\varphi \frac{\mathrm{d}P_\ell^1(\cos\theta)}{r\,\mathrm{d}\theta} Z_n(kr)\mathbf{e}_\theta - \sin\varphi \frac{P_\ell^1(\cos\theta)}{r\sin\theta} Z_\ell(kr)\mathbf{e}_\varphi , \qquad (5.110)
$$

and $Z_\ell(kr)$ represents $j_\ell(k_{1L}^{\pm}r)$. We note that no magnetic field is associated with the longitudinal modes.

The unknown expansion coefficients a_ℓ, b_ℓ, c_ℓ, d_ℓ, and $f_\ell^{\pm}$ can be determined by the BCs at the surface of the spherical EG. The usual three BCs at $r = a$ are that the tangential components of the electric and magnetic fields be continuous, as in the local case, plus the condition that the normal component of electric field is also continuous [20]. As mentioned before, in an EG sphere both the transverse and longitudinal (usually neglected) waves give a contribution to the value of electric field, we have

$$(E_{i\theta} + E_{s\theta})\,|_{r=a} = \left(E_{t\theta} + E_{l\theta}^+ + E_{l\theta}^-\right)|_{r=a} , \tag{5.111}$$

$$\left(E_{i\phi} + E_{s\varphi}\right)|_{r=a} = \left(E_{t\varphi} + E_{l\varphi}^+ + E_{l\varphi}^-\right)|_{r=a} , \tag{5.112}$$

$$(H_{i\theta} + H_{s\theta})\,|_{r=a} = H_{t\theta}|_{r=a} , \tag{5.113}$$

$$\left(H_{i\phi} + H_{s\phi}\right)|_{r=a} = H_{t\phi}|_{r=a} , \tag{5.114}$$

$$\varepsilon_2\,(E_{ir} + E_{sr})\,|_{r=a} = \varepsilon_b\left(E_{tr} + E_{lr}^+ + E_{lr}^-\right)|_{r=a} . \tag{5.115}$$

However, the presence of the Bohm potential in the longitudinal dielectric function of the system increases the order of k_{1L} in (5.107), when compared to the equation in the standard nonlocal model. Thus, it is necessary to impose a new additional boundary condition for the present system, as (5.100).

Solving the system of (5.111)–(5.115) and also (5.100), we find that the coefficients of the scattered wave are given by

$$a_\ell = \frac{\left(c_\ell + \varepsilon_2\left[(k_{1T}a)j_\ell(k_{1T}a)\right]'\right) j_\ell(k_2 a) - \varepsilon_{1T} j_\ell(k_{1T}a)\left[(k_2 a)j_\ell(k_2 a)\right]'}{\left(c_\ell + \varepsilon_2\left[(k_{1T}a)j_\ell(k_{1T}a)\right]'\right) h_\ell^{(1)}(k_2 a) - \varepsilon_{1T} j_\ell(k_{1T}a)\left[(k_2 a)h_\ell^{(1)}(k_2 a)\right]'} , \tag{5.116}$$

$$b_\ell = \frac{j_\ell(k_{1T}a)\left[(k_2 a)j_\ell(k_2 a)\right]' - j_\ell(k_2 a)\left[(k_{1T}a)j_\ell(k_{1T}a)\right]'}{j_\ell(k_{1T}a)\left[(k_2 a)h_\ell^{(1)}(k_2 a)\right]' - h_\ell^{(1)}(k_2 a)\left[(k_{1T}a)j_\ell(k_{1T}a)\right]'} , \tag{5.117}$$

where

$$c_\ell = \ell(\ell+1)j_\ell(k_{1T}a)\varepsilon_2\left(\frac{\varepsilon_{1T}}{\varepsilon_b} - 1\right)\frac{j_\ell(k_{1L}^+ a) - \Gamma_\ell j_\ell(k_{1L}^- a)}{(k_{1L}^+ a)j_\ell'(k_{1L}^+ a) - \Gamma_\ell(k_{1L}^- a)j_\ell'(k_{1L}^- a)} , \tag{5.118}$$

$$\Gamma_\ell = \frac{(k_{1L}^+ a)^4 \, j_\ell\left(k_{1L}^+ a\right)}{(k_{1L}^- a)^4 \, j_\ell\left(k_{1L}^- a\right)} . \tag{5.119}$$

The coefficients b_ℓ are exactly the same as in the classical Mie scattering theory. The coefficients a_ℓ reduce to those of the classical local theory when c_ℓ is equal to zero [18]. Also, the standard nonlocal response result can be obtained when $\Gamma_\ell = 0$ and $k_{1L}^+ = k_L$, where k_L is defined by (4.77).

The denominator of a_ℓ vanishes at the frequencies of the radiative TM^r SPP modes. For $\gamma = 0$, the radiative dispersion relation associated with these SPPs is

$$\left(c_\ell + \varepsilon_2\left[(k_{1T}a)\,j_\ell(k_{1T}a)\right]'\right)h_\ell^{(1)}(k_2a) - \varepsilon_{1T}\,j_\ell(k_{1T}a)\left[(k_2a)h_\ell^{(1)}(k_2a)\right]' = 0\,,$$
(5.120)

where in the absence of the Bohm potential or gradient correction, i.e., $\beta = 0$, this equation reduces to equation (14) in [21]. Also, the denominator of b_ℓ vanishes at the frequencies of the radiative TE^r SPP modes. For $\gamma = 0$, the radiative dispersion relation associated with these SPPs is

$$j_\ell(k_{1T}a)\left[(k_2a)h_\ell^{(1)}(k_2a)\right]' - h_\ell^{(1)}(k_2a)\left[(k_{1T}a)\,j_\ell(k_{1T}a)\right]' = 0\,.$$
(5.121)

Finally, we note that the extinction properties follow from the coefficients a_ℓ and b_ℓ in the usual way [18]. Thus, the extinction cross section of the system (in units of the geometric cross section) is given by (3.167).

References

1. M.A. Kaliteevski, S. Brand, J.M. Chamberlain, R.A. Abram, V.V. Nikolaev, Effect of longitudinal excitations on surface plasmons. Solid State Commun. **144**, 413–417 (2007)
2. W. Yan, M. Wubs, N.A. Mortensen, Hyperbolic metamaterials: nonlocal response regularizes broadband supersingularity. Phys. Rev. B **86**, 205429 (2012)
3. A. Moreau, C. Ciracì, D.R. Smith, Impact of nonlocal response on metallo-dielectric multilayers and optical patch antennas. Phys. Rev. B **87**, 045401 (2013)
4. J.R. Wait, Radiation from sources immersed in compressible plasma media. Can. J. Phys. **42**, 1760–1780 (1964)
5. E. Matsuo, M. Tsuji, Energy flow and group velocity of electromagnetic surface wave in hydrodynamic approximation. J. Phys. Soc. Jpn. **45**, 575–580 (1978)
6. A. Moradi, Bohm potential and inequality of group and energy transport velocities of plasmonic waves on metal-insulator waveguides. Phys. Plasmas **24**, 072104 (2017)
7. A.D. Boardman, *Electromagnetic Surface Modes* (Wiley, New York, 1982)
8. H.J. Lee, S.H. Cho, Dispersion of surface waves propagating along planar interfaces between plasmas or plasma and free space. Plasma Phys. Control. Fusion **37**, 989–1002 (1995)
9. S. Raza, T. Christensen, M. Wubs, S.I. Bozhevolnyi, N.A. Mortensen, Nonlocal response in thin-film waveguides: loss versus nonlocality and breaking of complementarity. Phys. Rev. B **88**, 115401 (2013)
10. K. Dharamvir, B. Singla, K.N. Pathak, V.V. Paranjape, Plasmon excitations in a metallic slab. Phys. Rev. B **48**, 12330–12333 (1992)
11. B.E. Sernelius, *Surface Modes in Physics* (Wiley-VCH, Berlin, 2001)
12. S. Raza, S.I. Bozhevolnyi, M. Wubs, N.A. Mortensen, Nonlocal optical response in metallic nanostructures. J. Phys. Condens. Matter **27**, 183204 (2015)
13. A. Moradi, Extinction properties of metallic nanowires: quantum diffraction and retardation effects. Phys. Lett. A **379**, 2379–2383 (2015)
14. G.C. Aers, A.D. Boardman, B.V. Paranjape, Non-radiative surface plasma-polariton modes of inhomogeneous metal circular cylinders. J. Phys. F Met. Phys. **10**, 53–65 (1980)

15. R. Ruppin, Extinction properties of thin metallic nanowires. Opt. Commun. **190**, 205–209 (2001)
16. A. Moradi, Oblique incidence scattering from single-walled carbon nanotubes. Phys. Plasmas **17**, 033504 (2010)
17. A. Moradi, Extinction properties of single-walled carbon nanotubes: two-fluid model. Phys. Plasmas **21**, 032106 (2014)
18. C.F. Bohren, D.R. Huffman, *Absorption and Scattering of Light by Small Particles* (Wiley, New York, 1983)
19. Y.-Y. Zhang, S.-B. An, Y.-H. Song, N. Kang, Z.L. Mišković, Y.-N. Wang, Plasmon excitation in metal slab by fast point charge: the role of additional boundary conditions in quantum hydrodynamic model. Phys. Plasmas **21**, 102114 (2014)
20. A. Moradi, Quantum nonlocal effects on optical properties of spherical nanoparticles. Phys. Plasmas **22**, 022119 (2015)
21. A.D. Boardman, B.V. Paranjape, The optical surface modes of metal spheres. J. Phys. F Met. Phys. **7**, 1935–1945 (1977)

Part II
Two-Dimensional Electron Gases

Chapter 6
Electrostatic Problems Involving Two-Dimensional Electron Gases in Planar Geometry

Abstract In this chapter, we study the electrostatic boundary-value problems involving planar two-dimensional electron gas layers. The main interest and the key first applications of presented boundary-value problems concern doped graphene, while keeping in mind the analysis can be applied to semiconductor inversion layers, high electron mobility transistor, and potentially other planar two-dimensional electron gas systems. For brevity, in many sections of this chapter the $\exp(-i\omega t)$ time factor is suppressed. Furthermore, all media under consideration are nonmagnetic and attention is only confined to the linear phenomena.

6.1 Plasmonic Properties of Monolayer Two-Dimensional Electron Gases

Plasmonic properties of planar 2DEG systems have been studied since 1960s. In 1967, Stern published a paper [1] in which the plasmonic spectrum of a planar 2DEG has been studied. However, despite this long history, there are few available theoretical account of many of the basic properties of SPs of planar 2DEG layers. The aim of the present chapter is to discuss such accounts.

6.1.1 Dispersion Relation

Let us assume a planar 2DEG in equilibrium that is uniformly distributed in the plane $z = 0$ in a Cartesian coordinate system with the position vector $\mathbf{r} = (x, y, z)$ and has the surface density (per unit area) of n_0. Furthermore, a substrate with relative dielectric ε_1 is supposed to occupy the region $z < 0$ underneath the 2DEG, whereas the region $z > 0$ is assumed to be a semi-infinite insulator with relative dielectric constant ε_2.

The original version of this chapter was revised. The correction to this chapter is available at https://doi.org/10.1007/978-3-030-43836-4_11

A. Moradi, *Canonical Problems in the Theory of Plasmonics*, Springer Series in Optical Sciences 230, https://doi.org/10.1007/978-3-030-43836-4_6

209

Now, assuming that $n(x, t)$ is the first-order perturbed density (per unit area) of the homogeneous electron fluid on the 2DEG layer, due to the propagation of an electrostatic surface wave parallel to the layer $z = 0$ along the x-direction. Based on the SHD theory (see Sect. 1.2), the electronic excitations of a 2DEG layer can be described by Newton's equation of motion, in conjunction with the continuity and Poisson equations. In the linear approximation, one has

$$\frac{\partial v_x}{\partial t} = \frac{e}{m_e} \frac{\partial \Phi}{\partial x}\Big|_{z=0} - \frac{\alpha^2}{n_0} \frac{\partial n}{\partial x} \,, \tag{6.1}$$

$$\frac{\partial n}{\partial t} + n_0 \frac{\partial v_x}{\partial x} = 0 \,, \tag{6.2}$$

$$\left(\frac{\partial^2}{\partial x^2} + \frac{\partial^2}{\partial z^2} \right) \Phi(x, z) = \begin{cases} 0 \,, & z \neq 0 \,, \\ en/\varepsilon_0 \,, & z = 0 \,, \end{cases} \tag{6.3}$$

where $v_x(x, t)$ is the first-order perturbed values of electrons velocity along the x-direction. In the right-hand side of (6.1), the first term is the force on electrons due to the tangential component of the electric field, i.e., $E_x = -\partial \Phi / \partial x$ evaluated at the 2DEG layer $z = 0$, where Φ is the self-consistent electrostatic potential. The second term is the force due to the internal interaction in the electron fluid, with $\alpha = v_F/\sqrt{2}$ that is the speed of propagation of density disturbances in a uniform 2D homogeneous electron fluid [2, 3]. We further note that the Laplacian in (6.3) is naturally of a 3D character. We obtain from (6.1) and (6.2) after the elimination of the velocity $v_x(x, t)$

$$\left(\omega^2 + \alpha^2 \frac{\partial^2}{\partial x^2} \right) n = \frac{en_0}{m_e} \frac{\partial^2}{\partial x^2} \Phi\Big|_{z=0} \,, \tag{6.4}$$

where ω is the angular frequency of a SP. Due to the planar symmetry of the present system, one can replace the quantities n and Φ in (6.3) and (6.4) by expressions of the form

$$\Phi(x, z) = \tilde{\Phi}(z) \exp(ik_x x) \,, \tag{6.5}$$

$$n(x) = \tilde{n} \exp(ik_x x) \,, \tag{6.6}$$

where k_x is the longitudinal wavenumber in the x-direction. After substitution, one finds

$$\left(\frac{d^2}{dz^2} - k_x^2 \right) \tilde{\Phi}(z) = 0 \,, \qquad z \neq 0 \,, \tag{6.7}$$

and

$$\tilde{n} = -\frac{en_0}{m_e} \frac{k_x^2}{\omega^2 - \alpha^2 k_x^2} \tilde{\Phi}(z=0) . \tag{6.8}$$

The solution of (6.7) has the form

$$\tilde{\Phi}(z) = \begin{cases} A_1 e^{+k_x z} , & z \le 0 , \\ A_2 e^{-k_x z} , & z \ge 0 , \end{cases} \tag{6.9}$$

where the relations between the coefficients A_1 and A_2 can be determined from the matching BCs at the separation surface. For the present case, the BCs are

$$\Phi_1\big|_{z=0} = \Phi_2\big|_{z=0} , \tag{6.10a}$$

$$\varepsilon_2 \frac{\partial \Phi_2}{\partial z}\Big|_{z=0} - \varepsilon_1 \frac{\partial \Phi_1}{\partial z}\Big|_{z=0} = \frac{en}{\varepsilon_0} , \tag{6.10b}$$

where subscript 1 denotes the region below the 2DEG layer and subscript 2 denotes the region above the layer. We note that (6.10b) means that, due to the polarization of the 2DEG layer, the z-component of the electric field is discontinuous at $z = 0$. On applying the electrostatic BCs, the dispersion relation for the system is given as

$$\omega^2 = \alpha^2 k_x^2 + \frac{n_0 e^2}{\varepsilon_0 m_e} \frac{k_x}{\varepsilon_1 + \varepsilon_2} , \tag{6.11}$$

where the dispersion curve is similar to that for $\alpha = 0$, except that for large ω, $\omega \simeq \alpha k_x$. By contrast, with $\alpha = 0$, the high-frequency behavior is $\omega \propto k_x^{1/2}$. Figure 6.1 shows the spectrum of SPs of a suspended doped graphene[1] [5] in vacuum ($\varepsilon_1 = 1 = \varepsilon_2$) in terms of the dimensionless variables, obtained from (6.11) when $\alpha = 0$. For this system, we have $m_e = \hbar k_F / v_F$ that is the effective electron mass in doped graphene, where $k_F = \sqrt{\pi n_0}$ and $v_F \approx c/300$ are the Fermi wavenumber and Fermi velocity in doped graphene, respectively [5]. Let us note that the validity of present model for graphene by using the hydrodynamic theory is limited by the relation $k v_F \ll \omega \ll \omega_F$, where $\omega_F = k_F v_F$. Therefore for typical doping densities of $n_0 \le 10^{13} \mathrm{cm}^{-2}$, the corresponding Fermi energy of $E_F = \hbar \omega_F \le 0.4 \, \mathrm{eV}$ guarantees that present fluid model is suitable for studying the plasmonic properties of graphene for applications in the THz to mid-infrared range of frequencies [6].

---

[1]Graphene, a single sheet of carbon atoms forming hexagonal lattice was first experimentally discovered by Novoselov *et al.* in 2004 [4]. Here a doped monolayer graphene is modeled as a 2D massless EG.

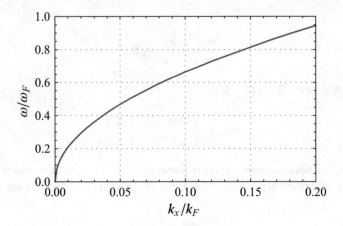

Fig. 6.1 Dispersion curve of SPs of a doped monolayer graphene as given by (6.11), when $\varepsilon_1 = 1 = \varepsilon_2$. Here $\omega_F = k_F v_F$, $k_F = \sqrt{\pi n_0}$, and $v_F \approx c/300$

6.1.2 Power Flow

For the power flow density associated with a SP of a planar 2DEG layer, we have, in the three media,

$$
S_x = - \begin{cases}
\varepsilon_0 \varepsilon_1 \Phi_1 \dfrac{\partial}{\partial t} \dfrac{\partial}{\partial x} \Phi_1 \,, & z \le 0 \,, \\[2mm]
e n_0 \Phi v_x - m_e \alpha^2 n v_x \,, & z = 0 \,, \\[2mm]
\varepsilon_0 \varepsilon_2 \Phi_2 \dfrac{\partial}{\partial t} \dfrac{\partial}{\partial x} \Phi_2 \,, & z \ge 0 \,,
\end{cases}
\tag{6.12}
$$

where $\Phi = \Phi_1|_{z=0}$ or $\Phi = \Phi_2|_{z=0}$. After the elimination of v_x, n, Φ_1, Φ_2, and Φ in the above equation by using (6.1)–(6.11), the cycle-averaged of (6.12) can be written as

$$
S_x = -\frac{\varepsilon_0 \omega k_x}{2} A_1^2
\begin{cases}
\varepsilon_1 e^{2k_x z} \,, & z \le 0 \,, \\[2mm]
-\dfrac{n_0 e^2}{\varepsilon_0 m_e} \dfrac{\omega^2}{\left[\omega^2 - \alpha^2 k_x^2\right]^2} \,, & z = 0 \,, \\[2mm]
\varepsilon_2 e^{-2k_x z} \,, & z \ge 0 \,,
\end{cases}
\tag{6.13}
$$

It can be seen that power flow densities are largest at the 2DEG layer, and their amplitudes decay exponentially with increasing distance into each medium from the interfaces. Also, it is evident that the power flow in a 2DEG layer occurs in the $+x$-direction, while in the dielectric regions, the power flow occurs in the $-x$-direction, i.e., opposite to the direction of phase propagation.

The total power flow associated with the SPs is determined by an integration over z. If the integrated Poynting vectors are denoted by angle brackets, the power flow through an area in the yz plane of infinite length in the z-direction and unit width in the y-direction is[2]

$$\langle S_x \rangle = -\frac{\varepsilon_0 \omega}{4} A_1^2 \left[\varepsilon_1 + \varepsilon_2 - 2\frac{n_0 e^2}{\varepsilon_0 m_e} \frac{\omega^2 k_x}{\left[\omega^2 - \alpha^2 k_x^2\right]^2} \right]. \tag{6.14}$$

This total power flow (per unit width) is positive, when k_x is positive.

6.1.3   Energy Distribution

We now consider the energy density distribution in the transverse direction. For the cycle-averaged of energy density distribution associated with the SPs of a planar 2DEG layer, we have, in the three media

$$U = \frac{\varepsilon_0}{4} \begin{cases} \varepsilon_1 |\nabla \Phi_1|^2 \,, & z \leq 0 \,, \\ \dfrac{m_e n_0}{\varepsilon_0} |v_x|^2 + \dfrac{m_e}{\varepsilon_0 n_0} \alpha^2 |n|^2 \,, & z = 0 \,, \\ \varepsilon_2 |\nabla \Phi_2|^2 \,, & z \geq 0 \,, \end{cases} \tag{6.15}$$

where we should note that the dimension of the second term is different from the first and third terms. After the elimination of v_x, n, Φ_1, Φ_2, and Φ in the above equation by using (6.1)–(6.11), we obtain

$$U = \frac{\varepsilon_0 k_x^2}{2} A_1^2 \begin{cases} \varepsilon_1 e^{2k_x z} \,, & z \leq 0 \,, \\ \dfrac{1}{2} \dfrac{n_0 e^2}{\varepsilon_0 m_e} \dfrac{\omega^2 + \alpha^2 k_x^2}{\left[\omega^2 - \alpha^2 k_x^2\right]^2} \,, & z = 0 \,, \\ \varepsilon_2 e^{-2k_x z} \,, & z \geq 0 \,, \end{cases} \tag{6.16}$$

The total energy density associated with the SPs is again determined by integration over z, the energy per unit surface area being

---

[2]The dimensional analysis of (6.12) shows this equation is more appropriate to be written as

$$S_x = - \begin{cases} \varepsilon_0 \varepsilon_1 \Phi_1 \dfrac{\partial}{\partial t} \dfrac{\partial}{\partial x} \Phi_1 \,, & z \leq 0 \,, \\ \varepsilon_0 \varepsilon_2 \Phi_2 \dfrac{\partial}{\partial t} \dfrac{\partial}{\partial x} \Phi_2 \,, & z \geq 0 \,, \end{cases}$$

where the second term in (6.12) should be added in the calculation of total power flow density as

$$\langle S_x \rangle = -\left[e n_0 \Phi v_x - m_e \alpha^2 n v_x \right] - \varepsilon_0 \varepsilon_1 \int_{-\infty}^{0} \Phi_1 \frac{\partial}{\partial t} \frac{\partial}{\partial x} \Phi_1 \, dz - \varepsilon_0 \varepsilon_2 \int_{0}^{+\infty} \Phi_2 \frac{\partial}{\partial t} \frac{\partial}{\partial x} \Phi_2 \, dz.$$

$$\langle U \rangle = \frac{\varepsilon_0 k_x}{4} A_1^2 \left[\varepsilon_1 + \varepsilon_2 + \frac{n_0 e^2}{\varepsilon_0 m_e} \frac{\omega^2 + \alpha^2 k_x^2}{\left[\omega^2 - \alpha^2 k_x^2 \right]^2} k_x \right]. \tag{6.17}$$

Also, the position of the center of energy along the z-axis can be written as

$$\langle z \rangle \equiv \frac{1}{\langle U \rangle} \int_{-\infty}^{+\infty} z \, U \, dz = \frac{1}{2k_x} \frac{\varepsilon_2 - \varepsilon_1}{\varepsilon_1 + \varepsilon_2 + \dfrac{n_0 e^2}{\varepsilon_0 m_e} \dfrac{\omega^2 + \alpha^2 k_x^2}{\left[\omega^2 - \alpha^2 k_x^2 \right]^2} k_x}. \tag{6.18}$$

Equation (6.18) shows that for $\varepsilon_1 = \varepsilon_2$ the center of energy is located in the 2DEG layer, i.e., $\langle z \rangle = 0$, while for $\varepsilon_2 > \varepsilon_1$ it moves into the region with relative dielectric constant ε_2, i.e., $\langle z \rangle > 0$ and for $\varepsilon_2 < \varepsilon_1$ it moves into the region with relative dielectric constant ε_1, i.e., $\langle z \rangle < 0$.

6.1.4   Energy Velocity

The energy velocity of the SPs is given as the ratio of the total power flow density (per unit width) and total energy density (per unit area), such as

$$v_e = \frac{\langle S_x \rangle}{\langle U \rangle} = -\frac{\omega}{k_x} \frac{\varepsilon_1 + \varepsilon_2 - 2 \dfrac{n_0 e^2}{\varepsilon_0 m_e} \dfrac{\omega^2 k_x}{\left[\omega^2 - \alpha^2 k_x^2 \right]^2}}{\varepsilon_1 + \varepsilon_2 + \dfrac{n_0 e^2}{\varepsilon_0 m_e} \dfrac{\omega^2 + \alpha^2 k_x^2}{\left[\omega^2 - \alpha^2 k_x^2 \right]^2} k_x}. \tag{6.19}$$

The expression on the right-hand side is in quantitative agreement with the group velocity found from the corresponding dispersion relation by means of the usual formula, i.e., (2.51). This agreement is quite a stringent test of the accuracy of the results. To show this point analytically, we restrict our attention to the case, where $\alpha = 0$. In this case, from (6.11) and (2.51) we find $v_g = \omega/2k_x$, where $\omega^2 = n_0 e^2 k_x / \varepsilon_0 (\varepsilon_1 + \varepsilon_2) m_e$. On the other hand, by using (6.19) we obtain $v_e = \omega/2k_x$ which is identical to the group velocity.

6.1.5   Dispersion Relation of Surface Magneto Plasmon: Perpendicular Configuration

Here we study the propagation of a SMP on the surface of a planar 2DEG layer in the perpendicular configuration [7], where a static magnetic field $\mathbf{B}_0 = B_0 \mathbf{e}_z$ is normal to the 2DEG layer. In this case, Newton's equation of motion for the electrons, i.e., (6.1) must be read as

$$\frac{\partial v_x}{\partial t} = \frac{e}{m_e}\frac{\partial \Phi}{\partial x}\Big|_{z=0} - \omega_c v_y - \frac{\alpha^2}{n_0}\frac{\partial n}{\partial x}, \tag{6.20a}$$

$$\frac{\partial v_y}{\partial t} = \omega_c v_x, \tag{6.20b}$$

where $\omega_c = eB_0/m_e$ is the cyclotron frequency of an electron, as mentioned before. We obtain from (6.20a) and (6.20b) after the elimination of the velocity $v_y(x, t)$

$$\left(\frac{\partial}{\partial t} + i\frac{\omega_c^2}{\omega}\right) v_x(x, t) = \frac{e}{m_e}\frac{\partial \Phi}{\partial x}\Big|_{z=0} - \frac{\alpha^2}{n_0}\frac{\partial n}{\partial x}, \tag{6.21}$$

and proceeding as in previous section, give a new dispersion relation as

$$\omega^2 = \omega_c^2 + \alpha^2 k_x^2 + \frac{n_0 e^2}{\varepsilon_0 m_e}\frac{k_x}{\varepsilon_1 + \varepsilon_2}. \tag{6.22}$$

This wave is known in the gas plasma physics as the upper hybrid mode. In Fig. 6.2, one can see that the dispersion of a SMP on the surface of a planar 2DEG layer is shifted toward higher frequencies with respect to the corresponding SP dispersion when $\mathbf{B}_0 = 0$. This is consequence of the hybridization between SPs of a planar 2DEG layer and cyclotronic excitations (with frequency ω_c) driven by the static magnetic field.

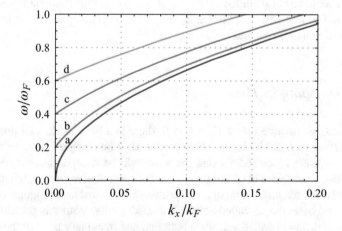

Fig. 6.2 Dispersion curves of SMPs of a doped monolayer graphene in the perpendicular configuration as given by (6.22), when $\varepsilon_1 = 1 = \varepsilon_2$. The different curves refer to (a) $\omega_c/\omega_F = 0$, (b) $\omega_c/\omega_F = 0.2$, (c) $\omega_c/\omega_F = 0.4$, and (d) $\omega_c/\omega_F = 0.6$

6.2   Plasmonic Properties of Monolayer Two-Dimensional Electron Gases: Effect of Collision

In the presence of the collision effect, the equations of drift motion of the electrons, i.e., (6.1) may be written as

$$\frac{\partial v_x}{\partial t} = \frac{e}{m_e}\frac{\partial \Phi}{\partial x}\Big|_{z=0} - \frac{\alpha^2}{n_0}\frac{\partial n}{\partial x} - \gamma v_x \ , \tag{6.23}$$

where γ is a phenomenological collision frequency. Then, we assume, as in previous section, that all of the perturbed physical quantities associated with the SPs have the form

$$A(x,t) = \tilde{A}\exp(ik_x x)\exp(-i\omega t) \ , \tag{6.24}$$

where $\tilde{A}$ is the amplitude of the physical quantity. This merely involves replacing $\partial/\partial t$ by $-i\omega$ and $\partial/\partial x$ by ik_x, leading to

$$-i\left(\omega + i\gamma\right)v_x = ik_x\frac{e}{m_e}\Phi\Big|_{z=0} - ik_x\frac{\alpha^2}{n_0}n \ , \tag{6.25}$$

and proceeding as in previous section, give a new dispersion relation as

$$\omega\left(\omega + i\gamma\right) = \alpha^2 k_x^2 + \frac{n_0 e^2}{\varepsilon_0 m_e}\frac{k_x}{\varepsilon_1 + \varepsilon_2} \ . \tag{6.26}$$

Until now we have not specified the nature of ω and k_x. In the following, we consider two cases. The first case will be that k_x is real and ω is complex, while the second case will be that ω is real and k_x is complex.

6.2.1   Damping in Time

In this case, we assume that at time $t = 0$, there is a SP with k_x real propagating on the surface of a 2DEG layer. Therefore, ω must be complex in order to satisfy (6.26). This means [see (6.24)] that the wave will be damped in time, but not in space. Thus, we assume in (6.26) that ω has the form $\omega = \omega_r + i\omega_i$, where ω_r and ω_i are real and imaginary parts of ω, respectively. Note that ω_r is angular frequency of wave, and ω_i is the so-called temporal *damping rate*. With this notation, (6.26) leads to two different equations, since both real and imaginary parts of the equation must be separately equal to zero. We have

$$\omega_r^2 - \omega_i^2 - \gamma \omega_i = \alpha^2 k_x^2 + \frac{n_0 e^2}{\varepsilon_0 m_e} \frac{k_x}{\varepsilon_1 + \varepsilon_2} \,, \tag{6.27a}$$

$$2\omega_i + \gamma = 0 \,, \tag{6.27b}$$

where (6.27b) gives the damping rate directly

$$\omega_i = -\frac{\gamma}{2} \,. \tag{6.28}$$

We can see that ω_i is a negative quantity, as we might expect for a damping rate.[3] Using (6.28) we see from (6.24) that all of the perturbed physical quantities associated with the SPs will have the form

$$A(x,t) = \tilde{A} \exp(-\gamma t/2) \exp[i(k_x x - \omega_r t)] \,, \tag{6.29}$$

which explicitly exhibits the damping effect due to collisions predicted by (6.28). This damping can modify the propagation properties of the waves in a number of other ways. For example, in contrast to collisionless propagation, a SP has a cutoff wavenumber, as we will now demonstrate. Substituting (6.28) into (6.27a) gives

$$\omega_r^2 = -\frac{\gamma^2}{4} + \alpha^2 k_x^2 + \frac{n_0 e^2}{\varepsilon_0 m_e} \frac{k_x}{\varepsilon_1 + \varepsilon_2} \,. \tag{6.30}$$

From (6.30) it can be seen that the range of possible wavelengths has been strongly modified. Because of collisions there is now a cutoff wavenumber k_c such that $k_x > k_c$, where k_c is given by

$$k_c = -\frac{n_0 e^2}{2\varepsilon_0 m_e \alpha^2} \frac{1}{\varepsilon_1 + \varepsilon_2} + \frac{1}{2}\sqrt{\left(\frac{n_0 e^2}{\varepsilon_0 m_e \alpha^2} \frac{1}{\varepsilon_1 + \varepsilon_2}\right)^2 + \frac{\gamma^2}{\alpha^2}} \,. \tag{6.31}$$

Thus, the wavelength of SPs is always smaller than a cutoff wavelength λ_c given by $\lambda_c = 2\pi/k_c$, which on might note in passing is the same as the wavelength of a SP of wave frequency $\omega_{rc} = \gamma/2$ in a collisionless 2DEG.

6.2.1.1 Power Flow

In this case, for the cycle-averaged of (6.12) we have

---

[3] A positive ω_i would have indicated wave growth.

$$
S_x = -\frac{\varepsilon_0 \omega_r k_x}{2} A_1^2 \begin{cases} \varepsilon_1 e^{2k_x z} e^{-\gamma t}\,, & z \leq 0\,, \\[2mm] -\dfrac{n_0 e^2}{\varepsilon_0 m_e} \dfrac{\omega_r^2 + \gamma^2/4}{\left[\omega_r^2 + \gamma^2/4 - \alpha^2 k_x^2\right]^2} e^{-\gamma t}\,, & z = 0\,, \\[2mm] \varepsilon_2 e^{-2k_x z} e^{-\gamma t}\,, & z \geq 0\,, \end{cases} \tag{6.32}
$$

and, therefore

$$
\langle S_x \rangle = -\frac{\varepsilon_0 \omega_r}{4} A_1^2 \left[\varepsilon_1 + \varepsilon_2 - 2\frac{n_0 e^2}{\varepsilon_0 m_e} \frac{\omega_r^2 + \gamma^2/4}{\left[\omega_r^2 + \gamma^2/4 - \alpha^2 k_x^2\right]^2} k_x \right] e^{-\gamma t}\,. \tag{6.33}
$$

6.2.1.2   Energy Distribution

After the elimination of v_x, n, Φ_1, Φ_2, and Φ in (6.15), we obtain

$$
U = \frac{\varepsilon_0 k_x^2}{2} A_1^2 \begin{cases} \varepsilon_1 e^{2k_x z} e^{-\gamma t}\,, & z \leq 0\,, \\[2mm] \dfrac{1}{2}\dfrac{n_0 e^2}{\varepsilon_0 m_e} \dfrac{\omega_r^2 + \gamma^2/4 + \alpha^2 k_x^2}{\left[\omega_r^2 + \gamma^2/4 - \alpha^2 k_x^2\right]^2} e^{-\gamma t}\,, & z = 0\,, \\[2mm] \varepsilon_2 e^{-2k_x z} e^{-\gamma t}\,, & z \geq 0\,, \end{cases} \tag{6.34}
$$

The total energy density (per unit area) associated with the SPs is again determined by integration over z, the energy per unit surface area being

$$
\langle U \rangle = \frac{\varepsilon_0 k_x}{4} A_1^2 \left[\varepsilon_1 + \varepsilon_2 + \frac{n_0 e^2}{\varepsilon_0 m_e} \frac{\omega_r^2 + \gamma^2/4 + \alpha^2 k_x^2}{\left[\omega_r^2 + \gamma^2/4 - \alpha^2 k_x^2\right]^2} k_x \right] e^{-\gamma t}\,. \tag{6.35}
$$

Also,

$$
P_d = n_0 m_e \gamma v_x^2\,, \tag{6.36}
$$

represents the rate of energy loss in the 2DEG layer by dissipation [8], where

$$
v_x = \frac{\omega_r - i\gamma/2}{n_0 k_x} n\,, \tag{6.37}
$$

and

$$
n = -\frac{e n_0}{m_e} \frac{k_x^2}{\omega_r^2 + \gamma^2/4 - \alpha^2 k_x^2}\,. \tag{6.38}
$$

The dissipation, however, is entirely associated with the frictional force, which acts on the moving charges.

6.2.1.3 Energy Velocity

Now, the energy velocity of the damped (in time) SPs is given as

$$
v_e = \frac{\langle S_x \rangle}{\langle U \rangle} = -\frac{\omega_r}{k_x} \frac{\varepsilon_1 + \varepsilon_2 - 2\dfrac{n_0 e^2}{\varepsilon_0 m_e} \dfrac{\omega_r^2 + \gamma^2/4}{\left[\omega_r^2 + \gamma^2/4 - \alpha^2 k_x^2\right]^2} k_x}{\varepsilon_1 + \varepsilon_2 + \dfrac{n_0 e^2}{\varepsilon_0 m_e} \dfrac{\omega_r^2 + \gamma^2/4 + \alpha^2 k_x^2}{\left[\omega_r^2 + \gamma^2/4 - \alpha^2 k_x^2\right]^2} k_x} ,
\tag{6.39}
$$

that remains positive at all frequencies.

6.2.2 Damping in Space

We now consider a SP is continuously propagates with real angular frequency ω, but due to collisions is spatially damped. Now k_x will be considered to be a complex quantity, with ω being real. We choose k_x to have the form $k_x = k_r + ik_i$, where k_r and k_i are the real and imaginary parts of the wavenumber, respectively. Substituting $k_x = k_r + ik_i$ into (6.26) and setting the real and imaginary parts of the resulting equation separately equal to zero gives the following two equations

$$
\omega^2 = \alpha^2 \left(k_r^2 - k_i^2 \right) + \frac{n_0 e^2}{\varepsilon_0 m_e} \frac{k_r}{\varepsilon_1 + \varepsilon_2} ,
\tag{6.40a}
$$

$$
\gamma \omega = 2\alpha^2 k_r k_i + \frac{n_0 e^2}{\varepsilon_0 m_e} \frac{k_i}{\varepsilon_1 + \varepsilon_2} .
\tag{6.40b}
$$

Using (6.40b) we can eliminate k_i from (6.40a). This gives the equation

$$
\omega^2 = \alpha^2 \left[k_r^2 - \left(\frac{\gamma \omega}{2\alpha^2 k_r + \dfrac{n_0 e^2}{\varepsilon_0 m_e} \dfrac{1}{\varepsilon_1 + \varepsilon_2}} \right)^2 \right] + \frac{n_0 e^2}{\varepsilon_0 m_e} \frac{k_r}{\varepsilon_1 + \varepsilon_2} ,
\tag{6.41}
$$

that is the dispersion equation for the present problem.

6.2.2.1 Power Flow

In this case, the cycle-averaged of (6.12) yields

$$S_x = -\frac{\varepsilon_0 \omega k_r}{2}$$

$$\times A_1^2 \begin{cases} \varepsilon_1 e^{2k_r z} e^{-2k_i x} , & z \le 0 , \\ -\dfrac{n_0 e^2}{\varepsilon_0 m_e} \dfrac{1}{k_r} \dfrac{\left[\omega^2 + 2\alpha^2 k_i^2\right] k_r + \left[\gamma\omega - 2\alpha^2 k_r k_i\right] k_i}{\left[\omega^2 - \alpha^2 \left(k_r^2 - k_i^2\right)\right]^2 + \left[\gamma\omega - 2\alpha^2 k_r k_i\right]^2} e^{-2k_i x} , & z = 0 , \\ \varepsilon_2 e^{-2k_r z} e^{-2k_i x} , & z \ge 0 , \end{cases}$$
(6.42)

and therefore

$$\langle S_x \rangle = -\frac{\varepsilon_0 \omega}{4} A_1^2 \left\{ \varepsilon_1 + \varepsilon_2 \right.$$

$$\left. -2\frac{n_0 e^2}{\varepsilon_0 m_e} \frac{\left[\omega^2 + 2\alpha^2 k_i^2\right] k_r + \left[\gamma\omega - 2\alpha^2 k_r k_i\right] k_i}{\left[\omega^2 - \alpha^2 \left(k_r^2 - k_i^2\right)\right]^2 + \left[\gamma\omega - 2\alpha^2 k_r k_i\right]^2} \right\} e^{-2k_i x} .$$
(6.43)

where

$$k_i = \gamma\omega \left[2\alpha^2 k_r + \frac{n_0 e^2}{\varepsilon_0 m_e} \frac{1}{\varepsilon_1 + \varepsilon_2} \right]^{-1} .$$
(6.44)

6.2.2.2   Energy Distribution

Using (6.15), we obtain

$$U = \frac{\varepsilon_0}{2} \left(k_r^2 + k_i^2\right)$$

$$\times A_1^2 \begin{cases} \varepsilon_1 e^{2k_r z} e^{-2k_i x} , & z \le 0 , \\ \dfrac{1}{2} \dfrac{n_0 e^2}{\varepsilon_0 m_e} \dfrac{\omega^2 + \alpha^2 \left(k_r^2 + k_i^2\right)}{\left[\omega^2 - \alpha^2 \left(k_r^2 - k_i^2\right)\right]^2 + \left[\gamma\omega - 2\alpha^2 k_r k_i\right]^2} e^{-2k_i x} , & z = 0 , \\ \varepsilon_2 e^{-2k_r z} e^{-2k_i x} , & z \ge 0 . \end{cases}$$
(6.45)

Thus, for the total energy density (per unit area) associated with the damped (in space) SPs, we find

$$\langle U \rangle = \frac{\varepsilon_0}{4} \left(k_r^2 + k_i^2 \right) A_1^2 \left\{ \frac{\varepsilon_1 + \varepsilon_2}{k_r} \right.$$

$$\left. + \frac{n_0 e^2}{\varepsilon_0 m_e} \frac{\omega^2 + \alpha^2 \left(k_r^2 + k_i^2 \right)}{\left[\omega^2 - \alpha^2 \left(k_r^2 - k_i^2 \right) \right]^2 + \left[\gamma \omega - 2\alpha^2 k_r k_i \right]^2} \right\} e^{-2k_i x} . \qquad (6.46)$$

Furthermore, using (6.47), the energy dissipated per cycle in a 2DEG layer is

$$\overline{P_d} = \pi \frac{n_0 e^2}{m_e} A_1^2 \frac{\omega \gamma \left(k_r^2 + k_i^2 \right)}{\left[\omega^2 - \alpha^2 \left(k_r^2 - k_i^2 \right) \right]^2 + \left[\gamma \omega - 2\alpha^2 k_r k_i \right]^2} e^{-2k_i x} , \qquad (6.47)$$

6.2.2.3 Energy Velocity

The energy velocity of the damped (in space) SPs is given as

$$v_e = -\frac{\omega}{k_r^2 + k_i^2} \frac{\varepsilon_1 + \varepsilon_2 - 2 \frac{n_0 e^2}{\varepsilon_0 m_e} \frac{\left[\omega^2 + 2\alpha^2 k_i^2 \right] k_r + \left[\gamma \omega - 2\alpha^2 k_r k_i \right] k_i}{\left[\omega^2 - \alpha^2 \left(k_r^2 - k_i^2 \right) \right]^2 + \left[\gamma \omega - 2\alpha^2 k_r k_i \right]^2}}{\frac{\varepsilon_1 + \varepsilon_2}{k_r} + \frac{n_0 e^2}{\varepsilon_0 m_e} \frac{\omega^2 + \alpha^2 \left(k_r^2 + k_i^2 \right)}{\left[\omega^2 - \alpha^2 \left(k_r^2 - k_i^2 \right) \right]^2 + \left[\gamma \omega - 2\alpha^2 k_r k_i \right]^2}} ,$$

$$(6.48)$$

that remains positive at all frequencies.

6.2.2.4 Attenuation Properties

The attenuation of the waves in the x-direction is $-k_i$ in nepers per meter. Thus $-k_i / k_r$ is a measure of the damping of the SP. If $\gamma = 0$, this quantity is zero since $k_x = k_r$ is a real quantity and $k_i = 0$. It is easy to find that the damping of the SP increases as γ becomes larger. Furthermore, the propagation length[4] of SPs of a planar 2DEG layer is given by

$$L_{sp} = \frac{1}{2k_i} , \qquad (6.49)$$

which decreases with increasing γ as expected.

---

[4]In a similar manner with previous description for SPPs in Sect. 3.2.1, the SPs propagation length is defined as the distance covered by SPs when its power/intensity falls to $1/e$ of its initial value, and can be determined by the imaginary part of the SPs wavenumber.

6.3 Quantization of Surface Plasmon Fields of Monolayer Two-Dimensional Electron Gases

Let us consider a planar 2DEG layer at the interface between media with relative dielectric constants ε_1 (for $z < 0$) and ε_2 (for $z > 0$). Now, following the canonical quantization [9], we start from the *Lagrangian* density of the present system, i.e., 2DEG+electrostatic field, which is the sum of four parts, as

$$\ell(\mathbf{r}) = \frac{1}{2}\varepsilon_0 (\nabla \varPhi)^2 + \ell_{sp}(\mathbf{r}_s) , \qquad (6.50)$$

where $\mathbf{r} = (x, y, z)$ and $\mathbf{r}_\parallel = (x, y, z = 0)$ is a position in the 2DEG layer, and

$$\ell_{sp}(\mathbf{r}_\parallel) = \frac{m_e n_0}{2} \parallel \dot{\mathbf{u}}^2(\mathbf{r}_\parallel) \parallel -\frac{m_e \alpha^2}{2n_0} n^2(\mathbf{r}_\parallel) + en(\mathbf{r}_\parallel)\varPhi(\mathbf{r})|_{z=0} . \qquad (6.51)$$

Also, the velocity of electrons $\mathbf{v}(\mathbf{r}_\parallel)$ is expressed in terms of the displacement vector of electrons $\mathbf{u}(\mathbf{r}_\parallel)$ as $\mathbf{v} = \dot{\mathbf{u}}$. We note that the term $\frac{1}{2}\varepsilon_0 (\nabla \varPhi)^2$ is density of electric potential energy of the system. Also, as shown in the previous section, the term $\frac{1}{2}m_e \parallel \mathbf{v}^2 \parallel$ is the kinetic energy density of the medium, while $m_e \alpha^2 n^2 / 2n_0$ shows the hydrodynamic compressional energy density or potential energy density. Finally, the term $en(\mathbf{r}_\parallel)\varPhi(\mathbf{r})|_{z=0}$ is the energy density of interaction part, which includes the interaction between the electrostatic field and the polarization field. Thus, the total Lagrangian $\mathscr{L}_{sp}$ for the present system is

$$\mathscr{L}_{sp} = \frac{1}{2}\varepsilon_0 \int d^3\mathbf{r}(\nabla \varPhi)^2 + \int d^2\mathbf{r}_s \ell_{sp}(\mathbf{r}_s) . \qquad (6.52)$$

The electrostatic potential fluctuations due to the SPs can be expanded by all eigenmodes as follows

$$\varPhi(\mathbf{r}, t) = \sum_{\mathbf{k}_\parallel} e^{i\left(\mathbf{k}_\parallel \cdot \mathbf{r}_\parallel - \omega_{\mathbf{k}_\parallel} t\right)} \begin{cases} \varPhi_{1\mathbf{k}_\parallel} e^{+k_\parallel z} , & z \leq 0 , \\ \varPhi_{2\mathbf{k}_s} e^{-k_\parallel z} , & z \geq 0 , \end{cases} \qquad (6.53)$$

where $k_\parallel = |\mathbf{k}_\parallel|$ and $\mathbf{k}_\parallel = k_x \mathbf{e}_x + k_y \mathbf{e}_y$. Now, we may express the electronic density corresponding to these modes as a localized surface density, namely

$$n(\mathbf{r}_\parallel, t) = \sum_{\mathbf{k}_\parallel} n_{\mathbf{k}_\parallel} e^{i\left(\mathbf{k}_\parallel \cdot \mathbf{r}_\parallel - \omega_{\mathbf{k}_\parallel} t\right)} , \qquad (6.54)$$

and we propose an expansion for the displacement field in the same manner, as

$$\mathbf{u}(\mathbf{r}_\parallel, t) = -i \sum_{\mathbf{k}_\parallel} u_{\mathbf{k}_\parallel} e^{i\left(\mathbf{k}_\parallel \cdot \mathbf{r}_\parallel - \omega_{\mathbf{k}_\parallel} t\right)} . \tag{6.55}$$

To determine the relations between the coefficients in (6.53)–(6.55), we have to use BCs, i.e., (6.10a) and (6.10b). By substituting (6.54) and (6.55) into 2D continuity equation, we obtain

$$n_{\mathbf{k}_\parallel} = -n_0 k_\parallel u_{\mathbf{k}_\parallel} . \tag{6.56}$$

Also, by substituting (6.53)–(6.55) into the BCs, we find the following expressions

$$\Phi_{1\mathbf{k}_\parallel} = \Phi_{2\mathbf{k}_\parallel} = \frac{e n_0}{\varepsilon_0 (\varepsilon_1 + \varepsilon_2)} u_{\mathbf{k}_\parallel} . \tag{6.57}$$

Since we are concerned with real valued fields in the total Lagrangian, $u_{\mathbf{k}_\parallel}$ must satisfy the Hermitian relation $u^*_{\mathbf{k}_\parallel} = u_{-\mathbf{k}_\parallel}$. In order to calculate the potential electrical energy portion of $\mathscr{L}_{sp}$, i.e., the first term of (6.50), we perform a partial integration as follows

$$\int_V d^3\mathbf{r}(\nabla\Phi)^2 = \oint_S ds\Phi \frac{\partial\Phi}{\partial z} - \int_V d^3\mathbf{r}\Phi\nabla^2\Phi , \tag{6.58}$$

where S is a closed surface enclosing the volume V. We note that as S recedes to infinity, the surface integral becomes arbitrarily small, and may consequently be neglected. Therefore, using $\nabla^2\Phi = en/\varepsilon_0$, the potential electrical energy portion of $\mathscr{L}_{sp}$, can be written as

$$\frac{1}{2}\varepsilon_0 \int d^3\mathbf{r}(\nabla\Phi)^2 = -\frac{1}{2}\varepsilon_0 \int d^2\mathbf{r}_\parallel \Phi|_{z=0} \nabla\Phi^2|_{z=0}$$

$$= -\frac{1}{2}e \int d^2\mathbf{r}_\parallel n(\mathbf{r}_\parallel)\Phi(\mathbf{r}_\parallel) = \frac{e^2 n_0^2 A}{2\varepsilon_0(\varepsilon_1 + \varepsilon_2)} \sum_{\mathbf{k}_\parallel} k_\parallel u_{\mathbf{k}_\parallel} u_{-\mathbf{k}_\parallel} , \tag{6.59}$$

where A is the area of the boundary surface $z = 0$. The kinetic energy portion of $\mathscr{L}_{sp}$ can be calculated by using (6.55) in the first term of (6.51) and integration over the 2DEG layer, and we find

$$\frac{m_e n_0}{2} \int d^2\mathbf{r}_\parallel \parallel \dot{\mathbf{u}}^2(\mathbf{r}_\parallel) \parallel = \frac{m_e n_0}{2} \sum_{\mathbf{k}_\parallel} \sum_{\mathbf{k}_\parallel'} \int d^2\mathbf{r}_\parallel \dot{u}_{\mathbf{k}_\parallel} \dot{u}_{\mathbf{k}_\parallel'} e^{i(\mathbf{k}_\parallel + \mathbf{k}_s') \cdot \mathbf{r}_s}$$

$$= \frac{m_e n_0 A}{2} \sum_{\mathbf{k}_\parallel} \dot{u}_{\mathbf{k}_\parallel} \dot{u}_{-\mathbf{k}_\parallel} . \tag{6.60}$$

The hydrodynamic compressional energy portion of $\mathscr{L}_{sp}$ can be calculated in the same manner. By using (6.54) in the second term of (6.51) and integration over the 2DEG layer, we find

$$\frac{m_e\alpha^2}{2n_0} \int d^2\mathbf{r}_\| n^2(\mathbf{r}_\|) = \frac{m_e n_0 A \alpha^2}{2} \sum_{\mathbf{k}_\|} k_\|^2 u_{\mathbf{k}_\|} u_{-\mathbf{k}_\|} . \tag{6.61}$$

Finally, the interaction portion of $\mathscr{L}_{sp}$ can be written as

$$e \int d^2\mathbf{r}_\| n(\mathbf{r}_\|) \Phi(\mathbf{r}_\|) = -\frac{e^2 n_0^2 A}{\varepsilon_0(\varepsilon_1 + \varepsilon_2)} \sum_{\mathbf{k}_\|} k_\| u_{\mathbf{k}_\|} u_{-\mathbf{k}_\|} . \tag{6.62}$$

Now the Lagrangian of the system is expressed in terms of $u_{\mathbf{k}_\|}$ and $\dot{u}_{\mathbf{k}_\|}$ as

$$\mathscr{L}_{sp} = \frac{m_e n_0 A}{2} \sum_{\mathbf{k}_\|} \dot{u}_{\mathbf{k}_\|} \dot{u}_{-\mathbf{k}_\|} - \frac{m_e n_0 A}{2} \sum_{\mathbf{k}_\|} \left[\alpha^2 k_\|^2 + \frac{e^2 n_0 k_\|}{m_e \varepsilon_0(\varepsilon_1 + \varepsilon_2)} \right] u_{\mathbf{k}_\|} u_{-\mathbf{k}_\|} . \tag{6.63}$$

The momentum conjugate to the field coordinate, $u_{\mathbf{k}_\|}$, is given by

$$p_{\mathbf{k}_\|} = \frac{\partial \mathscr{L}_{sp}}{\partial \dot{u}_{\mathbf{k}_\|}} = m_e n_0 A \dot{u}_{-\mathbf{k}_\|} , \tag{6.64}$$

and we find the Hamiltonian $\mathscr{H}_{sp}$ for the SP oscillations as

$$\mathscr{H}_{sp} = \sum_{\mathbf{k}_\|} p_{\mathbf{k}_\|} \dot{u}_{\mathbf{k}_\|} - \mathscr{L}_{sp} = \frac{1}{2 m_e n_0 A} \sum_{\mathbf{k}_\|} p_{\mathbf{k}_\|} p_{-\mathbf{k}_\|}$$

$$+ \frac{m_e n_0 A}{2} \sum_{\mathbf{k}_\|} \left[\alpha^2 k_\|^2 + \frac{e^2 n_0 k_\|}{m_e \varepsilon_0(\varepsilon_1 + \varepsilon_2)} \right] u_{\mathbf{k}_\|} u_{-\mathbf{k}_\|} . \tag{6.65}$$

The classical SP oscillations field is now quantized by the Bose–Einstein commutation relations

$$[u_{\mathbf{k}_\|}, p_{\mathbf{k}_\|'}] = i\hbar \delta_{\mathbf{k}_\| \mathbf{k}_\|'} . \tag{6.66}$$

Based on the Hamiltonian relation in (6.65), we introduce creation and annihilation operators, for both the positive and negative wavenumbers, as

$$\hat{a}_{\mathbf{k}_\|}^\dagger = -i(2 m_e n_0 A)^{-1/2} (\hbar \omega_{\mathbf{k}_\|})^{-1/2} p_{\mathbf{k}_\|}$$

$$+ \left(\frac{m_e n_0 A}{2 \hbar \omega_{\mathbf{k}_\|}} \right)^{1/2} \left[\alpha^2 k_\|^2 + \frac{e^2 n_0 k_\|}{m_e \varepsilon_0(\varepsilon_1 + \varepsilon_2)} \right]^{1/2} u_{-\mathbf{k}_\|} , \tag{6.67}$$

$$\hat{a}^{\dagger}_{-\mathbf{k}_{\|}} = -i(2m_e n_0 A)^{-1/2}(\hbar\omega_{\mathbf{k}_{\|}})^{-1/2} p_{-\mathbf{k}_{\|}}$$

$$+ \left(\frac{m_e n_0 A}{2\hbar\omega_{\mathbf{k}_{\|}}}\right)^{1/2} \left[\alpha^2 k_{\|}^2 + \frac{e^2 n_0 k_{\|}}{m_e \varepsilon_0 (\varepsilon_1 + \varepsilon_2)}\right]^{1/2} u_{\mathbf{k}_{\|}}, \qquad (6.68)$$

$$\hat{a}_{\mathbf{k}_{\|}} = i(2m_e n_0 A)^{-1/2}(\hbar\omega_{\mathbf{k}_{\|}})^{-1/2} p_{-\mathbf{k}_{\|}}$$

$$+ \left(\frac{m_e n_0 A}{2\hbar\omega_{\mathbf{k}_{\|}}}\right)^{1/2} \left[\alpha^2 k_{\|}^2 + \frac{e^2 n_0 k_{\|}}{m_e \varepsilon_0 (\varepsilon_1 + \varepsilon_2)}\right]^{1/2} u_{\mathbf{k}_{\|}}, \qquad (6.69)$$

$$\hat{a}_{-\mathbf{k}_{\|}} = i(2m_e n_0 A)^{-1/2}(\hbar\omega_{\mathbf{k}_{\|}})^{-1/2} p_{\mathbf{k}_{\|}}$$

$$+ \left(\frac{m_e n_0 A}{2\hbar\omega_{\mathbf{k}_{\|}}}\right)^{1/2} \left[\alpha^2 k_{\|}^2 + \frac{e^2 n_0 k_{\|}}{m_e \varepsilon_0 (\varepsilon_1 + \varepsilon_2)}\right]^{1/2} u_{-\mathbf{k}_{\|}}. \qquad (6.70)$$

By their definitions, they must have unit real commutators and the commutation rules are understood as (2.19) and (2.20). In terms of these operators, the Hamiltonian relation in (6.65) is reduced to the harmonic oscillation form

$$\mathscr{H}_{sp} = \sum_{\mathbf{k}_{\|}} \hbar\omega_{\mathbf{k}_{\|}} \left[\hat{a}^{\dagger}_{\mathbf{k}_{\|}} \hat{a}_{\mathbf{k}_{\|}} + \frac{1}{2}\right], \qquad (6.71)$$

provided that $\omega_{\mathbf{k}_{\|}}$ satisfies the relation,

$$\omega_{\mathbf{k}_{\|}}^2 = \alpha^2 k_{\|}^2 + \frac{e^2 n_0 k_{\|}}{m_e \varepsilon_0 (\varepsilon_1 + \varepsilon_2)}, \qquad (6.72)$$

which coincides with the one derived by the corresponding classical theory, i.e., (6.11). Now, the eigenstates of $\mathscr{H}_{sp}$ may be built starting from the vacuum state $|0\rangle$ in the usual way,

$$|n_{\mathbf{k}_{\|}}\rangle = \frac{\left(\hat{a}^{\dagger}_{\mathbf{k}_{\|}}\right)^{n_{\mathbf{k}_{\|}}}}{\sqrt{n_{\mathbf{k}_{\|}}!}}|0\rangle, \qquad (6.73)$$

corresponding to the excitation of $n_{\mathbf{k}_{\|}}$ SPs of a given mode $\mathbf{k}_{\|}$.[5] To conclude the quantization procedure, let us note that Φ, $\mathbf{v}$, and n are now operators and may be

---

[5] We note that $|0\rangle$ is the state which obeys $\hat{a}_{\mathbf{k}_{\|}}|0\rangle = 0$, and in (6.73) the $n_{\mathbf{k}_{\|}}!$ is for normalization.

expressed in terms of the $\hat{a}_{\mathbf{k}_\parallel}$ and $\hat{a}_{\mathbf{k}_\parallel}^\dagger$ as follows

$$
\hat{\Phi}(\mathbf{r}, t) = \frac{e}{\varepsilon_0(\varepsilon_1 + \varepsilon_2)} \left(\frac{\hbar n_0}{2m_e A} \right)^{1/2}
$$
$$
\sum_{\mathbf{k}_\parallel} \frac{1}{\omega_{\mathbf{k}_\parallel}^{1/2}} \left[\hat{a}_{\mathbf{k}_\parallel} e^{i\left(\mathbf{k}_\parallel \cdot \mathbf{r}_\parallel - \omega_{\mathbf{k}_\parallel} t\right)} + \hat{a}_{\mathbf{k}_\parallel}^\dagger e^{-i\left(\mathbf{k}_\parallel \cdot \mathbf{r}_\parallel - \omega_{\mathbf{k}_\parallel} t\right)} \right] \left\{ \begin{array}{l} e^{+k_\parallel z} , \; z \leq 0 , \\ e^{-k_\parallel z} , \; z \geq 0 , \end{array} \right. \tag{6.74}
$$

$$
\hat{\mathbf{v}}(\mathbf{r}_\parallel, t) = -\left(\frac{\hbar}{2m_e n_0 A} \right)^{1/2} \sum_{\mathbf{k}_\parallel} \omega_{\mathbf{k}_\parallel}^{1/2} \left[\hat{a}_{\mathbf{k}_\parallel} e^{i\left(\mathbf{k}_\parallel \cdot \mathbf{r}_\parallel - \omega_{\mathbf{k}_\parallel} t\right)} + \hat{a}_{\mathbf{k}_\parallel}^\dagger e^{-i\left(\mathbf{k}_\parallel \cdot \mathbf{r}_\parallel - \omega_{\mathbf{k}_\parallel} t\right)} \right] ,
$$
$$
\tag{6.75}
$$
$$
\hat{n}(\mathbf{r}_\parallel, t) = -\left(\frac{\hbar n_0}{2m_e A} \right)^{1/2} \sum_{\mathbf{k}_\parallel} \frac{k_\parallel}{\omega_{\mathbf{k}_\parallel}^{1/2}} \left[\hat{a}_{\mathbf{k}_\parallel} e^{i\left(\mathbf{k}_\parallel \cdot \mathbf{r}_\parallel - \omega_{\mathbf{k}_\parallel} t\right)} + \hat{a}_{\mathbf{k}_\parallel}^\dagger e^{-i\left(\mathbf{k}_\parallel \cdot \mathbf{r}_\parallel - \omega_{\mathbf{k}_\parallel} t\right)} \right] .
$$
$$
\tag{6.76}
$$

6.3.1   Interaction with External Probes

We now consider an external particle of charge Ze moving along a prescribed trajectory $\mathbf{r}(t) \equiv \{x(t), y(t), z(t)\}$. The electron–SP interaction Hamiltonian has the form

$$
\mathcal{H}_{int}(t) = Ze\hat{\Phi}[\mathbf{r}(t)] , \tag{6.77}
$$

where $\hat{\Phi}$ is given by (6.74). Using (6.74) for $z \geq 0$ in (6.77), we obtain the expression for the interaction energy for particles moving above the planar 2DEG layer along a trajectory $\mathbf{r}(t)$, as

$$
\mathcal{H}_{int}(t) = \sum_{\mathbf{k}_\parallel} \left[f_{\mathbf{k}_\parallel}(t) \hat{a}_{\mathbf{k}_\parallel} e^{-i\omega_{\mathbf{k}_\parallel} t} + f_{\mathbf{k}_\parallel}^*(t) \hat{a}_{\mathbf{k}_\parallel}^\dagger e^{i\omega_{\mathbf{k}_\parallel} t} \right] , \tag{6.78}
$$

with

$$
f_{\mathbf{k}_\parallel}(t) = -\frac{Ze^2}{\varepsilon_0(\varepsilon_1 + \varepsilon_2)} \left(\frac{\hbar n_0}{2m_e \omega_{\mathbf{k}_\parallel} A} \right)^{1/2} e^{-k_\parallel z(t)} e^{i\mathbf{k}_\parallel \cdot \mathbf{r}_\parallel(t)} , \tag{6.79}
$$

where now the values of (x, y, z) are those corresponding to the trajectory $\mathbf{r}(t)$ and $f_{\mathbf{k}_\parallel}^*(t)$ is the complex conjugate of $f_{\mathbf{k}_\parallel}(t)$. The total Hamiltonian for the SP field interacting with an external probe is given now in the second quantization formalism by

$$\mathcal{H} = \sum_{\mathbf{k}_\parallel} \hbar\omega_{\mathbf{k}_\parallel} \left[\hat{a}_{\mathbf{k}_\parallel}^\dagger \hat{a}_{\mathbf{k}_\parallel} + \frac{1}{2} \right] + \sum_{\mathbf{k}_\parallel} \left[f_{\mathbf{k}_\parallel}(t) \hat{a}_{\mathbf{k}_\parallel} e^{-i\omega_{\mathbf{k}_s} t} + f_{\mathbf{k}_\parallel}^*(t) \hat{a}_{\mathbf{k}_\parallel}^\dagger e^{i\omega_{\mathbf{k}_\parallel} t} \right].$$

(6.80)

We note that the interaction Hamiltonian term has a special form [10, 11], and therefore the time evolution of the SP field state may be found in the interaction picture from the Schrödinger equation, as

$$i\hbar \frac{\partial |\Psi(t)\rangle}{\partial t} = \mathcal{H}_{int} |\Psi(t)\rangle ,$$

(6.81)

where $|\Psi(t)\rangle$ is the quantum state of the SP field in the interaction picture and may be represented as a *coherent state* [12] with the following form

$$|\Psi(t)\rangle = \exp\left[-i \sum_{\mathbf{k}_\parallel} \left(\Upsilon_{\mathbf{k}_\parallel}(t) \hat{a}_{\mathbf{k}_s} + \Upsilon_{\mathbf{k}_\parallel}^*(t) \hat{a}_{\mathbf{k}_\parallel}^\dagger \right) \right] |0\rangle ,$$

(6.82)

where

$$\Upsilon_{\mathbf{k}_\parallel}(t) = -\frac{Ze}{\hbar} \int_{-\infty}^{t} f_{\mathbf{k}_\parallel}(t') e^{-i\omega_{\mathbf{k}_\parallel} t'} \, dt' .$$

(6.83)

and is assumed that the SP field is initially in the vacuum state $|\Psi(-\infty)\rangle = |0\rangle$. The solution of (6.82) can be expanded in the eigenstates $|n_{\mathbf{k}_\parallel}\rangle$ of the free Hamiltonian $\mathcal{H}_{sp}$, and we find

$$|\Psi(t)\rangle = \prod_{\mathbf{k}_\parallel} \exp\left(-\frac{1}{2} |\Upsilon_{\mathbf{k}_\parallel}(t)|^2 \right) \sum_{n=0}^{\infty} \left[-i\Upsilon_{\mathbf{k}_\parallel}^*(t) \right]^n \frac{\left(\hat{a}_{\mathbf{k}_\parallel}^\dagger \right)^{n_{\mathbf{k}_\parallel}}}{n_{\mathbf{k}_\parallel}!} |0\rangle ,$$

(6.84)

which has the complete time evolution of the SP field. Now, by using (6.84) the probability of excitation of $n_{\mathbf{k}_\parallel}$ SPs of a given mode $\mathbf{k}_\parallel$, after interacting with the medium between times $t = -\infty$ and $t = +\infty$, can be written as

$$P_{n_{\mathbf{k}_\parallel}} = |\langle n_{\mathbf{k}_\parallel} |\Psi(\infty)\rangle|^2 = \exp(-Q_{\mathbf{k}_\parallel}) \frac{(Q_{\mathbf{k}_\parallel})^{n_{\mathbf{k}_\parallel}}}{n_{\mathbf{k}_\parallel}!} ,$$

(6.85)

where $Q_{\mathbf{k}_\parallel} = |\Upsilon_{\mathbf{k}_\parallel}(\infty)|^2$. We note the parameter $Q_{\mathbf{k}_\parallel}$ shows the average total number of SPs excited in mode $\mathbf{k}_\parallel$, i.e., $\langle N_{\mathbf{k}_\parallel} \rangle = Q_{\mathbf{k}_\parallel}$. Furthermore, the composite probability to excite $n_{\mathbf{k}_{\parallel 1}}$ SPs of the mode $\mathbf{k}_{\parallel 1}$, $n_{\mathbf{k}_{\parallel 2}}$ SPs of the mode $\mathbf{k}_{\parallel 2}$, and so on, will be given by

$$P_{\left\{n_{\mathbf{k}_{\parallel}}\right\}} = \exp(-Q) \prod_{\mathbf{k}_{\parallel}} \frac{(Q_{\mathbf{k}_s})^{n_{\mathbf{k}_{\parallel}}}}{n_{\mathbf{k}_{\parallel}}!} , \qquad (6.86)$$

where $Q = \sum_{\mathbf{k}_{\parallel}} Q_{\mathbf{k}_{\parallel}}$. Now, the probability of exciting only one SP (of any mode) is given by

$$P_1 = Q \exp(-Q) , \qquad (6.87)$$

and the probability of no-SP excitation is simply

$$P_0 = \exp(-Q) . \qquad (6.88)$$

6.4   Plasmonic Properties of Bilayer Electron Gas Structures

In the previous sections, we investigated the main features of a SP in a single planar 2DEG layer. We note the presence of two 2DEG layers introduces new features which are the result of the combined effect of the geometry [13–15]. Therefore, for a bilayer 2DEG structure, we expect new physical behavior of the dispersion relation, energy, power flow, and energy velocity of plasmonic waves, in comparison with those obtained for a monolayer 2DEG. The aim of the present section is to provide such accounts.

6.4.1   Frequencies of Surface Plasmon Modes and Electrostatic Potential Distribution

We consider a structure with two parallel 2DEG layers with large area placed in the planes $z = z_j$ with $j = 1, 2$ in a 3D Cartesian coordinate system with coordinates (x, y, z), as shown in Fig. 6.3. We define (x, y, z_j) to represent the coordinates of a point in the jth 2DEG layer $z = z_j$. In the present structure, an insulator substrate with relative dielectric constant ε_1 is supposed to occupy the region $z < z_1$ underneath the two 2DEG layers. Also, 2DEG layers are separated by an insulator of thickness $z_2 - z_1 = d$ and relative dielectric constant ε_2, whereas the region $z > z_2$ is assumed to be a semi-infinite insulator with relative dielectric constant ε_3.

Again, we consider the propagation of a SP in the x-direction, so the electric potential is represented in the form of (6.5). For this geometry, the solution of (6.7), when $z \neq z_j$, has the form

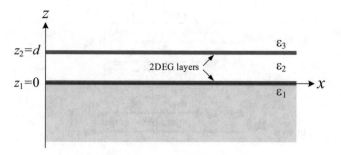

Fig. 6.3 Schematic representation of a structure with two parallel 2DEG layers capable of supporting SP modes at each of its layers

$$\tilde{\Phi}(z) = \begin{cases} A_1 e^{+k_x z} , & z \leq z_1 , \\ A_2 e^{+k_x z} + A_3 e^{-k_x z} , & z_1 \leq z \leq z_2 , \\ A_4 e^{-k_x z} , & z \geq z_2 , \end{cases} \tag{6.89}$$

where the relations between the coefficients A_1–A_4 can be determined from the matching BCs at the separation surfaces. The BCs at $z = z_1 = 0$, and $z = z_2 = d$, are

$$\Phi_j \big|_{z=z_j} = \Phi_{j+1} \big|_{z=z_j} , \tag{6.90a}$$

$$\varepsilon_{j+1} \frac{\partial \Phi_{j+1}}{\partial z} \Big|_{z=z_j} - \varepsilon_j \frac{\partial \Phi_j}{\partial z} \Big|_{z=z_j} = \frac{e n_j}{\varepsilon_0} , \tag{6.90b}$$

where Φ_j denotes the electric potential below the jth 2DEG layer and Φ_{j+1} denotes the electric potential above the layer and

$$n_j = -\frac{e n_0}{m_e} \frac{k_x^2}{\omega^2 - \alpha^2 k_x^2} \Phi_j |_{z=z_j} . \tag{6.91}$$

The application of the electrostatic BCs at $z = z_1 = 0$, and $z = z_2 = d$, gives rise to the following dispersion relation

$$\left(\frac{\varepsilon_1 + \varepsilon_2 - \dfrac{n_0 e^2}{\varepsilon_0 m_e} \dfrac{k_x}{\omega^2 - \alpha^2 k_x^2}}{\varepsilon_1 - \varepsilon_2 - \dfrac{n_0 e^2}{\varepsilon_0 m_e} \dfrac{k_x}{\omega^2 - \alpha^2 k_x^2}} \right) \left(\frac{\varepsilon_3 + \varepsilon_2 - \dfrac{n_0 e^2}{\varepsilon_0 m_e} \dfrac{k_x}{\omega^2 - \alpha^2 k_x^2}}{\varepsilon_3 - \varepsilon_2 - \dfrac{n_0 e^2}{\varepsilon_0 m_e} \dfrac{k_x}{\omega^2 - \alpha^2 k_x^2}} \right) = e^{-2k_x d} . \tag{6.92}$$

Also, we note that three unknown amplitudes A_2–A_4 can be related to the amplitude A_1 through the BCs. In this way, by applying the mentioned BCs we will be able to find the relationship between the constants A_1–A_4, as

$$A_1 = e^{-2k_x d} \frac{\varepsilon_2 - \varepsilon_3 + \dfrac{n_0 e^2}{\varepsilon_0 m_e} \dfrac{k_x}{\omega^2 - \alpha^2 k_x^2}}{\varepsilon_1 + \varepsilon_2 - \dfrac{n_0 e^2}{\varepsilon_0 m_e} \dfrac{k_x}{\omega^2 - \alpha^2 k_x^2}} A_4 , \qquad (6.93a)$$

$$A_2 = e^{-2k_x d} \frac{\varepsilon_2 - \varepsilon_3 + \dfrac{n_0 e^2}{\varepsilon_0 m_e} \dfrac{k_x}{\omega^2 - \alpha^2 k_x^2}}{2\varepsilon_2} A_4 , \qquad (6.93b)$$

$$A_3 = \frac{\varepsilon_2 + \varepsilon_3 - \dfrac{n_0 e^2}{\varepsilon_0 m_e} \dfrac{k_x}{\omega^2 - \alpha^2 k_x^2}}{2\varepsilon_2} A_4 . \qquad (6.93c)$$

For the simple case, when $\varepsilon_1 = \varepsilon_2 = \varepsilon_3 = 1$, we find two branches for ω defining the resonant frequencies of the SP excitations in a bilayer 2DEG, as

$$\omega_\pm^2 = \alpha^2 k_x^2 + \frac{n_0 e^2}{2\varepsilon_0 m_e} k_x \left(1 \pm e^{-k_x d}\right) . \qquad (6.94)$$

According to (6.94), the two SP modes of a bilayer graphene in vacuum are plotted in Fig. 6.4 for $d = 5/k_F$. It is apparent that the splitting between the two branches of SP frequencies in (6.94) is rather substantial owing to the strong electrostatic interaction between the two electron fluids. If we let the layers go to infinity, the modes decouple and there are two independent modes. Also, by using (6.89) and (6.93), we calculate the normalized profiles of electrostatic potentials in Fig. 6.5, for $d = 5/k_F$ and $k_x = 0.1 k_F$, corresponding to the two labeled points in Fig. 6.4.

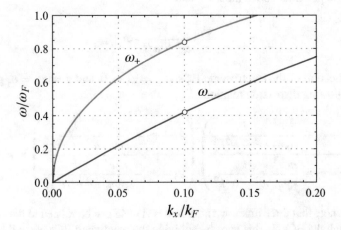

Fig. 6.4 Dispersion curves of SP modes of a bilayer graphene free-standing in vacuum, as given by (6.94) for $d = 5/k_F$. Here $\omega_F = k_F v_F$, $k_F = \sqrt{\pi n_0}$, and $v_F \approx c/300$

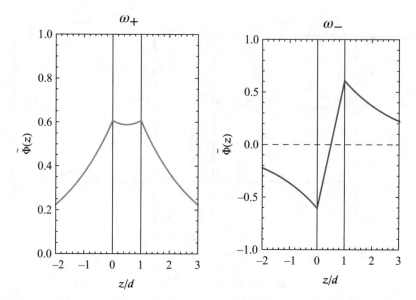

Fig. 6.5 Normalized profile $\tilde{\Phi}(z)$ of SP modes of a bilayer graphene free-standing in vacuum, when $d = 5/k_F$ and $k_x = 0.1k_F$ corresponding to the two labeled points in Fig. 6.4. Panels ω_+ and ω_- show the results for symmetric (even) and anti-symmetric (odd) modes, respectively

6.4.2   Induced Surface Charge

Now it is a straightforward matter to compute the surface charge σ_{ind} induced on the two layers of the system. Using (6.91), we find

$$\sigma_{ind}(x)\big|_{z=z_j} = -en_j = \frac{n_0 e^2}{m_e} \frac{k_x^2}{\omega^2 - \alpha^2 k_x^2} \tilde{\Phi}_j(z)\big|_{z=z_j} \cos k_x x , \qquad (6.95)$$

where $\Phi_j(x, z)$ is considered as $\tilde{\Phi}_j(z) \cos k_x x$ that shows a standing wave in the x-direction. Therefore, by substituting (6.89) into above equation and using (6.93), we arrive to

$$\sigma_{ind}(x)\big|_{z=d} = \frac{n_0 e^2}{m_e} \frac{k_x^2}{\omega^2 - \alpha^2 k_x^2} e^{-k_x d} A_4 \cos k_x x , \qquad (6.96a)$$

$$\sigma_{ind}(x)\big|_{z=0} = \frac{n_0 e^2}{m_e} \frac{k_x^2}{\omega^2 - \alpha^2 k_x^2} \frac{\varepsilon_2 - \varepsilon_3 + \dfrac{n_0 e^2}{\varepsilon_0 m_e} \dfrac{k_x}{\omega^2 - \alpha^2 k_x^2}}{\varepsilon_1 + \varepsilon_2 - \dfrac{n_0 e^2}{\varepsilon_0 m_e} \dfrac{k_x}{\omega^2 - \alpha^2 k_x^2}} e^{-2k_x d} A_4 \cos k_x x . $$

$$(6.96b)$$

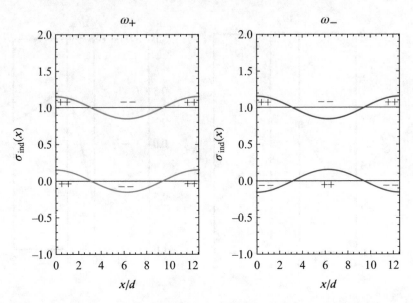

Fig. 6.6 Distribution of induced surface charge across a bilayer graphene free-standing in vacuum, as given by (6.96a) and (6.96b), when $d = 5/k_F$ and $k_x = 0.1k_F$ corresponding to the two labeled points in Fig. 6.4. Panels ω_+ and ω_- show the results for symmetric (even) and anti-symmetric (odd) modes, respectively

Figure 6.6 gives the normalized induced charge density σ_{ind} across a bilayer graphene free-standing in vacuum as a function x/d, when $d = 5/k_F$ and $k_x = 0.1k_F$, corresponding to the two labeled points in Fig. 6.4. The distribution of induced charge gives rise to two standing surface charge waves and shows that the high-frequency mode ω_+ is symmetric and low-frequency mode ω_- is anti-symmetric.

6.4.3   Power Flow

For the power flow density associated with the SP modes of a bilayer 2DEG, we have, in the five media,

$$S_x = - \begin{cases} \varepsilon_0 \varepsilon_1 \Phi_1 \dfrac{\partial}{\partial t} \dfrac{\partial}{\partial x} \Phi_1 , & z \leq z_1 , \\[2mm] e n_0 \Phi_1 v_{1x} - m_e \alpha^2 n_1 v_{1x} , & z = z_1 , \\[2mm] \varepsilon_0 \varepsilon_2 \Phi_2 \dfrac{\partial}{\partial t} \dfrac{\partial}{\partial x} \Phi_2 , & z_1 \leq z \leq z_2 , \\[2mm] e n_0 \Phi_3 v_{2x} - m_e \alpha^2 n_2 v_{2x} , & z = z_2 , \\[2mm] \varepsilon_0 \varepsilon_3 \Phi_3 \dfrac{\partial}{\partial t} \dfrac{\partial}{\partial x} \Phi_3 , & z \geq z_2 . \end{cases} \qquad (6.97)$$

After the elimination of v_{jx}, n_j, Φ_1, Φ_2, and Φ_3 in this equation, the cycle-averaged of (6.97) can be written as

$$
S_x = -\frac{\varepsilon_0 \omega k_x}{2}
\begin{cases}
\varepsilon_1 A_1^2 e^{2k_x z} , & z \leq z_1 , \\[2mm]
-\dfrac{n_0 e^2}{\varepsilon_0 m_e} \dfrac{\omega^2}{\left[\omega^2 - \alpha^2 k_x^2\right]^2} A_1^2 , & z = z_1 , \\[2mm]
\varepsilon_2 \left[A_2 e^{+k_x z} + A_3 e^{-k_x z} \right]^2 , & z_1 \leq z \leq z_2 , \\[2mm]
-\dfrac{n_0 e^2}{\varepsilon_0 m_e} \dfrac{\omega^2}{\left[\omega^2 - \alpha^2 k_x^2\right]^2} A_4^2 e^{-2k_x d} , & z = z_2 , \\[2mm]
\varepsilon_3 A_4^2 e^{-2k_x z} , & z \geq z_2 .
\end{cases}
\tag{6.98}
$$

The total flow of energy through an area in the yz plane of infinite length in the z-direction and unit width in the y-direction is

$$
\langle S_x \rangle = -\frac{\varepsilon_0 \omega}{4} \Bigg\{ \left(\varepsilon_1 - 2\frac{n_0 e^2}{\varepsilon_0 m_e} \frac{\omega^2 k_x}{\left[\omega^2 - \alpha^2 k_x^2\right]^2} \right) A_1^2
$$
$$
+ \left(\varepsilon_3 - 2\frac{n_0 e^2}{\varepsilon_0 m_e} \frac{\omega^2 k_x}{\left[\omega^2 - \alpha^2 k_x^2\right]^2} \right) A_4^2 e^{-2k_x d}
$$
$$
+ \varepsilon_2 \left[A_2^2 \left(e^{2k_x d} - 1 \right) - A_3^2 \left(e^{-2k_x d} - 1 \right) + 4k_x d A_2 A_3 \right] \Bigg\} .
\tag{6.99}
$$

This total power flow density (per unit width) is positive for symmetric and antisymmetric modes, when k_x is positive.

6.4.4 Energy Distribution

Here, for the cycle-averaged of energy density distribution associated with the SP modes of a bilayer 2DEG, we have, in the five media

$$
U = \frac{\varepsilon_0}{4}
\begin{cases}
\varepsilon_1 |\nabla \Phi_1|^2 , & z \leq z_1 , \\[2mm]
\dfrac{m_e n_0}{\varepsilon_0} |v_{1x}|^2 + \dfrac{m_e}{\varepsilon_0 n_0} \alpha^2 |n_1|^2 , & z = z_1 , \\[2mm]
\varepsilon_2 |\nabla \Phi_2|^2 , & z_1 \leq z \leq z_2 , \\[2mm]
\dfrac{m_e n_0}{\varepsilon_0} |v_{2x}|^2 + \dfrac{m_e}{\varepsilon_0 n_0} \alpha^2 |n_2|^2 , & z = z_2 , \\[2mm]
\varepsilon_3 |\nabla \Phi_3|^2 , & z \geq z_2 ,
\end{cases}
\tag{6.100}
$$

After the elimination of v_{jx}, n_j, Φ_1, Φ_2, and Φ_3 in (6.100), we obtain

$$U = \frac{\varepsilon_0 k_x^2}{2} \begin{cases} \varepsilon_1 A_1^2 e^{2k_x z}, & z \leq z_1, \\[2mm] \dfrac{1}{2} \dfrac{n_0 e^2}{\varepsilon_0 m_e} \dfrac{\omega^2 + \alpha^2 k_x^2}{\left[\omega^2 - \alpha^2 k_x^2\right]^2} A_1^2, & z = z_1, \\[2mm] \varepsilon_2 \left[A_2^2 e^{+2k_x z} + A_3^2 e^{-2k_x z}\right], & z_1 \leq z \leq z_2, \\[2mm] \dfrac{1}{2} \dfrac{n_0 e^2}{\varepsilon_0 m_e} \dfrac{\omega^2 + \alpha^2 k_x^2}{\left[\omega^2 - \alpha^2 k_x^2\right]^2} A_4^2 e^{-2k_x d}, & z = z_2, \\[2mm] \varepsilon_3 A_4^2 e^{-2k_x z}, & z \geq z_2, \end{cases} \tag{6.101}$$

It is clear that all contributions to the energy density are positive, as expected on physical grounds. The total energy density associated with the SPs is again determined by integration over z, the energy per unit surface area being

$$\langle U \rangle = \frac{\varepsilon_0 k_x}{4} \left\{ \left(\varepsilon_1 + \frac{n_0 e^2}{\varepsilon_0 m_e} \frac{\omega^2 + \alpha^2 k_x^2}{\left[\omega^2 - \alpha^2 k_x^2\right]^2} k_x \right) A_1^2 \right.$$
$$+ \left(\varepsilon_3 + \frac{n_0 e^2}{\varepsilon_0 m_e} \frac{\omega^2 + \alpha^2 k_x^2}{\left[\omega^2 - \alpha^2 k_x^2\right]^2} k_x \right) A_4^2 e^{-2k_x d}$$
$$\left. + \varepsilon_2 \left[A_2^2 \left(e^{2k_x d} - 1 \right) - A_3^2 \left(e^{-2k_x d} - 1 \right) \right] \right\}. \tag{6.102}$$

Also, the energy velocity (or group velocity) of the SPs is given as the ratio of the total power flow density (per unit width) and total energy density (per unit area), such as previous sections.

6.5   Plasmonic Properties of a Superlattice of Alternating Two-Dimensional Electron Gas Layers

We consider here the structure shown in Fig. 6.7; an infinite superlattice consisting of an alternating ABABAB structure, where media A and B are of thickness a and b, respectively, and the corresponding relative dielectric constants are ε_a and ε_b. At each interface it is assumed that there is a planar 2DEG layer, whose thickness is negligible compared with a and b. The unit cells of the structure are designated by the index n and $L = a + b$ is the length of a unit cell, as illustrated in the figure. Here we wish to obtain the dispersion relation of SPs of the whole structure which propagate in the x-direction and have angular frequency ω. The electrostatic potential $\Phi(x, t)$ then satisfies Laplace's equation everywhere and must obey appropriate BCs at each EG layer.

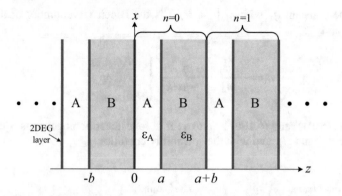

Fig. 6.7 Schematic representation of an infinitely extended superlattice consisting of 2DEG layers separated alternately by media A and B of different thicknesses and dielectric constants. The unit cells of the structure are indexed by n, as illustrated

Again, the electric potential is represented in the form of (6.5). For the present problem, the general solution of (6.7), when $z \neq nL$ and $z \neq nL + a$, has the form [16, 17]

$$\tilde{\Phi}(z) = \begin{cases} A_+ e^{+k_x(z-nL)} + A_- e^{-k_x(z-nL)} , & nL \leq z \leq nL + a , \\ B_+ e^{+k_x(z-nL-a)} + B_- e^{-k_x(z-nL-a)} , & nL + a \leq z \leq (n+1)L , \end{cases}$$
(6.103)

where the relations between the coefficients A_+, A_-, B_+, and B_- can be determined from the matching BCs at the separation surfaces. The BCs at $z = nL + a$ are

$$\Phi_A\big|_{z=nL+a} = \Phi_B\big|_{z=nL+a} ,$$
(6.104a)

$$\varepsilon_B \frac{\partial \Phi_B}{\partial z}\bigg|_{z=nL+a} - \varepsilon_A \frac{\partial \Phi_A}{\partial z}\bigg|_{z=nL+a} = \frac{en}{\varepsilon_0} ,$$
(6.104b)

where Φ_A and Φ_B denote the electric potential in the media A and B, respectively and

$$n = -\frac{en_0}{m_e} \frac{k_x^2}{\omega^2 - \alpha^2 k_x^2} \Phi|_{z=nL+a} .$$
(6.105)

Another appropriate BCs are

$$\Phi_A\big|_{z=nL} = e^{-iq_z L} \Phi_B\big|_{z=(n+1)L} ,$$
(6.106a)

$$\varepsilon_A \frac{\partial \Phi_A}{\partial z}\bigg|_{z=nL} - e^{-iq_z L} \varepsilon_B \frac{\partial \Phi_B}{\partial z}\bigg|_{z=(n+1)L} = \frac{en}{\varepsilon_0} ,$$
(6.106b)

where the constant q_z with $|q_z| < \pi/L$ is the Bloch wavenumber in the Bloch theorem and

$$
n = -\frac{en_0}{m_e} \frac{k_x^2}{\omega^2 - \alpha^2 k_x^2}
\begin{cases}
\Phi_A|_{z=nL}, \\
\quad or \\
\Phi_B|_{z=(n+1)L}.
\end{cases}
\tag{6.107}
$$

Equations (6.104) and (6.106) lead to a set of four linear homogeneous equations in A_+, A_-, B_+, and B_- and with the solvability condition

$$
\begin{vmatrix}
e^{k_x a} & e^{-k_x a} & 1 & 1 \\
\left(\varepsilon_A - \dfrac{n_0 e^2/\varepsilon_0 m_e}{\omega^2 - \alpha^2 k_x^2} k_x\right) e^{k_x a} & -\left(\varepsilon_A + \dfrac{n_0 e^2/\varepsilon_0 m_e}{\omega^2 - \alpha^2 k_x^2} k_x\right) e^{-k_x a} & \varepsilon_B & -\varepsilon_B \\
1 & 1 & \dfrac{e^{k_x b}}{e^{i q_z L}} & \dfrac{e^{-k_x b}}{e^{i q_z L}} \\
\left(\varepsilon_A + \dfrac{n_0 e^2/\varepsilon_0 m_e}{\omega^2 - \alpha^2 k_x^2} k_x\right) & -\left(\varepsilon_A - \dfrac{n_0 e^2/\varepsilon_0 m_e}{\omega^2 - \alpha^2 k_x^2} k_x\right) & \varepsilon_B \dfrac{e^{k_x b}}{e^{i q_z L}} & -\varepsilon_B \dfrac{e^{-k_x b}}{e^{i q_z L}}
\end{vmatrix}
= 0.
\tag{6.108}
$$

It is straightforward, albeit lengthy, to show that (6.108) is equivalent to

$$
\frac{n_0 e^2}{\varepsilon_0 m_e} \frac{k_x}{\omega^2 - \alpha^2 k_x^2} = \varepsilon_A \coth(k_x a) + \varepsilon_B \coth(k_x b)
$$

$$
\pm \frac{\left[\varepsilon_A^2 \sinh^2(k_x b) + \varepsilon_B^2 \sinh^2(k_x a) + 2\varepsilon_A \varepsilon_B \sinh(k_x a)\sinh(k_x b)\cos(q_z L)\right]^{1/2}}{\sinh(k_x a)\sinh(k_x b)}.
\tag{6.109}
$$

This dispersion relation[6] has two branches, corresponding to the two choices of sign on the right-hand side. If $\alpha = 0$, the result is equivalent to the previous result in [18, 19]. Also, (6.109) simplifies considerably when the media A and B are identical (i.e., the case of $b = a$ and $\varepsilon_B = \varepsilon_A$). In this case, we find

$$
\omega^2 = \alpha^2 k_x^2 + \frac{n_0 e^2}{2\varepsilon_0 \varepsilon_A m_e} \frac{\sinh(k_x a)}{\cosh(k_x a) - \cos(q_z a)} k_x,
\tag{6.110}
$$

in agreement with previous calculations using different methods [20–22], but in all other cases there is a stop band between the two branches. Figure 6.8 shows the spectra of plasmon oscillations of a superlattice of monolayer graphenes in terms of the dimensionless variables obtained from (6.110) when $\varepsilon_A = 1$ and $a = 5/k_F$.

---

[6]The recipe for reducing (6.108) to (6.109) is as follows: (1) First add column 1 to column 2 and column 3 to column 4. (2) Subtract 1/2 of column 2 from column 1 and 1/2 of column 4 from column 3. (3) Now expand the resulting determinant and finally get (6.109).

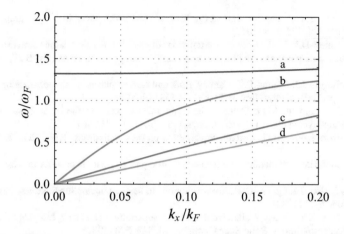

Fig. 6.8 Dispersion curves of SPs of a superlattice of monolayer graphenes as given by (6.110), when $\varepsilon_A = 1$ and $a = 5/k_F$. The different curves refer to (**a**) $q_z a = 0$, (**b**) $q_z a = \pi/6$, (**c**) $q_z a = \pi/2$, and (**d**) $q_z a = \pi$. Here $\omega_F = k_F v_F$ where $k_F = \sqrt{\pi n_0}$ and $v_F \approx c/300$

References

1. F. Stern, Polarizability of a two-dimensional electron gas. Phys. Rev. Lett. **18**, 546–548 (1967)
2. A.L. Fetter, Electrodynamics of a layered electron gas. I. single layer. Ann. Phys. **81**, 367–393 (1973)
3. A.D. Karsono, D.R. Tilley, Electron gas in one and two dimensions. J. Phys. C Solid State Phys. **10**, 2123–2129 (1977)
4. K.S. Novoselov, A.K. Geim, S.V. Morozov, D. Jiang, Y. Zhang, S.V. Dubonos, I.V. Grigorieva, A.A. Firsov, Electric field effect in atomically thin carbon films. Science **306**, 666–669 (2004)
5. P.A.D. Goncalves, N.M.R. Peres, *An Introduction to Graphene Plasmonic* (World Scientific, Singapore, 2016)
6. Z.L. Mišković, S. Segui, J.L. Gervasoni, N.R. Arista, Energy losses and transition radiation produced by the interaction of charged particles with a graphene sheet. Phys. Rev. B **94**, 125414 (2016)
7. A. Moradi, Energy density and energy flow of magnetoplasmonic waves on graphene. Solid State Commun. **253**, 63–66 (2017)
8. A. Moradi, Damping properties of plasmonic waves on graphene. Phys. Plasmas **24**, 072114 (2017)
9. B. Huttner, S.M. Barnett, Quantization of the electromagnetic field in dielectrics. Phys. Rev. A **46**, 4306–4322 (1992)
10. C. Denton, J.L. Gervasoni, R.O. Barrachina, N.R. Arista, Plasmon excitation by charged particles moving near a solid surface. Phys. Rev. A **57**, 4498–4511 (1998)
11. N.R. Arista, M.A. Fuentes, Interaction of charged particles with surface plasmons in cylindrical channels in solids. Phys. Rev. B **63**, 165401 (2001)
12. E. Merzbacher, *Quantum Mechanics* (Wiley, New York, 1970)
13. Y.V. Bludov, A. Ferreira, N.M.R. Peres, M.I. Vasilevskiy, A primer on surface plasmon-polaritons in graphene. Int. J. Mod. Phys. B **27**, 1341001 (2013)
14. C.Z. Li, Y.- N. Wang, Y.H. Song, Z.L. Mišković, Interactions of charged particle beams with double-layered two-dimensional quantum electron gases. Phys. Lett. A **378**, 1626–1631 (2014)

15. A. Moradi, Energy density and energy flow of plasmonic waves in bilayer graphene. Opt. Commun. **394**, 135–138 (2017)
16. R.E. Camley, D.L. Mills, Collective excitations of semi-infinite superlattice structures: surface plasmons, bulk plasmons, and the electron-energy-loss spectrum. Phys. Rev. B **29**, 1695–1706 (1984)
17. B.L. Johnson, J.T. Weiler, R.E. Camley, Bulk and surface plasmons and localization effects in finite superlattices. Phys. Rev. B **32**, 6544–6553 (1985)
18. N.C. Constantinou, M.G. Cottam, Bulk and surface plasmon modes in a superlattice of alternating layered electron gases. J. Phys. C **19**, 739–747 (1986)
19. E.L. Albuquerque, M.G. Cottam, Superlattice plasmon-polaritons. Phys. Rep. **233**, 67–135 (1993)
20. A.L. Fetter, Electrodynamics of a layered electron gas. II. periodic array. Ann. Phys. **88**, 1–25 (1974)
21. S. Das Sarma, J.J. Quinn, Collective excitations in semiconductor superlattices. Phys. Rev. B **25**, 7603–7618 (1982)
22. W.L. Bloss, E.M. Brody, Collective modes of a superlattice-plasmons, LO phonon-plasmons, and magnetoplasmons. Solid State Commun. **43**, 523–528 (1982)

Chapter 7
Electromagnetic Problems Involving Two-Dimensional Electron Gases in Planar Geometry

Abstract In this chapter, we study some optical properties of planar two-dimensional electron gas layers within the framework of classical electrodynamics. We use the standard hydrodynamic model to describe the dielectric response of a planar two-dimensional electron gas layer. Explicit results are given for a collection of electromagnetic boundary-value problems. For brevity, throughout the chapter the $\exp(-i\omega t)$ time factor is suppressed. Furthermore, all media under consideration are nonmagnetic and attention is only confined to the linear phenomena.

7.1 Plasmonic Properties of Monolayer Two-Dimensional Electron Gases

7.1.1 Fresnel Transmission and Reflection Coefficients

Let us consider an incident plane electromagnetic wave coming from below, i.e., $z < 0$ and propagating in an insulator medium with a relative real permittivity ε_1, reflected by a planar 2DEG layer located at $z = 0$, as indicated in Fig. 7.1. The region above the 2DEG layer being characterized by a relative real permittivity ε_2. As described earlier in Sect. 3.1, we choose a Cartesian coordinate system such that the 2DEG layer is perpendicular to the z-axis and the plane of incidence is perpendicular to the y-axis [see Fig. 7.1]. Following the same procedure used in Sect. 3.1, we can do the calculation of the reflection and transmission coefficients at the 2DEG surface $z = 0$ in two special wave polarizations, i.e., p- and s-polarized waves.

We consider a p-polarized wave, which has E_x, E_z, and H_y components. For this p-polarized wave, the total magnetic field H_{1y} in the first insulator space (i.e., $z < 0$)

The original version of this chapter was revised. The correction to this chapter is available at https://doi.org/10.1007/978-3-030-43836-4_11

A. Moradi, *Canonical Problems in the Theory of Plasmonics*, Springer Series in
Optical Sciences 230, https://doi.org/10.1007/978-3-030-43836-4_7

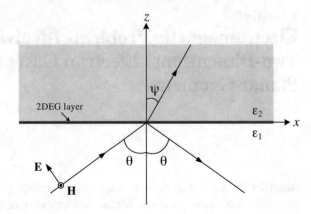

Fig. 7.1 Reflection and transmission of a p-polarized plane wave from a planar 2DEG layer

is taken to be in the form of a superposition of incident and reflected waves. Thus

$$H_{1y} = -H_0 \left[e^{ik_{1z}z} + r_p e^{-ik_{1z}z} \right] e^{ik_x x} , \qquad (7.1)$$

where r_p is the Fresnel reflection coefficient for p-polarized wave, H_0 represents the amplitude of the incident magnetic field, $k_{1z} = \sqrt{\varepsilon_1 k_0^2 - k_x^2} = n_1 k_0 \cos\theta$ with $n_1 = \sqrt{\varepsilon_1}$. Note that θ is the angle of incidence and $k_0 = \omega/c$ is the wavenumber in vacuum. Also, using (3.1b) and (3.1c) the total electric fields along the x- and z-directions in the first insulator space can be written as

$$E_{1x} = -\frac{k_{1z}}{\omega\varepsilon_0\varepsilon_1} H_0 \left[e^{ik_{1z}z} - r_p e^{-ik_{1z}z} \right] e^{ik_x x} , \qquad (7.2a)$$

$$E_{1z} = \frac{k_x}{\omega\varepsilon_0\varepsilon_1} H_0 \left[e^{ik_{1z}z} + r_p e^{-ik_{1z}z} \right] e^{ik_x x} . \qquad (7.2b)$$

In the region above the 2DEG layer, the magnetic field can be written as

$$H_{2y} = -t_p H_0 e^{ik_{2z}z} e^{ik_x x} , \qquad (7.3)$$

where t_p is the Fresnel transmission coefficient for p-polarized wave and $k_{2z} = \sqrt{\varepsilon_2 k_0^2 - k_x^2} = n_2 k_0 \cos\psi$ (ψ is the angle of refraction and $n_2 = \sqrt{\varepsilon_2}$). Also, we have

$$E_{2x} = -\frac{k_{2z}}{\omega\varepsilon_0\varepsilon_2} t_p H_0 e^{ik_{2z}z} e^{ik_x x} , \qquad (7.4a)$$

$$E_{2z} = \frac{k_x}{\omega\varepsilon_0\varepsilon_2} t_p H_0 e^{ik_{2z}z} e^{ik_x x} . \qquad (7.4b)$$

Now, starting with the relevant linearized equation of drift motion for a 2DEG in conjunction with the linearized continuity equation, the conditions relating the fields just above and just below the 2DEG layer can readily be obtained. We have

$$\frac{\partial v_x}{\partial t} = -\frac{e}{m_e} E_x \Big|_{z=0} - \frac{\alpha^2}{n_0} \frac{\partial n}{\partial x} - \gamma v_x \,, \tag{7.5}$$

$$\frac{\partial n}{\partial t} + n_0 \frac{\partial v_x}{\partial x} = 0 \,, \tag{7.6}$$

where $v_x(x,t)$, $n(x,t)$, γ, and α have the same meaning as in the previous chapter. Using $J_x = -en_0 v_x$ and assuming that all of the perturbed physical quantities associated with the surface wave have the form of (6.24), we obtain from (7.5) and (7.6) after the elimination of $n(x,t)$

$$\sigma_{xx} = \frac{ie^2 n_0}{m_e} \frac{\omega}{\omega(\omega + i\gamma) - \alpha^2 k_x^2} \,, \tag{7.7}$$

where σ_{xx} is the 2DEG optical conductivity [1] and for the present system we have $J_x = \sigma_{xx} E_x$ [see Chap. 1]. Using the subscript 1 to denote the region below the 2DEG layer and the subscript 2 to denote the region above the 2DEG, the BC

$$H_{2y}|_{z=0} - H_{1y}|_{z=0} = -J_x|_{z=0} \,, \tag{7.8}$$

immediately follow from Maxwell's curl equation for magnetic field. To within the same approximation

$$E_{1x}|_{z=0} = E_{2x}|_{z=0} \,. \tag{7.9}$$

So that, from the above BCs, we find

$$1 + r_p = t_p \left(1 + \frac{\sigma_{xx}}{\omega \varepsilon_0 \varepsilon_2} k_{2z}\right) \,, \tag{7.10a}$$

$$1 - r_p = \frac{\varepsilon_1}{\varepsilon_2} \frac{k_{2z}}{k_{1z}} t_p \,. \tag{7.10b}$$

Now, we can solve the system of (7.10a) and (7.10b) to obtain

$$r_p = \frac{\varepsilon_2 k_{1z} - k_{2z} \left(\varepsilon_1 - \dfrac{\sigma_{xx}}{\omega \varepsilon_0} k_{1z}\right)}{\varepsilon_2 k_{1z} + k_{2z} \left(\varepsilon_1 + \dfrac{\sigma_{xx}}{\omega \varepsilon_0} k_{1z}\right)} \,, \tag{7.11a}$$

$$t_p = \frac{2\varepsilon_2 k_{1z}}{\varepsilon_2 k_{1z} + k_{2z} \left(\varepsilon_1 + \dfrac{\sigma_{xx}}{\omega \varepsilon_0} k_{1z}\right)} \,. \tag{7.11b}$$

The *transmittance coefficient* is given by the ratio of the power flow densities in the two insulator media. Using the relations between E_x and H_y, we can express the time-averaged power flow densities perpendicular to the surface, given by (1.64), as

$$S_i = \frac{1}{2} \text{Re} \left[E_{ix} H_{iy}^* \right] = \frac{1}{2} \frac{\text{Re} \left[k_{1z} \right]}{\omega \varepsilon_0 \varepsilon_1} H_0^2 , \tag{7.12a}$$

$$S_t = \frac{1}{2} \text{Re} \left[E_{tx} H_{ty}^* \right] = \frac{1}{2} \frac{\text{Re} \left[k_{2z} \right]}{\omega \varepsilon_0 \varepsilon_2} |t_p|^2 H_0^2 , \tag{7.12b}$$

where H_0 is real. Therefore, the transmittance coefficient is

$$T_p = \frac{\varepsilon_1}{\varepsilon_2} \frac{\text{Re} \left[k_{2z} \right]}{\text{Re} \left[k_{1z} \right]} |t_p|^2 . \tag{7.13}$$

In a similar way the reflectance coefficient can be written as

$$R_p = |r_p|^2 . \tag{7.14}$$

Also, the *absorbance coefficient* follows from the other two via

$$A_p = 1 - R_p - T_p . \tag{7.15}$$

Figure 7.2 illustrates the behavior of the transmittance, reflectance, and absorbance coefficients as a function of the dimensionless frequency of the impinging radiation, at oblique incidence for a suspended ($\varepsilon_1 = 1 = \varepsilon_2$) doped graphene. The angle of incidence is $\theta = \pi/3$. In Fig. 7.2, we have used $E_F = 0.4\,\text{eV}$ (or $n_0 \approx 1 \times 10^{13}\,\text{cm}^{-2}$) and $\hbar \gamma = 5\,\text{meV}$ [2].

7.1.1.1   Fresnel Coefficients and Surface Plasmon Polaritons

The denominators of the TM Fresnel coefficients become zero at

$$\frac{\varepsilon_1}{k_{1z}} + \frac{\varepsilon_2}{k_{2z}} = -\frac{\sigma_{xx}}{\omega \varepsilon_0} . \tag{7.16}$$

This equation is identical with (7.27) when $\gamma = 0$ and $k_{jz} = i\kappa_j$ with $j = 1$ and 2. This means that the poles of the Fresnel coefficients provide the dispersion relation of SPPs confined to the interface [3, 4].

7.1.1.2   Regions of Ordinary and Total Reflection

As discussed earlier in Sect. 3.1, the matching of the phases of **E** and **H** at the interface $z = 0$ leads to the generalized Snell's law as

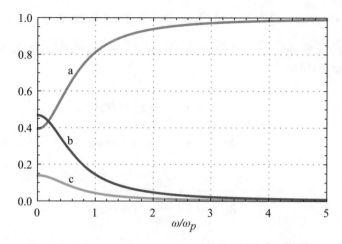

Fig. 7.2 Numerical results of electromagnetic radiation at oblique incidence through a suspended ($\varepsilon_1 = 1 = \varepsilon_2$) doped monolayer graphene, as a function of the dimensionless frequency of the impinging radiation. The different curves refer to (**a**) transmittance coefficient, (**b**) absorbance coefficient, (**c**) reflectance coefficient. Here $\omega_p = 4\alpha_{sc} E_F/\hbar$, where $E_F = \hbar k_F v_F$ and $\alpha_{sc} = e^2/4\pi\varepsilon_0\hbar c$ that is the *fine structure constant*. The angle of incidence is $\theta = \pi/3$, $E_F = 0.4\,\text{eV}$ (or $n_0 \approx 1 \times 10^{13}\text{cm}^{-2}$), and $\hbar\gamma = 5\,\text{meV}$

$$k_x = n_1 k_0 \sin\theta = n_2 k_0 \sin\psi \ . \tag{7.17}$$

In the case of a lossless planar 2DEG layer, i.e., when $\gamma = 0$ in (7.7), the region of ordinary reflection is defined by $0 < \sin^2\psi < 1$. If $\varepsilon_2 < \varepsilon_1$ (i.e., the incident medium has a refractive index larger than that of the second medium), the *critical angle* of incidence, where $\sin\psi = 1$ and $\psi = 90°$, is given by

$$\theta_c = \sin^{-1}\frac{n_2}{n_1}, \qquad n_2 < n_1 \ . \tag{7.18}$$

For waves incident from medium 1 at $\theta = \theta_c$, the refracted wave is propagating parallel to the interface. When $\sin^2\psi$ exceeds unity,[1] the field inside the region above the 2DEG layer would be evanescent and therefore $R_p = 1$, since r_p is the ratio of two complex numbers with the same absolute value [see (7.19a)]. This implies the total internal reflection of the incident wave.

---

[1]This means that ψ is a complex angle with a purely imaginary cosine, as $\cos\psi = -i\sqrt{\sin^2\psi - 1}$. The negative sign ensures that the transmitted wave decays exponentially as z approaches $+\infty$. In this case, the angle ψ loses its physical interpretation as the angle of refraction for the transmitted wave. It only means that $k_{2z} = n_2 k_0 \cos\psi$ and $k_{2x} = n_2 k_0 \sin\psi$.

7.1.1.3   Evanescent Waves and Phase Shift in Total Reflection

In the total reflection, the reflection and transmission coefficients given by (7.11) can be expressed as

$$
r_p = \frac{k_{1z}\left(\varepsilon_2 + i\dfrac{\sigma_{xx}}{\omega\varepsilon_0}\kappa_2\right) - i\kappa_2\varepsilon_1}{k_{1z}\left(\varepsilon_2 + i\dfrac{\sigma_{xx}}{\omega\varepsilon_0}\kappa_2\right) + i\kappa_2\varepsilon_1} ,
\tag{7.19a}
$$

$$
t_p = \frac{2\varepsilon_2 k_{1z}}{k_{1z}\left(\varepsilon_2 + i\dfrac{\sigma_{xx}}{\omega\varepsilon_0}\kappa_2\right) + i\kappa_2\varepsilon_1} ,
\tag{7.19b}
$$

where $\kappa_2 = \sqrt{k_x^2 - \varepsilon_2 k_0^2}$. Since $k_{2z} = i\kappa_2$ is purely imaginary in this case, and the Poynting vector is proportional to the real part of k_{2z}, from (7.13) we immediately conclude that no energy is transferred through the interface, i.e., $T_p = 0$ and all the light energies are totally reflected from the surface. However, we notice that t_p is not vanishing at total reflection. This means that even though the light energies are totally reflected, the electromagnetic fields still penetrate into the second medium [5, 6]. In fact, zero transmission of light energy only means that the normal component of the Poynting vector vanishes, and the power flow is parallel to the boundary surface.

Also, it is convenient to write the term $k_{1z}(\varepsilon_2 + i\sigma_{xx}\kappa_2/\omega\varepsilon_0) + i\kappa_2\varepsilon_1$ in the denominator of (7.19) in the polar form as

$$
k_{1z}\left(\varepsilon_2 + i\frac{\sigma_{xx}}{\omega\varepsilon_0}\kappa_2\right) + i\kappa_2\varepsilon_1 = \xi_p e^{i\phi_p} ,
\tag{7.20}
$$

where

$$
\xi_p = \left[k_{1z}^2\left(\varepsilon_2 + i\frac{\sigma_{xx}}{\omega\varepsilon_0}\kappa_2\right)^2 + \kappa_2^2\varepsilon_1^2 \right]^{1/2} ,
\tag{7.21a}
$$

$$
\phi_p = \tan^{-1}\left[\frac{\kappa_2\varepsilon_1}{k_{1z}\left(\varepsilon_2 + i\dfrac{\sigma_{xx}}{\omega\varepsilon_0}\kappa_2\right)} \right] .
\tag{7.21b}
$$

Therefore (7.19) can be rewritten as

$$
r_p = e^{-2i\phi_p} ,
\tag{7.22a}
$$

$$t_p = \frac{2\varepsilon_2 k_{1z}}{\xi_p} e^{-i\phi_p} . \tag{7.22b}$$

Equation (7.22a) indicates that the amplitude of the reflected wave is only different from the incident amplitude by a phase factor $2\phi_p$.

7.1.2 Power Flow in Total Reflection: Lateral Shift

As mentioned in the previous section, in the event of total reflection, there is a finite power flow parallel to the interface, i.e., in the 2DEG layer. This may appear to be surprising since, the reflected and incident powers being equal in the case of lossless total reflection, it is difficult to conceive of a power flow in the second medium as it would imply a violation of the law of conservation of energy. The logical solution to this apparent paradox has been discussed by Renard [7] and is explained in Fig. 7.3 for the present system, i.e., a planar 2DEG layer. The electromagnetic energy enters the second (reflecting) medium, where it propagates a certain distance parallel to the interface, and then returns to the first medium.[2] However, the *lateral shift*[3] evaluated by Renard [7] according to the energy-flux technique has been found to be different from that obtained by Artmann [9] using the *stationary phase method*. To remedy this issue, Yasumoto and Õishi [10] pointed out that the power flow along the interface is not entirely in the medium of lower refractive index and that

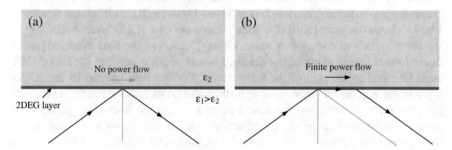

Fig. 7.3 The path of a ray in lossless total reflection: (**a**) geometrical reflection and (**b**) actual reflection. The ray propagation parallel to the interface in the reflection medium is a logical necessity in order to explain finite (non-zero) power flow parallel to the interface in the reflecting medium

---

[2]Evidently, there is no way to detect this lateral shift of rays in the case of reflection of a uniform plane wave of infinite extent. However, the lateral shift of a well-collimated beam (of finite transverse extent) should be observable.

[3]This lateral shift is associated with the names of two German scientists and is now referred to as the Goos–Hänchen shift [8].

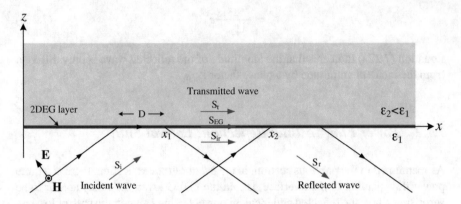

Fig. 7.4 Illustration of the lateral shift on a planar 2DEG layer for a finite plane wave as a consequence of total reflection. There exists an additional power flow S_{EG} parallel to the 2DEG layer in addition to the power flow S_{ir} produced by the interaction between the incident and reflected waves and the power flow inside the medium 2 carried by the evanescent wave in the x-direction

a part of the power flow along the interface is in the medium of higher refractive index.

Therefore, when the energy-flux method is implemented carefully with the inclusion of the necessary power flows in both media, the lateral shift deduced by the energy-flux method is identical to that obtained by the stationary phase method [11]. We now proceed to derive an expression for the lateral drift of an electromagnetic beam totally reflected at an insulator-2DEG-insulator structure. Fig. 7.4 shows the lateral shift of a finite plane wave on a 2DEG layer. According to the consideration of energy-flux conservation [7, 10, 11], the time-averaged power flow of the reflected plane across the strip whose width is equal to the lateral shift must be equal to the sum of the time-averaged power flow parallel to the 2DEG layer in the evanescent wave existing in the medium of lower refractive index and the power flow existing on the 2DEG layer and also the interaction power flow existing in the medium of higher refractive index in the region of overlap of the incident and the reflected wave packets. Thus, the lateral shift is given by

$$D = \frac{\frac{1}{2}m_e\alpha^2 nv_x + \lim_{(x_2-x_1)\to\infty}\int_{x_1}^{x_2}dx\int_{-z(x)}^{0}\frac{S_{xir}}{x_2-x_1}dz+\int_{0}^{+\infty}S_{xt}\,dz}{S_{zr}},$$

(7.23)

where S_{zr} is the z-component of the time-averaged power flow density vector of the reflected wave, S_{xt} is the x-component of the time-averaged power flow density vector of the transmitted wave, and $m_e\alpha^2 nv_x$ is equal with the x-component of the time-averaged power flow density vector in the 2DEG layer. Furthermore, S_{xir} represents the x-component of the time-averaged power flow density vector, produced in a triangular region [see Fig. 7.4] by the interaction between the two packets. The two wave packets overlap in this triangular region where $x_1 \leq x \leq x_2$ and $-z(x) \leq z \leq 0$. Note that for the second integral on the numerator of (7.23), we first calculate it over x from x_1 to x_2 and finally take the limit where $(x_2 - x_1) \to \infty$.

Similarly, the associated time delay T_d in the reflection process should be equal to the time needed for the total energy density (per unit area) (i.e., sum of the stored energy within the evanescent wave and stored energy of interaction between the two wave packets and also stored energy in the 2DEG layer) to travel along the interface for the distance D as

$$T_d = \frac{\frac{1}{4}\left(m_e n_0 |v_x|^2 + \frac{m_e}{n_0}\alpha^2 |n|^2\right) + \lim_{(x_2-x_1)\to\infty} \int_{x_1}^{x_2} dx \int_{-z(x)}^{0} \frac{U_{ir}}{x_2-x_1} dz + \int_{0}^{+\infty} U_t\, dz}{S_{zr}},$$

(7.24)

where in the above equations, U_t and U_{ir} denote the stored energy density of the evanescent wave and stored energy density of the interaction between the two wave packets, respectively. With the equations derived in previous sections for a p-polarized incident plane wave in conjunction with appropriate BCs, i.e., (7.8) and (7.8), the lateral shift and associated time delay evaluated as [11]

$$D = 2\frac{k_x}{k_{1z}\kappa_2}\frac{\varepsilon_1}{\xi_p^2}\left[\varepsilon_2 k_{1z}^2\left(1 + 2\frac{n_0 e^2}{\varepsilon_0 \varepsilon_2 m_e}\frac{\alpha^2 \kappa_2^3}{\left[\omega^2 - \alpha^2 k_x^2\right]^2}\right) + \kappa_2^2\left(\varepsilon_2 + i\frac{\sigma_{xx}}{\omega\varepsilon_0}\kappa_2\right)^{1/2}\right],$$

(7.25)

and

$$T_d = 2\frac{k_x^2}{\omega k_{1z}\kappa_2}\frac{\varepsilon_1}{\xi_p^2}\left[\varepsilon_2 k_{1z}^2\left(1 + \frac{n_0 e^2}{\varepsilon_0 \varepsilon_2 m_e}\frac{\omega^2 + \alpha^2 k_x^2}{\left[\omega^2 - \alpha^2 k_x^2\right]^2}\frac{\kappa_2^3}{k_x^2}\right) + \kappa_2^2\left(\varepsilon_2 + i\frac{\sigma_{xx}}{\omega\varepsilon_0}\kappa_2\right)^{1/2}\right].$$

(7.26)

The variation of lateral shift and associated time delay with respect to the angle of incidence θ is shown in Fig. 7.5 for a p-polarized incident plane wave packet on a monolayer graphene when $\alpha = 0$, $n_1 = 1.5$, and $n_2 = 1$.

7.1.3   Dispersion Relation

It is well-known that a planar 2DEG layer can support a p-polarized or TM SPP [12, 13]. We note that in general[4] a s-polarized SPP cannot propagate on the surface

---

[4]We do not discuss here the situation when a planar 2DEG layer such as graphene [14] supports the s-polarized or TE surface waves.

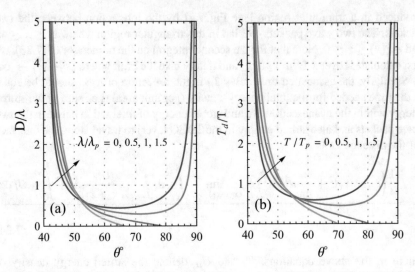

Fig. 7.5 Variation of (**a**) lateral shift for four different values of λ/λ_p and (**b**) associated time delay for four different values of T/T_p with respect to the angle of incidence θ of a p-polarized incident plane wave packet on a monolayer graphene when $\alpha = 0$, $n_1 = 1.5$, and $n_2 = 1$. Here $\lambda_p = 2\pi c/\omega_p$, $T_p = 2\pi/\omega_p$, and $\omega_p = 4\alpha_{sc} E_F/\hbar$, where $E_F = \hbar k_F v_F$ and $\alpha_{sc} = e^2/4\pi \varepsilon_0 \hbar c$

of such structures [14, 15]. A p-polarized SPP is assumed to propagate in the x-direction parallel to the surface of the system. The variation in the y-direction is taken to be zero, so that the disturbance is purely 2D in nature [see Sect. 3.1]. Thus in the region below the 2DEG layer and also the region above the 2DEG, this wave has the electric field given by $[E_x(z)\mathbf{e}_x + E_z(z)\mathbf{e}_z]\exp(ik_x x)$, and magnetic field given by $\mathbf{e}_y H_y(z)\exp(ik_x x)$.

As discussed in Sect. 3.1, H_y and E_z can be determined if the non-zero longitudinal component E_x is known [see (3.16a) and (3.16b)], where the x-component E_x can be obtained from the Helmholtz equation, i.e., (3.17), and the solution of (3.17) has the form of (3.18). On applying the electromagnetic BCs at $z = 0$, i.e., (7.8) and (7.9), and setting $\gamma = 0$, we find [12]

$$\frac{\varepsilon_1}{\kappa_1} + \frac{\varepsilon_2}{\kappa_2} = \frac{n_0 e^2}{\varepsilon_0 m_e}\frac{1}{\omega^2 - \alpha^2 k_x^2}, \qquad (7.27)$$

where $\kappa_j^2 = k_x^2 - \varepsilon_j \omega^2/c^2$ and $j = 1, 2$. With $\alpha = 0$ (7.27) is Nakayama's result [13], and with $\varepsilon_1 = 1 = \varepsilon_2$ it reduces to the result of Fetter [12]. The dispersion curve is similar to that for $\alpha = 0$, except that for large ω, $\omega \simeq \alpha k_x$. By contrast, with $\alpha = 0$, the high-frequency behavior is $\omega \propto k_x^{1/2}$ [13]. Figure 7.6 shows the spectrum of SPPs of a monolayer graphene, as given by (7.27), when $\alpha = 0$ and $\varepsilon_1 = 1 = \varepsilon_2$.

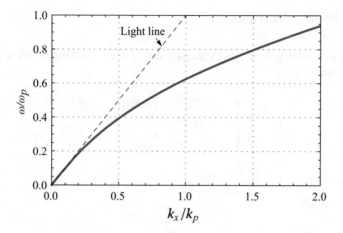

Fig. 7.6 Dispersion curve of SPPs of a monolayer graphene, as given by (7.27), when $\alpha = 0$, $\varepsilon_1 = 1 = \varepsilon_2$. Here $k_p = \omega_p/c$ and $\omega_p = 4\alpha_{sc}E_F/\hbar$, where $E_F = \hbar k_F v_F$ and $\alpha_{sc} = e^2/4\pi\varepsilon_0\hbar c$

7.1.4 Power Flow

For the power flow density associated with a SPP of a planar 2DEG layer, we have, in the three media,

$$\mathbf{S} = \begin{cases} \mathbf{E}_1 \times \mathbf{H}_1 \,, & z \leq 0 \,, \\ m_e\alpha^2 n v_x \,, & z = 0 \,, \\ \mathbf{E}_2 \times \mathbf{H}_2 \,, & z \geq 0 \,, \end{cases} \tag{7.28}$$

where subscripts 1 and 2 denote the regions below and above the 2DEG layer. After the elimination of v_x and n, the cycle-averaged x-components are

$$S_x = \frac{\varepsilon_0 k_x \omega}{2} A_1^2 \begin{cases} \dfrac{\varepsilon_1}{\kappa_1^2}e^{+2\kappa_1 z} \,, & z \leq 0 \,, \\[2mm] \dfrac{n_0 e^2}{\varepsilon_0 m_e}\dfrac{\alpha^2}{\left[\omega^2 - \alpha^2 k_x^2\right]^2} \,, & z = 0 \,, \\[2mm] \dfrac{\varepsilon_2}{\kappa_2^2}e^{-2\kappa_2 z} \,, & z \geq 0 \,. \end{cases} \tag{7.29}$$

The total power flow density associated with the SPPs is determined by an integration over z. The power flow through an area in the yz plane of infinite length in the z-direction and unit width in the y-direction is

$$\langle S_x \rangle = \frac{\varepsilon_0 k_x \omega}{4} A_1^2 \left[\frac{\varepsilon_1}{\kappa_1^3} + \frac{\varepsilon_2}{\kappa_2^3} + 2\frac{n_0 e^2}{\varepsilon_0 m_e}\frac{\alpha^2}{\left[\omega^2 - \alpha^2 k_x^2\right]^2} \right]. \tag{7.30}$$

This total power flow density (per unit width) is positive when k_x is positive.

7.1.5 Energy Distribution

For the cycle-averaged of energy density distribution associated with a SPP of a 2DEG layer, we have

$$
U = \frac{1}{4}
\begin{cases}
\varepsilon_0 \varepsilon_1 |\mathbf{E}_1|^2 + \mu_0 |\mathbf{H}_1|^2 , & z \le 0 , \\[2mm]
m_e n_0 |v_x|^2 + \dfrac{m_e}{n_0} \alpha^2 |n|^2 , & z = 0 , \\[2mm]
\varepsilon_0 \varepsilon_2 |\mathbf{E}_2|^2 + \mu_0 |\mathbf{H}_2|^2 , & z \ge 0 ,
\end{cases}
\tag{7.31}
$$

that yields

$$
U = \frac{\varepsilon_0}{2} A_1^2
\begin{cases}
\varepsilon_1 \dfrac{k_x^2}{\kappa_1^2} e^{+2\kappa_1 z} , & z \le 0 , \\[3mm]
\dfrac{1}{2} \dfrac{n_0 e^2}{\varepsilon_0 m_e} \dfrac{\omega^2 + \alpha^2 k_x^2}{\left[\omega^2 - \alpha^2 k_x^2 \right]^2} , & z = 0 , \\[3mm]
\varepsilon_2 \dfrac{k_x^2}{\kappa_2^2} e^{-2\kappa_2 z} , & z \ge 0 ,
\end{cases}
\tag{7.32}
$$

where all contributions to the energy density are positive. The total energy density associated with the SPPs is again determined by integration over z, the energy per unit surface area being

$$
\langle U \rangle = \frac{\varepsilon_0}{4} A_1^2 \left[k_x^2 \left(\frac{\varepsilon_1}{\kappa_1^3} + \frac{\varepsilon_2}{\kappa_2^3} \right) + \frac{n_0 e^2}{\varepsilon_0 m_e} \frac{\omega^2 + \alpha^2 k_x^2}{\left[\omega^2 - \alpha^2 k_x^2 \right]^2} \right] .
\tag{7.33}
$$

7.1.6 Energy Velocity

The energy velocity of the SPPs is given as the ratio of the total power flow density (per unit width) and total energy density (per unit area), such as

$$
v_e = \frac{\omega}{k_x} \frac{\dfrac{\varepsilon_1}{\kappa_1^3} + \dfrac{\varepsilon_2}{\kappa_2^3} + 2 \dfrac{n_0 e^2}{\varepsilon_0 m_e} \dfrac{\alpha^2}{\left[\omega^2 - \alpha^2 k_x^2 \right]^2}}{\dfrac{\varepsilon_1}{\kappa_1^3} + \dfrac{\varepsilon_2}{\kappa_2^3} + \dfrac{n_0 e^2}{\varepsilon_0 m_e} \dfrac{1}{k_x^2} \dfrac{\omega^2 + \alpha^2 k_x^2}{\left[\omega^2 - \alpha^2 k_x^2 \right]^2}} .
\tag{7.34}
$$

According to the above equation, the group velocity or energy velocity of SPPs on a suspended monolayer graphene in vacuum is plotted in Fig. 7.7.

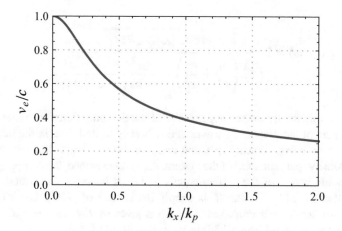

Fig. 7.7 Group (energy) velocity curve of SPPs of a suspended monolayer graphene in vacuum (when $\alpha = 0$) as given by (7.34). Here $k_p = \omega_p/c$ and $\omega_p = 4\alpha_{sc}E_F/\hbar$, where $E_F = \hbar k_F v_F$ and $\alpha_{sc} = e^2/4\pi\varepsilon_0\hbar c$

7.1.7 Damping Property

In the present section, we study the frequency dependence of the damping function of a SPP of a 2DEG layer. We wish to obtain an analytical expression for the damping function (which depends on the frequency and wavenumber) of a SPP of a 2DEG layer by using the perturbative method proposed by Loudon [16] and Nkoma et al. [17]. Such a procedure enables us to calculate the true SPPs damping rate to the first order in the damping parameter γ, introduced to describe the intrinsic damping of crystal oscillations. Also, this theory enables us to discuss both the propagation length and the lifetime of a surface plasmonic wave. The advantage of the perturbative method is that the damping properties result from the calculation of real dispersion relations. The plasmonic damping parameter or relaxation rate $\Gamma(k_x, \omega)$ of the present case may be determined by the following procedure. The kinetic and total energy densities (per unit area) U_k and $\langle U \rangle$ are first calculated in the absence of damping. If a small amount of damping is now reintroduced, the plasmonic energy relaxation rate to the lowest order in γ is

$$\Gamma(k_x, \omega) = 2\gamma \frac{U_k}{\langle U \rangle} , \tag{7.35}$$

where from (7.33) we have

$$U_k = A_1^2 \frac{n_0 e^2}{4m_e} \frac{\omega^2}{\left[\omega^2 - \alpha^2 k_x^2\right]^2} . \tag{7.36}$$

Therefore, we get

$$\Gamma(k_x, \omega) = 2\gamma \frac{\dfrac{n_0 e^2}{\varepsilon_0 m_e} \dfrac{\omega^2}{\left[\omega^2 - \alpha^2 k_x^2\right]^2}}{k_x^2 \left(\dfrac{\varepsilon_1}{\kappa_1^3} + \dfrac{\varepsilon_2}{\kappa_2^3}\right) + \dfrac{n_0 e^2}{\varepsilon_0 m_e} \dfrac{\omega^2 + \alpha^2 k_x^2}{\left[\omega^2 - \alpha^2 k_x^2\right]^2}}. \tag{7.37}$$

Let us note that the frequency dependence of the damping function comes from the retarded part of the plasmonic waves and it is easy to find that, in the nonretarded limit, the total energy density (per unit area) becomes twice as large as the kinetic energy density (per unit area) of the system. As a consequence, the damping function of plasmonic waves of the system equals γ, i.e., it becomes a constant. Also, let us note that the SPPs lifetime T is simply the inverse of (7.37), i.e., $T(k_x, \omega) = \Gamma^{-1}(k_x, \omega)$, while their propagation length is given by $L(k_x, \omega) = v_g \Gamma^{-1}(k_x, \omega)$, where v_g can be found from (7.34) or $v_g = \partial\omega/\partial k_x$ [18, 19].

By using (7.37), we can now calculate the SPPs damping rate and hence the SPPs propagation length as a function of frequency. The damping rate of the long-wavelength SPPs on a monolayer graphene in terms of the dimensionless variables is presented in Fig. 7.8 when $\varepsilon_1 = 1 = \varepsilon_2$. It is clear that the damping function of long-wavelength SPPs is equal to γ for $\omega \gg \omega_p$, but that the rate vanishes at the minimum frequency of zero. In panels (a) and (b) of Fig. 7.9, the frequency dependence of long-wavelength SPPs lifetime and propagation length of the system are shown, respectively, when $\varepsilon_1 = 1 = \varepsilon_2$ and $L_p = c/\gamma$. From panel (a) one can see that for $\omega \ll \omega_p$, by decreasing values of ω, the SPPs lifetime increases sharply and the SPPs lifetime reaches its minimum value, i.e., $1/\gamma$, for $\omega \gg \omega_p$. On the

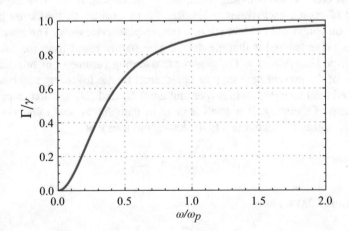

Fig. 7.8 Frequency dependence of relaxation rate of the long-wavelength SPPs of a doped monolayer graphene when $\alpha = 0$ and $\varepsilon_1 = 1 = \varepsilon_2$. Here $\omega_p = 4\alpha_{sc} E_F/\hbar$, where $E_F = \hbar k_F v_F$ and $\alpha_{sc} = e^2/4\pi\varepsilon_0\hbar c$

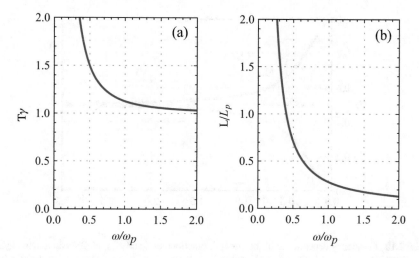

Fig. 7.9 Frequency dependence of (**a**) lifetime and (**b**) propagation length of the long-wavelength SPPs of a doped monolayer graphene when $\alpha = 0$ and $\varepsilon_1 = 1 = \varepsilon_2$. Here $\omega_p = 4\alpha_{sc} E_F/\hbar$, where $E_F = \hbar k_F v_F$ and $\alpha_{sc} = e^2/4\pi\varepsilon_0\hbar c$

other hand, from panel (b) it is obvious that for $\omega \ll \omega_p$ one can provide SPPs with large propagation length [1].

7.1.8 Surface Plasmon and Surface Electromagnetic Field Strength Functions

According to the results of Sect. 3.2.8, also for present system we have (3.44), where

$$\langle U_{ph}\rangle = \frac{\varepsilon_0}{4} A_1^2 k_x^2 \left(\frac{\varepsilon_1}{\kappa_1^3} + \frac{\varepsilon_2}{\kappa_2^3} \right) , \qquad (7.38)$$

$$\langle U_{sp}\rangle = A_1^2 \frac{n_0 e^2}{4m_e} \frac{\omega^2 + \alpha^2 k_x^2}{\left[\omega^2 - \alpha^2 k_x^2\right]^2} . \qquad (7.39)$$

Thus, for the SPs strength function Θ_{sp} and the surface electromagnetic strength function Θ_{ph} we obtain

$$\Theta_{ph} = \frac{k_x^2 \left(\dfrac{\varepsilon_1}{\kappa_1^3} + \dfrac{\varepsilon_2}{\kappa_2^3} \right)}{k_x^2 \left(\dfrac{\varepsilon_1}{\kappa_1^3} + \dfrac{\varepsilon_2}{\kappa_2^3} \right) + \dfrac{n_0 e^2}{\varepsilon_0 m_e} \dfrac{\omega^2 + \alpha^2 k_x^2}{\left[\omega^2 - \alpha^2 k_x^2\right]^2}} , \qquad (7.40)$$

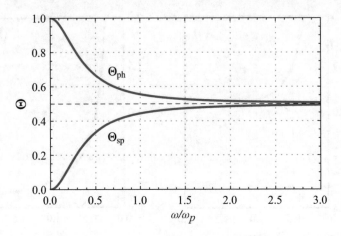

Fig. 7.10 Frequency dependence of the strength functions Θ_{sp} and Θ_{ph} of SPPs on a monolayer graphene, as given by (7.40) and (7.41), when $\alpha = 0$ and $\varepsilon_1 = 1 = \varepsilon_2$. Here $\omega_p = 4\alpha_{sc}E_F/\hbar$, where $E_F = \hbar k_F v_F$ and $\alpha_{sc} = e^2/4\pi\varepsilon_0\hbar c$

$$\Theta_{sp} = \frac{\dfrac{n_0 e^2}{\varepsilon_0 m_e}\dfrac{\omega^2 + \alpha^2 k_x^2}{\left[\omega^2 - \alpha^2 k_x^2\right]^2}}{k_x^2\left(\dfrac{\varepsilon_1}{\kappa_1^3} + \dfrac{\varepsilon_2}{\kappa_2^3}\right) + \dfrac{n_0 e^2}{\varepsilon_0 m_e}\dfrac{\omega^2 + \alpha^2 k_x^2}{\left[\omega^2 - \alpha^2 k_x^2\right]^2}}. \tag{7.41}$$

Figure 7.10 shows the variation of Θ_{ph} and Θ_{sp} with respect to the frequency of SPPs of a monolayer graphene. One can see that for very low values of ω/ω_p, a SPP of the system is totally photon-like and the role of the SP is negligibly small compared with that of the surface electromagnetic field. Only for high values of ω/ω_p the strength functions of the SP and surface electromagnetic field are comparable. Indeed, they are equal for high values of ω/ω_p.

7.1.9   Penetration Depth

To investigate the spatial extension of the electromagnetic field associated with a SPP, when $\alpha = 0$ and $\varepsilon_1 = \varepsilon = \varepsilon_2$, we use the SPPs decay constant κ perpendicular to the interface which allows to define the penetration depth $\delta = 1/\kappa$ at which the electromagnetic field falls to $1/e$. Figure 7.11 shows the penetration depth δ as a function of the frequency of the SPPs of a monolayer graphene. For large ω/ω_p (nonretarded condition), the penetration depth is $\delta = 1/k_x$ thereby leading to a strong concentration of the surface wave field on the system.

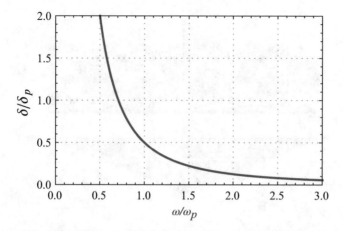

Fig. 7.11 Frequency dependence of dimensionless penetration depth δ/δ_p (with $\delta_p = c/\omega_p$) of SPPs of a monolayer graphene, when $\alpha = 0$ and $\varepsilon_1 = 1 = \varepsilon_2$. Here $\omega_p = 4\alpha_{sc}E_F/\hbar$, where $E_F = \hbar k_F v_F$ and $\alpha_{sc} = e^2/4\pi\varepsilon_0\hbar c$

7.2 Excitation of Surface Plasmon Polaritons on Monolayer Two-Dimensional Electron Gases

In Sect. 3.3, we learned that a SPP may be excited only with the evanescent wave and by the ATR method, where we related the amplitude of a SPP to the exciting incident radiation. We shall consider the ATR method in Otto's configuration to achieve light-SPP interaction in a planar 2DEG layer.[5] The scheme is illustrated in Fig. 7.12, where a monochromatic p-polarized wave in a glass prism is incident on a thin insulator spacing at the large face of the prism and at an angle of incidence greater than the critical angle for the total internal reflection. In the ATR regime, the magnetic fields of p-polarized electromagnetic waves in the three insulators with relative dielectric constants ε_1, ε_2, and ε_3 are of the form

$$H_y = -e^{ik_x x}\begin{cases} A_1 e^{ik_{1z}z} + A_2 e^{-ik_{1z}z} \,, & z \leq 0\,, \\ A_3 e^{\kappa_2 z} + A_4 e^{-\kappa_2 z}\,, & 0 \leq z \leq d\,, \\ A_5 e^{-\kappa_3 z}\,, & z \geq d\,, \end{cases} \tag{7.42}$$

where $k_{1z} = \sqrt{\varepsilon_1}k_0\cos\theta$, $\kappa_2^2 = k_x^2 - \varepsilon_2 k_0^2$, $\kappa_3^2 = k_x^2 - \varepsilon_3 k_0^2$ and one can express E_x in term of the corresponding magnetic field component H_y, using (3.1b). The four unknown amplitudes A_2–A_5 can be related to the incident amplitude A_1 through the appropriate BCs. Continuity of the tangential components of **E** and **H** at the $z = 0$ boundary and continuity and discontinuity of the tangential components of **E** and

---

[5]The Kretschmann configuration does not work for a 2DEG layer owing to its intrinsic two dimensionality [2].

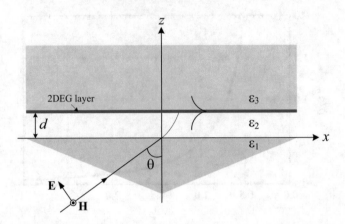

Fig. 7.12 Otto's configuration with a planar 2DEG layer for excitation of p-polarized SPPs

H, respectively, at the $z = d$ boundary lead to a linear system of four equations, which may be casted as a standard $\mathbf{Mb} = \mathbf{c}$ matrix equation, namely

$$\mathbf{M} \begin{pmatrix} A_2 \\ A_3 \\ A_4 \\ A_5 \end{pmatrix} = \begin{pmatrix} A_1 \\ A_1 \\ 0 \\ 0 \end{pmatrix}, \tag{7.43}$$

where the matrix $\mathbf{M}$ is given by

$$\mathbf{M} = \begin{pmatrix} -1 & 1 & 1 & 0 \\ 1 & -i\dfrac{\varepsilon_1}{\varepsilon_2}\dfrac{\kappa_2}{k_{1z}} & i\dfrac{\varepsilon_1}{\varepsilon_2}\dfrac{\kappa_2}{k_{1z}} & 0 \\ 0 & e^{\kappa_2 d} & e^{-\kappa_2 d} & -\left(1 + i\dfrac{\sigma_{xx}}{\omega\varepsilon_0\varepsilon_3}\kappa_3\right)e^{-\kappa_3 d} \\ 0 & e^{\kappa_2 d} & -e^{-\kappa_2 d} & \dfrac{\varepsilon_2}{\varepsilon_3}\dfrac{\kappa_3}{\kappa_2}e^{-\kappa_3 d} \end{pmatrix}. \tag{7.44}$$

Now, using Cramer's rule, we can find the reflection coefficient from the following quotient of determinants:

$$r_p = \frac{A_2}{A_1} = \frac{1}{A_1}\frac{\det \mathbf{m}}{\det \mathbf{M}}, \tag{7.45}$$

where the matrix $\mathbf{m}$ is built by replacing the first column of $\mathbf{M}$ by the column vector $(A_1, A_1, 0, 0)^T$, that is, the vector on the right-hand side of (7.43). After some algebra one arrives to the following expression for the reflection coefficient,

as (3.71) where the Fresnel reflection coefficients, with the 12 and 23 subscripts for glass-insulator and insulator-2DEG-insulator boundaries, respectively, are given by

$$r_{12} = \frac{\varepsilon_2 k_{1z} - i\varepsilon_1 \kappa_2}{\varepsilon_2 k_{1z} + i\varepsilon_1 \kappa_2} , \tag{7.46a}$$

$$r_{23} = \frac{\varepsilon_3 \kappa_2 - \left(\varepsilon_2 - i\dfrac{\sigma_{xx}}{\omega \varepsilon_0} \kappa_2\right) \kappa_3}{\varepsilon_3 \kappa_2 + \left(\varepsilon_2 + i\dfrac{\sigma_{xx}}{\omega \varepsilon_0} \kappa_2\right) \kappa_3} , \tag{7.46b}$$

and the expression for the reflectance coefficient across the 2DEG layer is given by

$$R_p = |r_p|^2 . \tag{7.47}$$

Let us note that if $\mathrm{Re}[\sigma_{xx}(\omega)] = 0$, that is, if the conductivity of the 2DEG layer is purely imaginary, then r_{23} is real, so that the denominator in (7.47) is simply the complex conjugate of the numerator, and therefore one would obtain $R_p = 1$, meaning that all the impinging electromagnetic radiation is reflected. An example of excitation and detection of SPPs of a monolayer graphene by means of prism coupling in Otto's configuration is depicted in Fig. 7.13 by varying the impinging angle at different fixed frequencies.

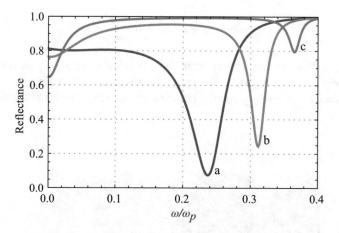

Fig. 7.13 Excitation and detection of SPPs on a monolayer graphene surface using the prism coupling technique in Otto's configuration by sweeping the light frequency at a fixed impinging angle θ. The different curves refer to (**a**) $\theta = 50°$, (**b**) $\theta = 60°$, and (**c**) $\theta = 72°$. The minimum in the reflectance coefficient, as given by (7.47), pinpoints the SPPs resonant frequency. The parameters are $E_F = 0.4\,\mathrm{eV}$, $\hbar\gamma = 0.1\,\mathrm{meV}$, $\varepsilon_1 = 14$, $\varepsilon_2 = 4 = \varepsilon_3$, and $d = 40\,\mu\mathrm{m}$

7.3   Plasmonic Properties of Monolayer Two-Dimensional Electron Gases: Static Magnetic Field Effect

7.3.1   Dispersion Relation

We consider an infinitesimally thin 2DEG layer in equilibrium with the surface density (per unit area) of n_0 that is uniformly distributed in the xy plane in the usual Cartesian coordinate system. We assume an external static magnetic field $\mathbf{B}_0 = B_0 \mathbf{e}_z$ that is normal to the 2DEG layer (perpendicular configuration). Referring to Fig. 7.14, we suppose that the 2DEG layer is sandwiched between two semi-infinite insulator media, characterized by the real relative dielectric constants ε_1 and ε_2. The medium 1 occupies $z < 0$ half-space and $z > 0$ half-space is occupied by medium 2.

We assume a SMPP that propagates parallel to the interface $z = 0$ along the x-direction. As mentioned in the Sect. 3.4, SMPPs are no longer p-polarized or s-polarized waves in general. Based on the linearized magneto-hydrodynamic theory, that is usually valid in the low-frequency and long-wavelength limits, the electronic excitations of a 2DEG layer may be described by the following linearized equations [20]:

$$\left(\omega^2 - \alpha^2 k_x^2\right) J_x + i\omega\omega_c J_y = i\omega \frac{n_0 e^2}{m_e} E_x , \tag{7.48a}$$

$$- i\omega\omega_c J_x + \omega^2 J_y = i\omega \frac{n_0 e^2}{m_e} E_y , \tag{7.48b}$$

where k_x and ω are the wavenumber and the angular frequency of the SMPPs, respectively, $\omega_c = eB_0/m_e$ is the cyclotron frequency of an electron and $J_x = -en_0 v_x$ and $J_y = -en_0 v_y$ are the polarization current densities along the x- and

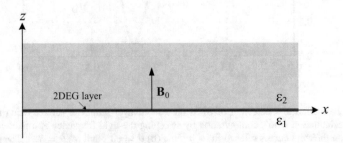

Fig. 7.14 Illustration of a planar 2DEG layer sandwiched between two semi-infinite insulators with relative dielectric constants ε_1 and ε_2 referring, respectively, to the top and bottom insulators. The 2DEG layer is located at the $z = 0$ plane and static magnetic field $\mathbf{B}_0 = B_0 \mathbf{e}_z$ that is normal to the 2DEG

y-directions, respectively, due to the motion of 2DEG. Using (7.48a) and (7.48b), we can find the conductivity of the system that is a 2D tensor, consisting of four factors or components as

$$\underline{\sigma}(\omega, k_x) = \frac{i n_0 e^2/m_e}{\omega^2 - \omega_c^2 - \alpha^2 k_x^2} \begin{pmatrix} \omega^2 & -i\omega\omega_c \\ i\omega\omega_c & \omega^2 - \alpha^2 k_x^2 \end{pmatrix}. \tag{7.49}$$

The SMPPs are solutions of Maxwell's equations with standard BCs at $z = 0$. The explicit forms for the E_y, E_z, H_y, and H_z components of the SMPPs can be expressed in terms of E_x and H_x components. The relevant solutions of E_x and H_x are

$$E_x = e^{ik_x x} \begin{cases} A_1 e^{+\kappa_1 z}, & z \le 0, \\ A_2 e^{-\kappa_2 z}, & z \ge 0, \end{cases} \tag{7.50}$$

$$H_x = e^{ik_x x} \begin{cases} B_1 e^{+\kappa_1 z}, & z \le 0, \\ B_2 e^{-\kappa_2 z}, & z \ge 0, \end{cases} \tag{7.51}$$

where $\kappa_j^2 = k_x^2 - \varepsilon_j \omega^2/c^2$ (with $j = 1, 2$). Also A_1 and B_1 (A_2 and B_2) are the amplitudes of the electric and magnetic field components in the region 1 (region 2), respectively. All the other components may then be determined using Maxwell's equations as

$$E_y = (-1)^j \frac{i\omega\mu_0}{\kappa_j} H_x, \qquad E_z = (-1)^j \frac{ik_x}{\kappa_j} E_x, \tag{7.52}$$

$$H_y = (-1)^{j+1} \frac{i\omega\varepsilon_0 \varepsilon_j}{\kappa_j} E_x, \qquad H_z = (-1)^j \frac{ik_x}{\kappa_j} H_x. \tag{7.53}$$

The field components have to satisfy the usual BCs at $z = 0$, that is, continuity of E_x and E_y and discontinuity of H_x and H_y due to the polarization of the 2DEG layer. Imposing the mentioned BCs gives

$$A_2 = A_1, \tag{7.54}$$

$$B_2 = -\frac{\kappa_2}{\kappa_1} B_1, \tag{7.55}$$

$$\frac{A_1}{B_1} = \Upsilon \frac{\eta_0/\kappa_1}{\dfrac{\varepsilon_1}{\kappa_1} + \dfrac{\varepsilon_2}{\kappa_2} - \dfrac{1}{\eta_0\omega\varepsilon_0} \Xi}, \tag{7.56}$$

$$\frac{B_1}{A_1} = -\Upsilon \frac{\kappa_1/\eta_0}{\kappa_1 + \kappa_2 + \dfrac{\omega\mu_0}{\eta_0}\Pi}, \tag{7.57}$$

where

$$\Xi = \frac{\omega \omega_p}{\omega^2 - \alpha^2 k_x^2 - \omega_c^2} ,$$

$$\Pi = \frac{\left[\omega^2 - \alpha^2 k_x^2\right] \omega_p}{\omega \left(\omega^2 - \alpha^2 k_x^2\right) - \omega \omega_c^2} ,$$

$$\Upsilon = \frac{\omega_c \omega_p}{\omega^2 - \alpha^2 k_x^2 - \omega_c^2} ,$$

with $\eta_0 = \sqrt{\mu_0/\varepsilon_0}$ that is the vacuum impedance and $\omega_p = 4\alpha_{sc} E_F/\hbar$, where $E_F = \hbar k_F v_F$ and $\alpha_{sc} = e^2/4\pi\varepsilon_0\hbar c$ that is the fine structure constant [21]. By combining (7.56) and (7.57), it is straightforward to obtain the general dispersion relation as

$$\left[\frac{\varepsilon_1}{\kappa_1} + \frac{\varepsilon_2}{\kappa_2} - \frac{\Xi}{\eta_0\omega\varepsilon_0}\right] \left[\kappa_1 + \kappa_2 + \Pi \frac{\omega\mu_0}{\eta_0}\right] = -\Upsilon^2 , \qquad (7.58)$$

that is a transcendental equation and needs to be solved numerically in order to obtain the spectrum of SMPPs. We note that the above dispersion relation gives us information on the SMPPs in order to obtain the power flow, energy density, energy velocity, wave polarization, transverse and longitudinal fields strength functions, and penetration depth of SMPPs. Clearly in a particular limit, by considering the case $\omega \gg \omega_c$ [22], from (7.58) we may find

$$\frac{\varepsilon_1}{\kappa_1} + \frac{\varepsilon_2}{\kappa_2} = \frac{c\omega_p}{\omega^2 - \omega_c^2 - \alpha^2 k_x^2} \qquad (7.59)$$

that shows the result obtained in [22]. However, for $\omega_c \geq \omega$ the dispersion relation of SMPPs in the present system becomes more intricate, obeying (7.58). In the following, we solve (7.58) in the case where the two insulator media are equal, i.e., $\varepsilon_1 = \varepsilon = \varepsilon_2$ and $\kappa^2 = k_x^2 - \varepsilon\omega^2/c^2$. We obtain

$$\kappa = \frac{\omega}{c} \frac{\left(4\varepsilon + \Upsilon^2 - \Xi\Pi\right) \pm \sqrt{\left(4\varepsilon + \Upsilon^2 - \Xi\Pi\right)^2 + 16\varepsilon\Xi\Pi}}{4\Xi} \qquad (7.60)$$

that shows the hybridization between plasmonic oscillations of the 2DEG system and cyclotronic excitations with frequency ω_c, driven by the static magnetic field. We note that for having SMPPs, the symbols $+$ and $-$ should be chosen for $\omega \geq \omega_c$ and $\omega \leq \omega_c$, respectively. The dispersion curves in terms of the dimensionless variables for the SMPPs on a monolayer graphene are presented in Fig. 7.15 for different values of ω_c/ω_p. These results could indicate that SMPPs propagating along a monolayer graphene may be profoundly influenced by the presence of an impressed magnetic field.

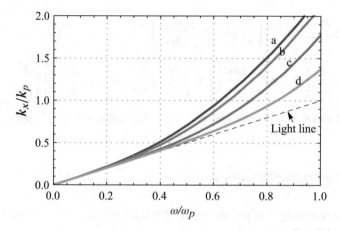

Fig. 7.15 Dispersion curves of SMPPs of a monolayer graphene as given by (7.58) in the perpendicular configuration when $\varepsilon_1 = 1 = \varepsilon_2$ and $n_0 = 9 \times 10^{13} \text{cm}^{-2}$. The different curves refer to (**a**) $\omega_c/\omega_p = 0$, (**b**) $\omega_c/\omega_p = 0.3$, (**c**) $\omega_c/\omega_p = 0.6$, and (**d**) $\omega_c/\omega_p = 0.9$. Here $k_p = \omega_p/c$ and $\omega_p = 4\alpha_{sc} E_F/\hbar$, where $E_F = \hbar k_F v_F$ and $\alpha_{sc} = e^2/4\pi\varepsilon_0\hbar c$

7.3.2 Power Flow

For the cycle-averaged x-component of power flow density associated with a SMPP on a 2DEG layer, in the three media, we have

$$
S_x = \frac{1}{2}\text{Re}\left\{
\begin{array}{ll}
E_{1y}H_{1z}^* - E_{1z}H_{1y}^*, & z \leq 0, \\
m_e\alpha^2 n v_x, & z = 0, \\
E_{2y}H_{2z}^* - E_{2z}H_{2y}^*, & z \geq 0,
\end{array}
\right.
\tag{7.61}
$$

where subscripts 1 and 2 denote the regions below and above the 2DEG layer, respectively. After the elimination of v_x, n, and electric and magnetic fields, we obtain

$$
S_x = \frac{\varepsilon_0\omega k_x}{2}|A_1|^2\left\{
\begin{array}{ll}
\dfrac{1}{\kappa_1^2}\left(\varepsilon_1 + \eta_0^2\dfrac{|B_1|^2}{|A_1|^2}\right)e^{2\kappa_1 z}, & z \leq 0, \\[3mm]
\alpha^2\dfrac{m_e}{\mu_0 n_0 e^2}\dfrac{1}{\omega^2}\left|\Xi + \Upsilon\dfrac{\omega\mu_0}{\kappa_1}\dfrac{B_1}{A_1}\right|^2, & z = 0, \\[3mm]
\dfrac{1}{\kappa_2^2}\left(\varepsilon_2 + \eta_0^2\dfrac{\kappa_2^2}{\kappa_1^2}\dfrac{|B_1|^2}{|A_1|^2}\right)e^{-2\kappa_2 z}, & z \geq 0.
\end{array}
\right.
\tag{7.62}
$$

The total power flow density associated with the SMPPs is determined by an integration over z. The power flow through an area in the yz plane of infinite length in the z-direction and unit width in the y-direction is

$$\langle S_x \rangle = \frac{\varepsilon_0 \omega k_x}{4} |A_1|^2 \left\{ \frac{1}{\kappa_1^3} \left(\varepsilon_1 + \eta_0^2 \frac{|B_1|^2}{|A_1|^2} \right) \right.$$

$$\left. + \frac{1}{\kappa_2^3} \left(\varepsilon_2 + \eta_0^2 \frac{\kappa_2^2}{\kappa_1^2} \frac{|B_1|^2}{|A_1|^2} \right) + \frac{m_e}{\mu_0 n_0 e^2} \frac{2\alpha^2}{\omega^2} \left| \Xi + \Upsilon \frac{\omega \mu_0}{\kappa_1} \frac{B_1}{A_1} \right|^2 \right\}. \qquad (7.63)$$

This total power flow density (per unit width) is positive when k_x is positive.

7.3.3 Energy Distribution

For the cycle-averaged of energy density distribution associated with the SMPPs of a 2DEG layer, we have

$$U = \frac{1}{4} \begin{cases} \varepsilon_0 \varepsilon_1 |\mathbf{E}_1|^2 + \mu_0 |\mathbf{H}_1|^2 \,, & z \leq 0 \,, \\ m_e n_0 \left(|v_x|^2 + |v_y|^2 \right) + \dfrac{m_e}{n_0} \alpha^2 |n|^2 \,, & z = 0 \,, \\ \varepsilon_0 \varepsilon_2 |\mathbf{E}_2|^2 + \mu_0 |\mathbf{H}_2|^2 \,, & z \geq 0 \,, \end{cases} \qquad (7.64)$$

that yields

$$U = \frac{\varepsilon_0}{4} |A_1|^2$$

$$\begin{cases} \left[\varepsilon_1 \left(1 + \dfrac{\omega^2 \mu_0^2}{\kappa_1^2} \dfrac{|B_1|^2}{|A_1|^2} + \dfrac{k_x^2}{\kappa_1^2} \right) + \eta_0^2 \left(\dfrac{|B_1|^2}{|A_1|^2} + \dfrac{\omega^2 \varepsilon_0^2 \varepsilon_1^2}{\kappa_1^2} + \dfrac{k_x^2}{\kappa_1^2} \dfrac{|B_1|^2}{|A_1|^2} \right) \right] e^{+2\kappa_1 z} \,, \\[4mm] \dfrac{m_e}{n_0 e^2 \mu_0^2} \left[\left| \Upsilon + \Pi \dfrac{\omega \mu_0}{\kappa_1} \dfrac{B_1}{A_1} \right|^2 + \left(1 + \dfrac{\alpha^2 k_x^2}{\omega^2} \right) \left| \Xi + \Upsilon \dfrac{\omega \mu_0}{\kappa_1} \dfrac{B_1}{A_1} \right|^2 \right] \,, \\[4mm] \left[\varepsilon_2 \left(1 + \dfrac{\omega^2 \mu_0^2}{\kappa_1^2} \dfrac{|B_1|^2}{|A_1|^2} + \dfrac{k_x^2}{\kappa_2^2} \right) + \eta_0^2 \left(\dfrac{\kappa_2^2}{\kappa_1^2} \dfrac{|B_1|^2}{|A_1|^2} + \dfrac{\omega^2 \varepsilon_0^2 \varepsilon_2^2}{\kappa_2^2} + \dfrac{k_x^2}{\kappa_1^2} \dfrac{|B_1|^2}{|A_1|^2} \right) \right] e^{-2\kappa_2 z} \,, \end{cases}$$

$$(7.65)$$

for $z \leq 0$, $z = 0$ and $z \geq 0$, respectively. The total energy density associated with the SMPPs is again determined by integration over z, the energy per unit surface area being

$$\langle U \rangle = \frac{1}{8} \varepsilon_0 |A_1|^2 \left\{ \frac{\varepsilon_1}{\kappa_1} \left(1 + \frac{\omega^2 \mu_0^2}{\kappa_1^2} \frac{|B_1|^2}{|A_1|^2} + \frac{k_x^2}{\kappa_1^2} \right) \right.$$

$$+\frac{1}{\kappa_1}\eta_0^2\left(\frac{|B_1|^2}{|A_1|^2}+\frac{\omega^2\varepsilon_0^2\varepsilon_1^2}{\kappa_1^2}+\frac{k_x^2}{\kappa_1^2}\frac{|B_1|^2}{|A_1|^2}\right)$$

$$+\frac{1}{\kappa_2}\left[\varepsilon_2\left(1+\frac{\omega^2\mu_0^2}{\kappa_1^2}\frac{|B_1|^2}{|A_1|^2}+\frac{k_x^2}{\kappa_2^2}\right)+\eta_0^2\left(\frac{\kappa_2^2}{\kappa_1^2}\frac{|B_1|^2}{|A_1|^2}+\frac{\omega^2\varepsilon_0^2\varepsilon_2^2}{\kappa_2^2}+\frac{k_x^2}{\kappa_1^2}\frac{|B_1|^2}{|A_1|^2}\right)\right]$$

$$+2\frac{m_e}{n_0e^2\mu_0^2}\left[\left|\Upsilon+\Pi\frac{\omega\mu_0}{\kappa_1}\frac{B_1}{A_1}\right|^2+\left(1+\frac{\alpha^2k_x^2}{\omega^2}\right)\left|\Xi+\Upsilon\frac{\omega\mu_0}{\kappa_1}\frac{B_1}{A_1}\right|^2\right]\right\}.$$

$$(7.66)$$

7.3.4 Energy Velocity

Again, the energy velocity of the SMPPs is given as the ratio of the total power flow density (per unit width) and total energy density (per unit area), such as $v_e = \langle S_x\rangle / \langle U\rangle$. The computed frequency variation of the phase and energy (group) velocities of SMPPs of a monolayer graphene is depicted in the dimensionless variables in Fig. 7.16 for different values of ω_c/ω_p. In general, it can be seen that the phase and energy velocities are less than the velocity of light, and it can be observed that by increasing values of ω/ω_p, the phase and energy velocities decrease sharply. It is also interesting to note that in the nonretarded condition (large values of ω/ω_p) the phase velocity is twice the energy velocity.

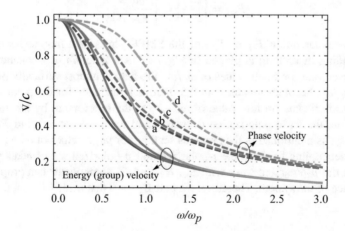

Fig. 7.16 Frequency variation of the phase velocity (dashed lines) and energy (group) velocity (solid lines) of SMPPs of a monolayer graphene in the perpendicular configuration when $\varepsilon_1 = 1 = \varepsilon_2$. The different curves refer to (**a**) $\omega_c/\omega_p = 0$, (**b**) $\omega_c/\omega_p = 0.3$, (**c**) $\omega_c/\omega_p = 0.6$, and (**d**) $\omega_c/\omega_p = 0.9$

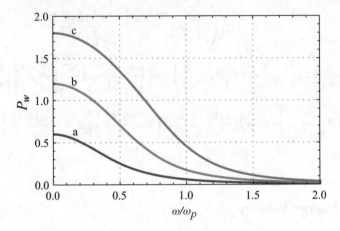

Fig. 7.17 Frequency variation of SMPP polarization of a monolayer graphene in the perpendicular configuration when $\varepsilon_1 = 1 = \varepsilon_2$. The different curves refer to (**a**) $\omega_c/\omega_p = 0.3$, (**b**) $\omega_c/\omega_p = 0.6$, and (**c**) $\omega_c/\omega_p = 0.9$

7.3.5 Wave Polarization

In the presence of a static external magnetic field, the wave polarization of the system may be discussed in a different way from previous formula in Sect. 3.2.4. Here, we may define

$$P_{wj} = \left| \frac{E_{jy}}{E_{jz}} \right| = \left| \frac{\mu_0 \omega}{k_x} \frac{B_j}{A_j} \right| . \tag{7.67}$$

The wave polarization $P_w = P_{wj}$ of the SMPP waves of a monolayer graphene free-standing in vacuum is plotted in Fig. 7.17 as a function of frequency. It can be observed that for small values of ω_c/ω_p the field is almost vertically polarized. When ω_c/ω_p becomes greater than 0.5 there is a tendency for the polarization to change to horizontal in low value of ω/ω_p. On the other hand, by increasing the value of ω, the wave polarization decreases and for $\omega \to \infty$ we find $P_w \to 0$. Physically, as mentioned in the previous section, for large values of ω, where $\omega \gg \omega_c$, the decoupling between the s-polarized and p-polarized waves takes place. In this case the p-polarized wave with E_x, H_x, and E_z components can propagate on the surface of the system. Therefore, we find $E_y \to 0$ that means $P_w \to 0$.

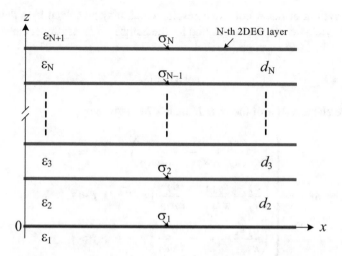

Fig. 7.18 Schematic representation of a N-layer 2DEG. The SPPs are propagating in the x-direction. The jth layer (with $1 \leq j \leq N$) sandwiched between two insulators with relative dielectric constants ε_j and ε_{j+1} referring, respectively, to the top and bottom dielectrics. Each insulator medium with relative dielectric constant ε_j has a thickness d_j

7.4   Dispersion Relation of Multilayer Two-Dimensional Electron Gases

Assume that a multilayer planar 2DEG system consists of N-layer 2DEG, as shown schematically in Fig. 7.18. Each layer such as j (with $1 \leq j \leq N$) is described by a conductivity σ_j. Also, the jth layer sandwiched between two insulators with relative dielectric constants ε_j and ε_{j+1} referring, respectively, to the bottom and top dielectrics. Each insulator medium with relative dielectric constant ε_j has a thickness d_j. Now we consider the propagation of a p-polarized SPP in the x-direction, so that the propagation along the x-direction is described by the multiplier $\exp(ik_x x)$, where k_x is the wavenumber. The solution of Maxwell's equations in a insulator layer with relative dielectric constant ε_j for the p-polarized SPPs yields

$$E_x^{(j)} = \left[A_+^{(j)} \exp(\kappa_j z) + A_-^{(j)} \exp(-\kappa_j z) \right] e^{ik_x x} , \qquad (7.68)$$

$$H_y^{(j)} = \frac{i\omega\varepsilon_0\varepsilon_j}{\kappa_j} \left[A_+^{(j)} \exp(\kappa_j z) - A_-^{(j)} \exp(-\kappa_j z) \right] e^{ik_x x} , \qquad (7.69)$$

where $\kappa_j^2 = k_x^2 - \varepsilon_j k_0^2$ and the meaning of the coefficients $A_\pm^{(j)}$ is different for the different media. The field components have to satisfy the usual BCs at each EG layer, that are, continuity of $E_x^{(j)}$ and discontinuity of $H_y^{(j)}$ due to the polarization of the 2DEG on the jth layer. Matching these BCs to a multilayer 2DEG system, we obtain a system of linear equations results which may be solved by a determinant method to yield the dispersion relation for the number of layers under consideration. For a N-layer 2DEG system, we find

$$\det \mathbf{M} = 0 , \tag{7.70}$$

where the element M_{ji} of the $N \times N$ matrix $\mathbf{M}$ is given by

$$
M_{ji} =
\begin{cases}
\left(\dfrac{\varepsilon_j}{\kappa_j} + \dfrac{\varepsilon_{j+1}}{\kappa_{j+1}} - \dfrac{\sigma_j}{i\omega\varepsilon_0} \right), & j = i, \\[2ex]
\left(\dfrac{\varepsilon_j}{\kappa_j} - \dfrac{\varepsilon_{j+1}}{\kappa_{j+1}} - \dfrac{\sigma_j}{i\omega\varepsilon_0} \right) e^{-\sum\limits_{j=2}^{i} \delta_j}, & j < i, \\[3ex]
\left(\dfrac{\varepsilon_{j+1}}{\kappa_{j+1}} - \dfrac{\varepsilon_j}{\kappa_j} - \dfrac{\sigma_j}{i\omega\varepsilon_0} \right) e^{-\sum\limits_{i=2}^{j} \delta_i}, & j > i,
\end{cases}
\tag{7.71}
$$

where $\delta_j = \kappa_j d_j$, $\delta_i = \kappa_i d_i$. It should be mentioned that (7.70) and (7.71) are of forms which easily lend themselves to programming of the general equation on a computer. This equation allows one to vary the number of layers at will. From (7.70) we obtain N positive roots for ω defining the SPPs dispersions which are clearly separated into a high-frequency ω_+, and a low-frequency, ω_- group.

7.4.1 Special Cases

7.4.1.1 Monolayer Electron Gas

For a 2DEG layer (see Fig. 7.18, for $N = 1$) with conductivity σ_1 that is sandwiched between two semi-infinite insulator media, characterized by the real relative dielectric constants ε_1 and ε_2, from (7.70) and (7.71) we find

$$\frac{\varepsilon_1}{\kappa_1} + \frac{\varepsilon_2}{\kappa_2} - \frac{\sigma_1}{i\omega\varepsilon_0} = 0 . \tag{7.72}$$

7.4.1.2 Bilayer Electron Gas

For a two-layer 2DEG (bilayer 2DEG, where $N = 2$) with conductivities σ_1 and σ_2 that are separated by an insulator material of thickness d_2 and relative dielectric constant ε_2, we have [23–25]

$$
\begin{vmatrix}
\left(\dfrac{\varepsilon_1}{\kappa_1} + \dfrac{\varepsilon_2}{\kappa_2} - \dfrac{\sigma_1}{i\omega\varepsilon_0}\right) & \left(\dfrac{\varepsilon_1}{\kappa_1} - \dfrac{\varepsilon_2}{\kappa_2} - \dfrac{\sigma_1}{i\omega\varepsilon_0}\right)e^{-\delta_2} \\[3mm]
\left(\dfrac{\varepsilon_3}{\kappa_3} - \dfrac{\varepsilon_2}{\kappa_2} - \dfrac{\sigma_2}{i\omega\varepsilon_0}\right)e^{-\delta_2} & \left(\dfrac{\varepsilon_2}{\kappa_2} + \dfrac{\varepsilon_3}{\kappa_3} - \dfrac{\sigma_2}{i\omega\varepsilon_0}\right)
\end{vmatrix} = 0 , \qquad (7.73)
$$

where in the case of $\sigma_2 = 0$, we find the result in [26]. From the above dispersion relation, we obtain two branches for ω defining the resonant frequencies of p-polarized SPPs with one branch having higher frequency than the bare spectrum for the 2DEG layer and other branch that is below the dispersion curve of a planar 2DEG layer.

7.4.1.3   Triple-Layer Electron Gas

For triple-layer 2DEG (see Fig. 7.18, for $N = 3$) with conductivities σ_1, σ_2, and σ_3 that are separated by insulator materials of thickness d_2 and d_2, we find

$$
\begin{vmatrix}
\left(\dfrac{\varepsilon_1}{\kappa_1}+\dfrac{\varepsilon_2}{\kappa_2}-\dfrac{\sigma_1}{i\omega\varepsilon_0}\right) & \left(\dfrac{\varepsilon_1}{\kappa_1}-\dfrac{\varepsilon_2}{\kappa_2}-\dfrac{\sigma_1}{i\omega\varepsilon_0}\right)e^{-\delta_2} & \left(\dfrac{\varepsilon_1}{\kappa_1}-\dfrac{\varepsilon_2}{\kappa_2}-\dfrac{\sigma_1}{i\omega\varepsilon_0}\right)e^{-\delta_2-\delta_3} \\[3mm]
\left(\dfrac{\varepsilon_3}{\kappa_3}-\dfrac{\varepsilon_2}{\kappa_2}-\dfrac{\sigma_2}{i\omega\varepsilon_0}\right)e^{-\delta_2} & \left(\dfrac{\varepsilon_2}{\kappa_2}+\dfrac{\varepsilon_3}{\kappa_3}-\dfrac{\sigma_2}{i\omega\varepsilon_0}\right) & \left(\dfrac{\varepsilon_2}{\kappa_2}-\dfrac{\varepsilon_3}{\kappa_3}-\dfrac{\sigma_2}{i\omega\varepsilon_0}\right)e^{-\delta_3} \\[3mm]
\left(\dfrac{\varepsilon_4}{\kappa_4}-\dfrac{\varepsilon_3}{\kappa_3}-\dfrac{\sigma_3}{i\omega\varepsilon_0}\right)e^{-\delta_2-\delta_3} & \left(\dfrac{\varepsilon_4}{\kappa_4}-\dfrac{\varepsilon_3}{\kappa_3}-\dfrac{\sigma_3}{i\omega\varepsilon_0}\right)e^{-\delta_3} & \left(\dfrac{\varepsilon_3}{\kappa_3}+\dfrac{\varepsilon_4}{\kappa_4}-\dfrac{\sigma_3}{i\omega\varepsilon_0}\right)
\end{vmatrix} = 0 ,
$$

$$(7.74)$$

where in the case of $\sigma_1 = \sigma_2 = \sigma_3 = 0$, we find the result in [27, 28].

7.4.1.4   Quadruple-Layer Electron Gas

For quadruple-layer 2DEG (i.e., $N = 4$) with conductivities σ_1–σ_4 that are separated by insulator materials of thickness d_2–d_3, from (7.70) and (7.71) we find

$$
\begin{vmatrix}
\left(\dfrac{\varepsilon_1}{\kappa_1}+\dfrac{\varepsilon_2}{\kappa_2}-\dfrac{\sigma_1}{i\omega\varepsilon_0}\right) & \left(\dfrac{\varepsilon_1}{\kappa_1}-\dfrac{\varepsilon_2}{\kappa_2}-\dfrac{\sigma_1}{i\omega\varepsilon_0}\right)e^{-\delta_2} & \left(\dfrac{\varepsilon_1}{\kappa_1}-\dfrac{\varepsilon_2}{\kappa_2}-\dfrac{\sigma_1}{i\omega\varepsilon_0}\right)e^{-\delta_2-\delta_3} & \left(\dfrac{\varepsilon_1}{\kappa_1}-\dfrac{\varepsilon_2}{\kappa_2}-\dfrac{\sigma_1}{i\omega\varepsilon_0}\right)e^{-\delta_2-\delta_3-\delta_4} \\[3mm]
\left(\dfrac{\varepsilon_3}{\kappa_3}-\dfrac{\varepsilon_2}{\kappa_2}-\dfrac{\sigma_2}{i\omega\varepsilon_0}\right)e^{-\delta_2} & \left(\dfrac{\varepsilon_2}{\kappa_2}+\dfrac{\varepsilon_3}{\kappa_3}-\dfrac{\sigma_2}{i\omega\varepsilon_0}\right) & \left(\dfrac{\varepsilon_2}{\kappa_2}-\dfrac{\varepsilon_3}{\kappa_3}-\dfrac{\sigma_2}{i\omega\varepsilon_0}\right)e^{-\delta_3} & \left(\dfrac{\varepsilon_2}{\kappa_2}-\dfrac{\varepsilon_3}{\kappa_3}-\dfrac{\sigma_2}{i\omega\varepsilon_0}\right)e^{-\delta_3-\delta_4} \\[3mm]
\left(\dfrac{\varepsilon_4}{\kappa_4}-\dfrac{\varepsilon_3}{\kappa_3}-\dfrac{\sigma_3}{i\omega\varepsilon_0}\right)e^{-\delta_2-\delta_3} & \left(\dfrac{\varepsilon_4}{\kappa_4}-\dfrac{\varepsilon_3}{\kappa_3}-\dfrac{\sigma_3}{i\omega\varepsilon_0}\right)e^{-\delta_3} & \left(\dfrac{\varepsilon_3}{\kappa_3}+\dfrac{\varepsilon_4}{\kappa_4}-\dfrac{\sigma_3}{i\omega\varepsilon_0}\right) & \left(\dfrac{\varepsilon_3}{\kappa_3}-\dfrac{\varepsilon_4}{\kappa_4}-\dfrac{\sigma_3}{i\omega\varepsilon_0}\right)e^{-\delta_4} \\[3mm]
\left(\dfrac{\varepsilon_5}{\kappa_5}-\dfrac{\varepsilon_4}{\kappa_4}-\dfrac{\sigma_4}{i\omega\varepsilon_0}\right)e^{-\delta_2-\delta_3-\delta_4} & \left(\dfrac{\varepsilon_5}{\kappa_5}-\dfrac{\varepsilon_4}{\kappa_4}-\dfrac{\sigma_4}{i\omega\varepsilon_0}\right)e^{-\delta_3-\delta_4} & \left(\dfrac{\varepsilon_5}{\kappa_5}-\dfrac{\varepsilon_4}{\kappa_4}-\dfrac{\sigma_4}{i\omega\varepsilon_0}\right)e^{-\delta_4} & \left(\dfrac{\varepsilon_4}{\kappa_4}+\dfrac{\varepsilon_5}{\kappa_5}-\dfrac{\sigma_4}{i\omega\varepsilon_0}\right)
\end{vmatrix} = 0 .
$$

$$(7.75)$$

7.5　Surface Plasmon Polaritons of a Superlattice of Alternating Two-Dimensional Electron Gas Layers

Again, we consider here the structure shown in Fig. 6.7, i.e., an infinite superlattice consisting of an alternating ABABAB structure. At each interface it is assumed that there is a planar 2DEG layer, whose thickness is negligible compared with a and b. For the present BVP when $z \neq nL$ and $z \neq nL + a$, the solutions of x- and y-components of electric and magnetic fields of a p-polarized SPP, respectively, have the form

$$
E_x(z) = \begin{cases} A_+ e^{+\kappa_A(z-nL)} + A_- e^{-\kappa_A(z-nL)} \,, & nL \leq z \leq nL + a \,, \\ B_+ e^{+\kappa_A(z-nL-a)} + B_- e^{-\kappa_A(z-nL-a)} \,, & nL + a \leq z \leq (n+1)L \,, \end{cases}
$$
(7.76)

$$
H_y(z) = i\varepsilon_0\omega \begin{cases} \dfrac{\varepsilon_A}{\kappa_A}\left[A_+ e^{+\kappa_A(z-nL)} - A_- e^{-\kappa_A(z-nL)}\right] \,, & nL \leq z \leq nL + a \,, \\ \dfrac{\varepsilon_B}{\kappa_B}\left[B_+ e^{+\kappa_A(z-nL-a)} - B_- e^{-\kappa_A(z-nL-a)}\right] \,, & nL+a \leq z \leq (n+1)L \,, \end{cases}
$$
(7.77)

where $\kappa_A^2 = k_x^2 - \varepsilon_A k_0^2$, $\kappa_B^2 = k_x^2 - \varepsilon_B k_0^2$, and the relations between the coefficients A_+, A_-, B_+, and B_- can be determined from the matching BCs at the separation surfaces. The BCs at $z = nL + a$ are

$$
E_{xA}\big|_{z=nL+a} = E_{xB}\big|_{z=nL+a} \,,
$$
(7.78a)

$$
H_{yB}\big|_{z=nL+a} - H_{yA}\big|_{z=nL+a} = -\sigma_{xx}E_{xB}\big|_{z=nL+a} \,,
$$
(7.78b)

where E_{xA} (H_{yA}) and E_{xB} (H_{yB}) denote the $x-$ ($y-$) component of electric (magnetic) fields in the media A and B, respectively. Also, we have

$$
E_{xA}\big|_{z=nL} = e^{-iq_zL} E_{xB}\big|_{z=(n+1)L} \,,
$$
(7.79a)

$$
H_{yA}\big|_{z=nL} - e^{-iq_zL} H_{yB}\big|_{z=(n+1)L} = -\sigma_{xx}E_{xA}\big|_{z=nL} \,,
$$
(7.79b)

where the constant q_z is the Bloch wavenumber in the Bloch theorem. Equations (7.78) and (7.79) lead to a set of four linear homogeneous equations in A_+, A_-, B_+, and B_- and with the solvability condition

$$
\begin{vmatrix} e^{\kappa_A a} & e^{-\kappa_A a} & 1 & 1 \\[2mm] \dfrac{\varepsilon_A}{\kappa_A}e^{\kappa_A a} & -\dfrac{\varepsilon_A}{\kappa_A}e^{-\kappa_A a} & \left(\dfrac{\varepsilon_B}{\kappa_B}+\dfrac{\sigma_{xx}}{i\omega\varepsilon_0}\right) & -\left(\dfrac{\varepsilon_B}{\kappa_B}-\dfrac{\sigma_{xx}}{i\omega\varepsilon_0}\right) \\[4mm] 1 & 1 & \dfrac{e^{\kappa_B b}}{e^{iq_zL}} & \dfrac{e^{-\kappa_B b}}{e^{iq_zL}} \\[4mm] \left(\dfrac{\varepsilon_A}{\kappa_A}+\dfrac{\sigma_{xx}}{i\omega\varepsilon_0}\right) & -\left(\dfrac{\varepsilon_A}{\kappa_A}-\dfrac{\sigma_{xx}}{i\omega\varepsilon_0}\right) & \dfrac{\varepsilon_B}{\kappa_B}\dfrac{e^{\kappa_B b}}{e^{iq_zL}} & -\dfrac{\varepsilon_B}{\kappa_B}\dfrac{e^{-\kappa_B b}}{e^{iq_zL}} \end{vmatrix} = 0 \,.
$$
(7.80)

The solution of this equation for ω gives the bulk plasmon dispersion relation for the alternating superlattice. It is straightforward, albeit lengthy, to show that (7.80) is equivalent to

$$\cos(q_z L) + \cosh(\kappa_A a) \left[\frac{\sigma_{xx}}{i\omega\varepsilon_0} \frac{\kappa_B}{\varepsilon_B} \sinh(\kappa_B b) - \cosh(\kappa_B b) \right]$$

$$+\sinh(\kappa_A a) \left[\frac{\sigma_{xx}}{i\omega\varepsilon_0} \frac{\kappa_A}{\varepsilon_A} \cosh(\kappa_B b) - \frac{\kappa_A^2 \varepsilon_B^2 + \kappa_B^2 \varepsilon_A^2 - \frac{\sigma_{xx}^2}{\omega^2\varepsilon_0^2}\kappa_A^2\kappa_B^2}{2\varepsilon_A\varepsilon_B\kappa_A\kappa_B} \sinh(\kappa_B b) \right] = 0.$$

$$(7.81)$$

If $\alpha = 0$ and $\gamma = 0$ in the conductivity formula of the system, i.e., (7.7), the result is equivalent to the previous result [29]. Also (7.81) simplifies considerably when the insulator media A and B are identical (i.e., the case of $b = a$ and $\varepsilon_B = \varepsilon_A$). In this case, we find

$$\cos(q_z L) - \cosh(2\kappa_A a) + \frac{\sigma_{xx}}{i\omega\varepsilon_0} \frac{\kappa_A}{\varepsilon_A} \sinh(2\kappa_A a) + \frac{\sigma_{xx}^2}{2\omega^2\varepsilon_0^2} \frac{\kappa_A^2}{\varepsilon_A^2} \sinh^2(\kappa_A a) = 0 .$$

$$(7.82)$$

References

1. A. Moradi, Damping properties of plasmonic waves on graphene. Phys. Plasmas **24**, 072114 (2017)
2. P.A.D. Goncalves, N.M.R. Peres, *An Introduction to Graphene Plasmonic* (World Scientific, Singapore, 2016)
3. M. Cardona, Fresnel reflection and surface plasmons. Am. J. Phys. **39**, 1277 (1971)
4. U. Hohenester, *Nano and Quantum Optics, an Introduction to Basic Principles and Theory* (Springer, Basel, 2020)
5. P. Yeh, *Optical Waves in Layered Media* (Wiley, New Jersey, 1998)
6. M.S. Sodha, N.C. Srivastava, *Microwave Propagation in Ferrimagnetics* (Plenum Press, New York, 1981)
7. R.H. Renard, Total reflection: a new evaluation of the Goos-Hänchen shift. J. Opt. Soc. Am. **54**, 1190–1197 (1964)
8. F. Goos, H. Hänchen, Ein neuer und fundamentaler Versuch zur Totalreflexion. Ann. Phys. **436**, 333–346 (1947)
9. K. Artmann, Berechnung der Seitenversetzung des totalreflektierten Strahles. Ann. Phys. **2**, 87–102 (1948)
10. K. Yasumoto, Y. Õishi, A new evaluation of the Goos-Hänchen shift and associated time delay. J. Appl. Phys. **54**, 2170–2176 (1983)
11. A. Moradi, Theory of Goos-Hänchen shift in graphene: energy-flux method. EPL **120**, 67002 (2017)
12. A.L. Fetter, Electrodynamics of a layered electron gas. I. single layer. Ann. Phys **81**, 367–393 (1973)

13. M. Nakayama, Theory of surface waves coupled to surface carriers. J. Phys. Soc. Jpn. **36**, 393–398 (1974)
14. S.A. Mikhailov, K. Ziegler, New electromagnetic mode in graphene. Phys. Rev. Lett. **99**, 016803 (2007)
15. A.V. Chaplik, Possible crystallization of charge carriers in low-density inversion layers. Sov. Phys. JETP **35**, 395–398 (1972)
16. R. Loudon, The propagation of electromagnetic energy through an absorbing dielectric. J. Phys. A: Gen. Phys. **3**, 233–245 (1970)
17. J. Nkoma, R. Loudon, D.R. Tille, Elementary properties of surface polaritons. J. Phys. C: Solid State Phys. **7**, 3547–3559 (1974)
18. M.S. Tomas, Z. Lenac, Thickness dependence of the surface-polariton relaxation rates in a crystal slab. Solid State Commun. **44**, 937–939 (1982)
19. L. Wendler, R. Haupt, Long-range surface plasmon-phonon-polaritons. J. Phys. C: Solid State Phys. **19**, 1871–1896 (1986)
20. A. Moradi, Energy density and energy flow of surface waves in a strongly magnetized graphene. J. Appl. Phys. **123**, 043103 (2018)
21. Y.V. Bludov, A. Ferreira, N.M.R. Peres, M.I. Vasilevskiy, A primer on surface plasmon-polaritons in graphene. Int. J. Mod. Phys. B **27**, 1341001 (2013)
22. A. Moradi, Energy density and energy flow of magnetoplasmonic waves on graphene. Solid State Commun. **253**, 63–66 (2017)
23. V.B. Jovanovic, I. Radovic, D. Borka, Z.L. Mišković, High-energy plasmon spectroscopy of freestanding multilayer graphene. Phys. Rev. B **84**, 155416 (2011)
24. P.I. Buslaev, I.V. Iorsh, I.V. Shadrivov, P.A. Belov, Y.S. Kivshar, Plasmons in waveguide structures formed by two graphene layers. JETP Lett. **97**, 535–539 (2013)
25. A. Moradi, Energy density and energy flow of plasmonic waves in bilayer graphene. Opt. Commun. **394**, 135–138 (2017)
26. A. Moradi, Plasmonic waves of graphene on a conducting substrate. J. Mod. Opt **66**, 353–357 (2019)
27. C.A. Ward, K. Bhasin, R.J. Bell, R.W. Alexander, I. Tyler, Multimedia dispersion relation for surface electromagnetic waves. J. Chem. Phys. **62**, 1674–1676 (1975)
28. F. Tao, H.F. Zhang, X.H. Yang, D. Cao, Surface plasmon polaritons of the metamaterial four-layered structures. J. Opt. Soc. Am. B **26**, 50–59 (2009)
29. N.C. Constantinou, M.G. Cottam, Bulk and surface plasmon modes in a superlattice of alternating layered electron gases. J. Phys. C **19**, 739–747 (1986)

Chapter 8
Electrostatic Problems Involving Two-Dimensional Electron Gases in Cylindrical Geometry

Abstract In this chapter, some electrostatic boundary-value problems involving two-dimensional electron gas layers in cylindrical geometry are studied. In this way, the dielectric response of a cylindrical electron gas layer is described by the standard hydrodynamic model. The main interest and the key first applications of presented boundary-value problems concern carbon nanotubes, while keeping in mind that the analysis can be applied to the other two-dimensional electron gas tubes. For brevity, in many sections of this chapter the $\exp(-i\omega t)$ time factor is suppressed. Furthermore, all media under consideration are nonmagnetic and attention is only confined to the linear phenomena.

8.1 Plasmonic Properties of Cylindrical Two-Dimensional Electron Gas Layers

8.1.1 Dispersion Relation

Let us consider a cylindrical 2DEG layer with a radius a, and assume that 2DEG distributed uniformly over the cylindrical surface, with the equilibrium density (per unit area) n_0. We take cylindrical polar coordinates $\mathbf{r} = (\rho, \phi, z)$ for an arbitrary point in space. We assume that $n(\phi, z, t)$ is the first-order perturbed density (per unit area) of the homogeneous electron fluid on the cylindrical 2DEG layer, due to the propagation of an electrostatic surface wave parallel to the cylindrical surface $\rho = a$.

Based on the SHD theory (see Sect. 1.2), the electronic excitations of the 2DEG surface can be described by Newton's equation of motion for the electrons, in conjunction with the continuity and Poisson equations. In the linear approximation, one has

The original version of this chapter was revised. The correction to this chapter is available at https://doi.org/10.1007/978-3-030-43836-4_11

A. Moradi, *Canonical Problems in the Theory of Plasmonics*, Springer Series in Optical Sciences 230, https://doi.org/10.1007/978-3-030-43836-4_8

$$\frac{\partial^2 \mathbf{u}}{\partial t^2} = \frac{e}{m_e} \nabla_\parallel \Phi \big|_{\rho=a} - \frac{\alpha^2}{n_0} \nabla_\parallel n \,, \tag{8.1}$$

$$n + n_0 \nabla_\parallel \cdot \mathbf{u} = 0 \,, \tag{8.2}$$

$$\nabla^2 \Phi(\mathbf{r}) = \begin{cases} 0 \,, & \rho \neq a \,, \\ en/\varepsilon_0 \,, & \rho = a \,, \end{cases} \tag{8.3}$$

where $\mathbf{u}(\phi, z, t) = a\phi \mathbf{e}_\phi + z\mathbf{e}_z$ is the first-order perturbed values of electrons displacement parallel to the cylindrical 2DEG layer $\rho = a$, and $\nabla_\parallel = \mathbf{e}_z(\partial/\partial z) + a^{-1}\mathbf{e}_\phi(\partial/\partial \phi)$ differentiates only tangentially to the surface of the 2DEG layer. In the right-hand side of (8.1) the first term is the force on electrons due to the tangential component of the electric field, i.e., $\mathbf{E} = -\nabla \Phi_\parallel$ evaluated at the 2DEG layer $\rho = a$, where Φ is the self-consistent electrostatic potential. The second term is the force due to the internal interaction in the 2DEG, with $\alpha = v_F/\sqrt{2}$ (v_F is the Fermi velocity as before) that is the propagation speed of density disturbances in a uniform 2D homogeneous electron fluid.

Now, after elimination of the displacement $\mathbf{u}(\phi, z, t)$ from (8.1) and (8.2) we obtain

$$\left(\omega^2 + \alpha^2 \nabla_\parallel^2\right) n = \frac{en_0}{m_e} \nabla_\parallel^2 \Phi \big|_{\rho=a} \,. \tag{8.4}$$

Due to the cylindrical symmetry of the present system, one can replace the quantities n and Φ in (8.3) and (8.4) by expressions of the form

$$\Phi(\rho, \phi, z) = \tilde{\Phi}(\rho) \exp(iqz) \exp(im\phi) \,, \tag{8.5}$$

$$n(\phi, z) = \tilde{n} \exp(iqz) \exp(im\phi) \,, \tag{8.6}$$

for a component of the above quantities. Also, m is the azimuthal quantum number, and q is the longitudinal wavenumber in the z-direction. Furthermore, for simplicity, in (8.5) and (8.6) we have dropped the explicit dependencies of Φ and n on q and m. This means that the eigenmodes can be classified by a continuous index q and a discrete index m. After substitution (8.5) and (8.6) in (8.3) and (8.4), one finds

$$\frac{d^2 \tilde{\Phi}(\rho)}{d\rho^2} + \frac{1}{\rho} \frac{d\tilde{\Phi}(\rho)}{d\rho} - \left(q^2 + \frac{m^2}{\rho^2}\right) \tilde{\Phi}(\rho) = 0 \tag{8.7}$$

and

$$\tilde{n} = -\frac{en_0}{m_e} \frac{\left(q^2 + \dfrac{m^2}{a^2}\right)}{\omega^2 - \alpha^2 \left(q^2 + \dfrac{m^2}{a^2}\right)} \tilde{\Phi} \big|_{\rho=a} \,. \tag{8.8}$$

The solution of (8.7) has the form

$$
\tilde{\Phi}(\rho) = \begin{cases} A_1 \dfrac{I_m(q\rho)}{I_m(qa)} \, , & \rho \leq a \, , \\[2ex] A_2 \dfrac{K_m(q\rho)}{K_m(qa)} \, , & \rho \geq a \, , \end{cases}
\tag{8.9}
$$

where A_1 and A_2 are constants. For the present case, the BCs are

$$
\Phi_1\big|_{\rho=a} = \Phi_2\big|_{\rho=a} \, ,
\tag{8.10a}
$$

$$
\varepsilon_2 \frac{\partial \Phi_2}{\partial \rho}\bigg|_{\rho=a} - \varepsilon_1 \frac{\partial \Phi_1}{\partial \rho}\bigg|_{\rho=a} = \frac{en}{\varepsilon_0} \, ,
\tag{8.10b}
$$

where subscript 1 denotes the region inside the cylindrical 2DEG layer and subscript 2 denotes the region outside the system. In writing (8.10b), we assumed that the relative dielectric constant of the region inside the cylindrical 2DEG layer is ε_1 and the relative dielectric constant of the surrounding medium is ε_2. On applying the electrostatic BCs at $\rho = a$, the SP dispersion relation for the system is given as

$$
\omega^2 = \alpha^2 \left(q^2 + \frac{m^2}{a^2} \right) - \frac{a\omega_p^2}{q} \left(q^2 + \frac{m^2}{a^2} \right) \frac{I_m(qa)K_m(qa)}{\varepsilon_2 I_m(qa)K_m'(qa) - \varepsilon_1 I_m'(qa)K_m(qa)} \, ,
\tag{8.11}
$$

where $\omega_p^2 = e^2 n_0 / \varepsilon_0 m_e a$. We may consider the case where the 2DEG tube is in a vacuum, i.e., $\varepsilon_1 = 1 = \varepsilon_2$. In this case (8.11) becomes [1]

$$
\omega^2 = \alpha^2 \left(q^2 + \frac{m^2}{a^2} \right) + a^2 \omega_p^2 \left(q^2 + \frac{m^2}{a^2} \right) I_m(qa)K_m(qa) \, .
\tag{8.12}
$$

Here the Wronskian property $I_m'(x)K_m(x) - I_m(x)K_m'(x) = 1/x$ has been used. It can be seen from (8.12) that the dispersion relation depends on radius of the tube a and the surface electron density n_0. From (8.12), two different dimensionality regimes can be distinguished depending on the limiting cases $qa \gg |m|$ and $qa \ll |m|$. For $qa \gg |m|$, Eq. (8.12) can be written approximately as

$$
\omega^2 = \alpha^2 q^2 + \frac{1}{2} qa\omega_p^2 \, ,
\tag{8.13}
$$

where we have used the asymptotic expressions of the Bessel functions $I_m(x) = e^x / \sqrt{2\pi x}$ and $K_m(x) = e^{-x}\sqrt{\pi/2x}$ (with the finite m) [2]. The second term of (8.13) overwhelms the first term for large radius of 2DEG tubes, and dispersion relation in this case becomes

$$
\omega^2 = \frac{1}{2} qa\omega_p^2 \, ,
\tag{8.14}
$$

which corresponds to a proper 2D behavior (namely, the plasmon energy is proportional to the square root of q), since the longitudinal momentum transfer $\hbar q$ is a continuous variable [3].

On the other hand, for $qa \ll |m|$, we use the well-known expressions of Bessel functions, $I_m(x) = a_m x^m$, $K_0(x) = \ln(1.123/x)$, and $K_m(x) = b_m x^{-m}$ ($m \neq 0$), where here $a_m = 2^{-m}/\Gamma(m+1)$ and $b_m = 2^{m-1}\Gamma(m)$. Then we get for $m \neq 0$

$$\omega^2 = \alpha^2 \frac{m^2}{a^2} + \frac{\omega_p^2}{2}m , \qquad m \neq 0 , \tag{8.15}$$

which may be viewed as 2D in character when $m/a \gg 1$, since m/a is then a quasi-continuous effective wavenumber along the perimeter of the cylinder. However, for small values of the azimuthal quantum number m, the right-hand side of (8.15) depends strongly on the radius of the 2DEG tube, unlike the case of (8.13) where the SP frequency is independent of the dimensions of the system. This latter case cannot be properly characterized as 2D, and we will adopt the convention of referring to it as a 1D case. For $m = 0$ and $qa \approx 0$, the SPs excitation has a traditional 1D character, namely, it exhibits (up to a slowly varying factor of the square root of a logarithmic term) a linear dependence on q,

$$\omega^2 = \omega_p^2 (qa)^2 \ln\left(\frac{1.123}{qa}\right) , \qquad m = 0 . \tag{8.16}$$

Figure 8.1 shows the spectrum of SP modes of a single-walled carbon nanotube (CNT)[1] in terms of the dimensionless variables in the $\sigma + \pi$ electron fluid model.[2] From this figure one can see that the frequency of the SP modes of a 2DEG tube increases continuously with increasing the value of qa_0 for all m and approach the SP frequency of a planar 2DEG layer, i.e., $\omega_p\sqrt{qa/2}$ for $qa \to \infty$. Figure 8.2 shows the influence of the tube radius on the dispersion curves for $m = 1$ and different values of a. It is clear that the dispersion curves will approach one for large radii.

Let us note that the validity of present theory for CNTs by using the hydrodynamic theory is limited by the relation $k_F a \gg 1$ that indicates such classical models give results in agreement with *random phase approximation* theory, as shown by Bose and Longe [6].

---

[1]Carbon nanotubes, often refers to multi-walled CNTs, were first synthesized in 1991 by Iijima [4] as graphitic carbon needles, ranging from 4 to 30 nm in diameter and up to 1 μm in length. Using the atomic density of a graphene sheet 38 nm^{-2}, the $\sigma + \pi$ surface electron density of a single-walled CNT can be approximated by $n_0 = n_{0\pi} + n_{0\sigma}$, where $n_{0\pi} = 38$ nm^{-2} and $n_{0\sigma} = 3 \times 38$ nm^{-2}. Here, we have $v_F = \hbar k_F/m_e$ and $k_F = \sqrt{2\pi n_0}$.

[2]The two electron fluid model divides both σ and π electron fluids [5]. We note that the investigation of plasmonics properties of single-walled CNTs by using the two fluid model makes an interesting BVP, but we leave it for the reader (see [5]).

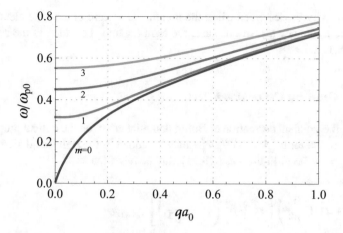

Fig. 8.1 Dispersion curves of SP modes of an isolated single-walled CNT with radius $a = 5a_0$, as given by (8.12) for different values of the parameter m. Here $\omega_{p0} = \left(n_0 e^2/\varepsilon_0 m_e a_0\right)^{1/2}$, $a_0 = 1\,\mathrm{nm}$, and $n_0 = 4 \times 38\,\mathrm{nm}^{-2}$

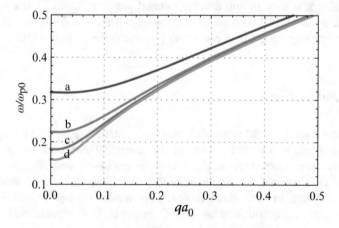

Fig. 8.2 Dispersion curves of SP modes of an isolated single-walled CNT with radius a, as given by (8.12), when $m = 1$. The different curves refer to (**a**) $a = 5a_0$, (**b**) $a = 10a_0$, (**c**) $a = 15a_0$, (**d**) $a = 20a_0$. Here $\omega_{p0} = \left(n_0 e^2/\varepsilon_0 m_e a_0\right)^{1/2}$, $a_0 = 1\,\mathrm{nm}$ and $n_0 = 4 \times 38\,\mathrm{nm}^{-2}$

8.1.1.1   Interband Transition Effect

To study the interband transitions effect on the optical properties of the system under consideration, the term $-\omega_0^2 \mathbf{u}$ may be added to the right-hand side of (8.1). This term with phenomenological *restoring frequency* ω_0 represents the restoring effects on the electron fluid displacement at $\rho = a$ which take into account the band structure of the electrons in a manner analogous to that invoked in devising the *Drude–Lorentz dielectric function* for carbon materials [7–10]. Thus, the interband transitions may be described using the mentioned picture of a bound electron with resonance frequency ω_0. In this case, the term ω_0^2 must be added to the right-hand side of (8.11).

8.1.1.2   Gradient Correction Effect

To study the gradient correction or Bohm potential effect on the optical properties of the system, the term $(\beta^2/n_0) \nabla_\parallel \nabla_\parallel^2 n$ may be added to the right-hand side of (8.1), where $\beta = v_F/2k_F$. In this case, (8.11) may be rewritten as

$$
\omega^2 = \left(q^2 + \frac{m^2}{a^2}\right)\left[\alpha^2 + \beta^2\left(q^2 + \frac{m^2}{a^2}\right)\right]
$$
$$
- \frac{a\omega_p^2}{q}\left(q^2 + \frac{m^2}{a^2}\right)\frac{I_m(qa)K_m(qa)}{\varepsilon_2 I_m(qa)K_m'(qa) - \varepsilon_1 I_m'(qa)K_m(qa)} . \tag{8.17}
$$

Numerically, it is easy to find that the internal pressure force of the electron [the term with α^2 in (8.17)] increases the frequency, while the dispersion behavior is almost affected by the gradient correction [the term with β^2 in (8.17)].

8.1.2   Power Flow

Let us now consider a SP mode that propagates along the z-axis of a cylindrical 2DEG layer and is a periodic wave in the azimuthal direction. In the previous section, we have assumed that all physical quantities to vary according to $\exp(iqz + im\phi - i\omega t)$. However, for applications to power flow and energy density, it is convenient to work in terms of real functions $\cos m\phi$ and $\sin m\phi$ [11, 12]. Therefore, we may replace the quantities n and Φ in (8.3) and (8.4) by expressions of the form

$$
\Phi(\mathbf{r}) = \tilde{\Phi}(\rho)\exp(iqz)\cos m\phi , \tag{8.18}
$$

$$
n(\phi, z) = \tilde{n}\exp(iqz)\cos m\phi . \tag{8.19}
$$

Now, for the power flow density associated with a SP on a 2DEG tube, we find

$$
S_z = -\begin{cases} \varepsilon_0\varepsilon_1\Phi_1\dfrac{\partial}{\partial t}\dfrac{\partial}{\partial z}\Phi_1 , & \rho \le a , \\[2mm] en_0\Phi v_z - m_e\alpha^2 n v_z , & \rho = a , \\[2mm] \varepsilon_0\varepsilon_2\Phi_2\dfrac{\partial}{\partial t}\dfrac{\partial}{\partial z}\Phi_2 , & \rho \ge a , \end{cases} \tag{8.20}
$$

where $\Phi = \Phi_1|_{\rho=a}$ or $\Phi = \Phi_2|_{\rho=a}$. After the elimination of $v_z = \partial z/\partial t$, n, Φ_1, Φ_2, and Φ in the above equation, the cycle-averaged of (8.20) can be written as

$$
S_z = -\frac{\varepsilon_0 \omega q}{2} A_1^2 \cos^2 m\phi
\begin{cases}
\varepsilon_1 \dfrac{I_m^2(q\rho)}{I_m^2(qa)}, & \rho \leq a, \\[3mm]
-\dfrac{n_0 e^2}{\varepsilon_0 m_e} \dfrac{\omega^2}{\left[\omega^2 - \alpha^2 \left(q^2 + \dfrac{m^2}{a^2}\right)\right]^2}, & \rho = a, \\[3mm]
\varepsilon_2 \dfrac{K_m^2(q\rho)}{K_m^2(qa)}, & \rho \geq a.
\end{cases}
\tag{8.21}
$$

The total power flow associated with the SP modes is determined by an integration over ϕ and ρ. We find

$$
\langle S_z \rangle = -\frac{\varepsilon_0 \omega q}{2} A_1^2 \left\{ \int_0^a \int_0^{2\pi} \varepsilon_1 \frac{I_m^2(q\rho)}{I_m^2(qa)} \cos^2 m\phi \rho \, d\rho \, d\phi \right.
$$

$$
+ \int_a^\infty \int_0^{2\pi} \varepsilon_2 \frac{K_m^2(q\rho)}{K_m^2(qa)} \cos^2 m\phi \rho \, d\rho \, d\phi
$$

$$
\left. - a \frac{n_0 e^2}{\varepsilon_0 m_e} \omega^2 \left[\omega^2 - \alpha^2 \left(q^2 + \frac{m^2}{a^2}\right)\right]^{-2} \int_0^{2\pi} \cos^2 m\phi \, d\phi \right\},
\tag{8.22}
$$

that yields

$$
\langle S_z \rangle = \frac{\varepsilon_0 \omega q a^2}{4} \left(\pi + \frac{\sin 4m\pi}{4m}\right) A_1^2 \left\{ \varepsilon_1 \left[\left[\frac{I_m'(qa)}{I_m(qa)}\right]^2 - \left(1 + \frac{m^2}{q^2 a^2}\right) \right] \right.
$$

$$
\left. - \varepsilon_2 \left[\left[\frac{K_m'(qa)}{K_m(qa)}\right]^2 - \left(1 + \frac{m^2}{q^2 a^2}\right) \right] + \frac{2\omega^2 \omega_p^2}{\left[\omega^2 - \alpha^2 \left(q^2 + \frac{m^2}{a^2}\right)\right]^2} \right\},
\tag{8.23}
$$

where in (8.23) the following relations:

$$
\int_0^a I_m^2(q\rho)\rho \, d\rho = -\frac{a^2}{2} \left[[I_m'(qa)]^2 - \left(1 + \frac{m^2}{q^2 a^2}\right) I_m^2(qa) \right],
\tag{8.24a}
$$

$$\int_a^\infty K_m^2(q\rho)\rho\,\mathrm{d}\rho = \frac{a^2}{2}\left[\left[K_m'(qa)\right]^2 - \left(1+\frac{m^2}{q^2a^2}\right)K_m^2(qa)\right], \qquad (8.24b)$$

have been used.

8.1.3 Energy Distribution

For the cycle-averaged of energy density distribution associated with the SP modes of a 2DEG tube, we have, in the three media,

$$U = \frac{\varepsilon_0}{4}\begin{cases} \varepsilon_1|\nabla\Phi_1|^2\,, & \rho \le a\,, \\[4pt] \dfrac{m_e n_0}{\varepsilon_0}\left(|v_\phi|^2+|v_z|^2\right)+\dfrac{m_e}{\varepsilon_0 n_0}\alpha^2|n|^2\,, & \rho = a\,, \\[4pt] \varepsilon_2|\nabla\Phi_2|^2\,, & \rho \ge a \end{cases} \qquad (8.25)$$

After elimination of v_ϕ, v_z, n, Φ_1, Φ_2, and Φ in the above equation, we obtain

$$U = \frac{\varepsilon_0}{4}A_1^2$$

$$\begin{cases} \varepsilon_1\left\{q^2\left[\left[\dfrac{I_m(q\rho)}{I_m(qa)}\right]^2+\left[\dfrac{I_m'(q\rho)}{I_m(qa)}\right]^2\right]\cos^2 m\phi + \dfrac{m^2}{\rho^2}\left[\dfrac{I_m(q\rho)}{I_m(qa)}\right]^2\sin^2 m\phi\right\}\,, & \rho \le a\,, \\[10pt] \dfrac{n_0 e^2}{\varepsilon_0 m_e}\dfrac{\omega^2\left(q^2\cos^2 m\phi + \dfrac{m^2}{a^2}\sin^2 m\phi\right)+\alpha^2\left(q^2+\dfrac{m^2}{a^2}\right)^2\cos^2 m\phi}{\left[\omega^2-\alpha^2\left(q^2+\dfrac{m^2}{a^2}\right)\right]^2}\,, & \rho = a\,, \\[10pt] \varepsilon_2\left\{q^2\left[\left[\dfrac{K_m(q\rho)}{K_m(qa)}\right]^2+\left[\dfrac{K_m'(q\rho)}{K_m(qa)}\right]^2\right]\cos^2 m\phi + \dfrac{m^2}{\rho^2}\left[\dfrac{K_m(q\rho)}{K_m(qa)}\right]^2\sin^2 m\phi\right\}\,, & \rho \ge a\,. \end{cases}$$
$$(8.26)$$

The total energy density associated with the SP modes is again determined by integration over ϕ and ρ. For general m the integrated energy density $\langle U\rangle$ can only be evaluated numerically. For $m = 0$, however, analytic expressions can be found. The total energy density (per unit length of cylinder) of fundamental SP mode is given by

$$\langle U\rangle = \frac{\pi\varepsilon_0}{4}q^2a^2A_1^2\left\{-\varepsilon_1\left[\left[\frac{I_1'(qa)}{I_0(qa)}\right]^2 - \frac{1}{q^2a^2}\left[\frac{I_1(qa)}{I_0(qa)}\right]^2 - 1\right]\right.$$

$$\left. +\varepsilon_2\left[\left[\frac{K_1'(qa)}{K_0(qa)}\right]^2 - \frac{1}{q^2a^2}\left[\frac{K_1(qa)}{K_0(qa)}\right]^2 - 1\right]+2\omega_p^2\frac{\omega^2+\alpha^2\omega^2}{\left[\omega^2-\alpha^2q^2\right]^2}\right\}\,.$$
$$(8.27)$$

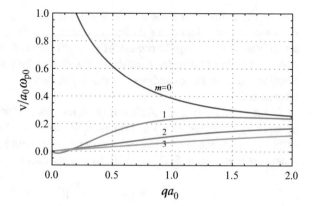

Fig. 8.3 Group (energy) velocity curves of SP modes of an isolated single-walled CNT with radius $a = a_0$ for different values of the parameter m. Here $\omega_{p0} = \left(n_0 e^2/\varepsilon_0 m_e a_0\right)^{1/2}$, $a_0 = 1$ nm and $n_0 = 4 \times 38$ nm^{-2}

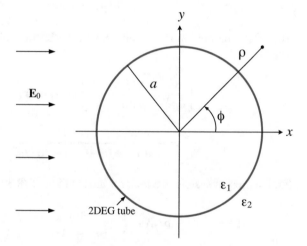

Fig. 8.4 Cross section of an infinitely long cylindrical 2DEG layer placed into an electrostatic field

Also, the energy velocity of the SP modes is given as the ratio of the total power flow and total energy density (per unit length of cylinder), that is in quantitative agreement with the group velocity found from the corresponding dispersion relation by means of the usual formula, i.e., (2.58). Figure 8.3 shows the energy (group) velocity of several SP modes of an isolated single-walled CNT with radius $a = a_0$.

8.1.4 Surface Plasmon Resonance

This BVP which is sketched in Fig. 8.4 deals with a cylindrical 2DEG layer of radius a and infinite length in a uniform quasi-static electric field $\mathbf{E} = \mathbf{e}_x E_0$ that is normal to its axis. We choose a cylindrical coordinate system with the origin taken at the axis of the cylinder and the x-axis along the electric field. The relative dielectric constants of inner and outer media are denoted by ε_1 and ε_2, respectively. Again, we assume that the equilibrium density (per unit area) of the system is n_0, while $n(\phi, z, t)$ is the first-order perturbed density (per unit area) of the homogeneous electron fluid on the cylindrical 2DEG layer, due to the presence of the external electric field.

In order to study the polarizability of the system, we use the SHD model by using (8.1)–(8.3), where we add the damping term $-\gamma\,\partial\mathbf{u}/\partial t$ to the right-hand side of (8.1). Due to the cylindrical symmetry of the present system, one can replace the quantities n and Φ in (8.1) and (8.3), by expressions of the form

$$\Phi(\rho, \phi) = \tilde{\Phi}(\rho)\cos\phi\,, \tag{8.28}$$

$$n(\phi) = \tilde{n}\cos\phi\,. \tag{8.29}$$

Therefore, the normal mode solutions of the SP frequency of the present problem satisfy

$$\left[\frac{\mathrm{d}^2}{\mathrm{d}\rho^2} + \frac{1}{\rho}\frac{\mathrm{d}}{\mathrm{d}\rho} - \frac{1}{\rho^2}\right]\tilde{\Phi}(\rho) = 0\,, \qquad \rho \neq a\,, \tag{8.30}$$

$$\tilde{n} = -\frac{en_0}{m_e a^2}\frac{1}{\omega\,(\omega + i\gamma) - \left(\alpha^2/a^2\right)}\tilde{\Phi}\Big|_{\rho=a}\,. \tag{8.31}$$

For the present system, we look for the solution of (8.30) of the form

$$\tilde{\Phi}(\rho) = \begin{cases} A_1\rho\,, & \rho \leq a\,, \\ -E_0\rho + A_2\rho^{-1}\,, & \rho \geq a\,. \end{cases} \tag{8.32}$$

To determine the unknown coefficients A_1 and A_2, the BCs, i.e., (8.10a) and (8.10b), must be enforced. The BCs yield

$$A_1 = \frac{-2\varepsilon_2}{\varepsilon_1 + \varepsilon_2 - \dfrac{\omega_p^2}{\omega\,(\omega + i\gamma) - \left(\alpha^2/a^2\right)}}E_0\,, \tag{8.33a}$$

$$A_2 = a^2\frac{\varepsilon_1 - \varepsilon_2 - \dfrac{\omega_p^2}{\omega\,(\omega + i\gamma) - \left(\alpha^2/a^2\right)}}{\varepsilon_1 + \varepsilon_2 - \dfrac{\omega_p^2}{\omega\,(\omega + i\gamma) - \left(\alpha^2/a^2\right)}}E_0\,. \tag{8.33b}$$

Substituting the magnitude of coefficients A_1 and A_2 in (8.28) and (8.32) gives

$$\Phi(\rho, \phi) = E_0 \rho \cos \phi \begin{cases} -\dfrac{2\varepsilon_2}{\varepsilon_1 + \varepsilon_2 - \dfrac{\omega_p^2}{\omega(\omega + i\gamma) - (\alpha^2/a^2)}}, & \rho \leq a, \\[4mm] -1 + \dfrac{a^2}{\rho^2} \dfrac{\varepsilon_1 - \varepsilon_2 - \dfrac{\omega_p^2}{\omega(\omega + i\gamma) - (\alpha^2/a^2)}}{\varepsilon_1 + \varepsilon_2 - \dfrac{\omega_p^2}{\omega(\omega + i\gamma) - (\alpha^2/a^2)}}, & \rho \geq a. \end{cases}$$

(8.34)

The electric field inside the system $\mathbf{E}_1 = -\nabla \Phi$ is uniform and parallel to the applied field and is given by

$$\mathbf{E}_1 = \frac{2\varepsilon_2}{\varepsilon_1 + \varepsilon_2 - \dfrac{\omega_p^2}{\omega(\omega + i\gamma) - (\alpha^2/a^2)}} E_0 \mathbf{e}_x$$

$$= E_0 \mathbf{e}_x - \frac{\varepsilon_1 - \varepsilon_2 - \dfrac{\omega_p^2}{\omega(\omega + i\gamma) - (\alpha^2/a^2)}}{\varepsilon_1 + \varepsilon_2 - \dfrac{\omega_p^2}{\omega(\omega + i\gamma) - (\alpha^2/a^2)}} E_0 \mathbf{e}_x .$$

(8.35)

In the simplest case, where $\gamma = 0$, $\varepsilon_1 = 1 = \varepsilon_2$, and $\omega^2 < \alpha^2/a^2 + \omega_p^2/2$, then $E_1 < E_0$.[3] The reduction of E_0 inside the 2DEG layer is attributed to the induced charge density on its surface as

$$\sigma_{ind}(\phi) = -en(\phi) = -\varepsilon_0 \frac{\omega_p^2}{\omega^2 - (\alpha^2/a^2) - (\omega_p^2/2)} E_0 \cos \phi .$$

(8.36)

As expected, it is positive in the region $0 \leq \phi \leq \pi/2$ and negative in the region $\pi/2 \leq \phi \leq \pi$. This induced charge, which has a $\cos \phi$ dependence, produces a uniform depolarization field in opposite direction with the external field. Also, the electric field outside the system is

$$\mathbf{E}_2 = E_0 \mathbf{e}_x + E_0 \frac{\varepsilon_1 - \varepsilon_2 - \dfrac{\omega_p^2}{\omega(\omega + i\gamma) - (\alpha^2/a^2)}}{\varepsilon_1 + \varepsilon_2 - \dfrac{\omega_p^2}{\omega(\omega + i\gamma) - (\alpha^2/a^2)}} \frac{a^2}{\rho^2} \left(1 - 2\sin^2 \phi\right) \mathbf{e}_x$$

---

[3] We note that if $\omega^2 > \alpha^2/a^2 + \omega_p^2/2$, then E_1 is greater than E_0. In this case the field inside the cylindrical 2DEG layer is strengthened.

$$+ 2E_0 \frac{\varepsilon_1 - \varepsilon_2 - \dfrac{\omega_p^2}{\omega(\omega + i\gamma) - (\alpha^2/a^2)}}{\varepsilon_1 + \varepsilon_2 - \dfrac{\omega_p^2}{\omega(\omega + i\gamma) - (\alpha^2/a^2)}} \frac{a^2}{\rho^2} \sin\phi \cos\phi\, \mathbf{e}_y \,. \qquad (8.37)$$

Now, using (2.78), we can get the corresponding normalized transversal polarizability, as

$$\alpha_{pol} = 2 \frac{\varepsilon_1 - \varepsilon_2 - \dfrac{\omega_p^2}{\omega(\omega + i\gamma) - (\alpha^2/a^2)}}{\varepsilon_1 + \varepsilon_2 - \dfrac{\omega_p^2}{\omega(\omega + i\gamma) - (\alpha^2/a^2)}} \,. \qquad (8.38)$$

Figure 8.5 shows the absolute value of α_{pol} with respect to dimensionless variable ω/ω_{p0} for a single-walled CNT with radius $a = a_0$. It is apparent that transversal polarizability experiences a resonant enhancement under the condition that the denominator of (8.38) is a minimum. Also, using the normalized transversal polarizability formula of the system, we may write the scattering and extinction absorption widths (in units of the geometric width $2a$) of the system, as [13]

$$Q_{sca} = \frac{\pi^2}{4}\varepsilon_2^{3/2}\left(\frac{a\omega}{c}\right)^3 \left| \frac{\varepsilon_1 - \varepsilon_2 - \dfrac{\omega_p^2}{\omega(\omega + i\gamma) - (\alpha^2/a^2)}}{\varepsilon_1 + \varepsilon_2 - \dfrac{\omega_p^2}{\omega(\omega + i\gamma) - (\alpha^2/a^2)}} \right|^2 , \qquad (8.39a)$$

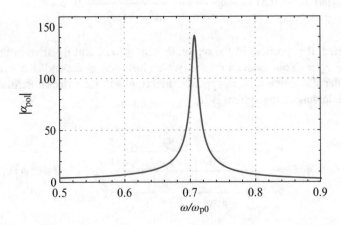

Fig. 8.5 Variation of the absolute value of the normalized transversal polarizability α_{pol} of a single-walled CNT with respect to the dimensionless variable ω/ω_{p0}, when $a = a_0$, $\varepsilon_1 = 1 = \varepsilon_2$, and $\gamma = 0.01\omega_p$. Here $\omega_{p0} = \left(n_0 e^2/\varepsilon_0 m_e a_0\right)^{1/2}$, $a_0 = 1\,\text{nm}$, and $n_0 = 4 \times 38\,\text{nm}^{-2}$

$$Q_{ext} = -\pi \varepsilon_2^{1/2} \left(\frac{a\omega}{c}\right) \mathrm{Re} \left[i \frac{\varepsilon_1 - \varepsilon_2 - \dfrac{\omega_p^2}{\omega\,(\omega + i\gamma) - \left(\alpha^2/a^2\right)}}{\varepsilon_1 + \varepsilon_2 - \dfrac{\omega_p^2}{\omega\,(\omega + i\gamma) - \left(\alpha^2/a^2\right)}} \right]. \tag{8.39b}$$

8.1.5   Effective Permittivity of a Composite of Cylindrical Two-Dimensional Electron Gas Layers

In dipolar approximation, the effective dielectric function ε_{eff} of a composite of aligned 2DEG tubes embedded in a host matrix with relative dielectric constant ε_2 may be derived, as (2.82), where f is the volume fraction of the embedded 2DEG tubes and α_{pol} is the normalized transversal polarizability of a 2DEG tube. By substituting (8.38) into (2.82), we find

$$\frac{\varepsilon_{eff} - \varepsilon_2}{\varepsilon_{eff} + \varepsilon_2} = f \frac{\varepsilon_1 - \varepsilon_2 - \dfrac{\omega_p^2}{\omega\,(\omega + i\gamma) - \left(\alpha^2/a^2\right)}}{\varepsilon_1 + \varepsilon_2 - \dfrac{\omega_p^2}{\omega\,(\omega + i\gamma) - \left(\alpha^2/a^2\right)}}. \tag{8.40}$$

This equation is the (complex) effective dielectric function of a composite of aligned 2DEG tubes, in the Maxwell-Garnett [14] approximation.

8.2   Quantization of Surface Plasmon Fields of Cylindrical Two-Dimensional Electron Gas Layers

We consider a cylindrical 2DEG layer free-standing in vacuum which has a radius a. As shown in Sect. 1.1, the electrostatic SP modes of the system are solutions of Laplace's equation in conjunction with the linearized hydrodynamic equations and standard BCs at $\rho = a$. We note that the energy of SP modes partly resides in the electric fields, partly in the kinetic energy of the electron motion, and partly in the hydrodynamic compressional energy.

The electric field energy density can be considered as a potential energy density $-\frac{1}{2}en\Phi$ of the oscillating 2DEG.[4] Also, the kinetic energy density is $\frac{1}{2}m_e\|\dot{\mathbf{u}}^2\|$.

---

[4]This term is equal with the term $\frac{1}{2}\varepsilon_0(\nabla\Phi)^2$ that is the density of electric potential energy of the system plus the term $en\Phi|_{\rho=a}$ that is the interaction part of energy density, i.e., the interaction between the electrostatic field and the polarization field (see Sect. 6.3).

Finally, for the hydrodynamic compressional energy density, we have $m_e\alpha^2 n^2/2n_0$ [12]. Now, we start with the Lagrangian density of the system which is the sum of three parts, as

$$\ell_{sp}(\phi, z) = \frac{m_e n_0}{2} \parallel \dot{\mathbf{u}}^2 \parallel - \frac{m_e \alpha^2}{2n_0} n^2 + \frac{e}{2} n\Phi|_{\rho=a} , \tag{8.41}$$

where $\mathbf{u}$ may be expressed in terms of the displacement potential, $\mathbf{u} = -\nabla_\parallel \Psi$ (the displacement field is irrotational). Thus, the total Lagrangian $\mathscr{L}_{sp}$ for the present system is

$$\mathscr{L}_{sp} = a \int_0^{2\pi} d\phi \int_{-L/2}^{L/2} dz \, \ell_{sp}(\phi, z) , \tag{8.42}$$

where L is the length of 2DEG tube. Now, the electrostatic potential fluctuations due to the SPs can be expanded by all eigenmodes as follows:

$$\Phi(\rho, \phi, z, t) = \sum_{q,m} e^{i(qz+m\phi)} e^{-i\omega_{q,m} t} \begin{cases} \Phi_{1q,m} I_m(q\rho) , & \rho \leq a , \\ \Phi_{2q,m} K_m(q\rho) , & \rho \geq a , \end{cases} \tag{8.43}$$

where q, m, and $\omega_{q,m}$ are wavenumber, azimuthal quantum number, and angular frequency of a SP mode, respectively. Also, we express the electronic density corresponding to these modes as a localized surface density, namely

$$n(\phi, z, t) = \sum_{q,m} n_{q,m} e^{i(qz+m\phi)} e^{-i\omega_{q,m} t} , \tag{8.44}$$

and we propose an expansion for the displacement potential in the same manner, as

$$\Psi(\phi, z, t) = \sum_{q,m} \Psi_{q,m} e^{i(qz+m\phi)} e^{-i\omega_{q,m} t} . \tag{8.45}$$

To determine the relations between the coefficients in (8.43)–(8.45), we have to use appropriate BCs, i.e., (8.10). Furthermore, by substituting (8.44) and (8.45) into the equation of continuity, we obtain

$$n_{q,m} = -n_0 \left(q^2 + \frac{m^2}{a^2} \right) \Psi_{q,m} , \tag{8.46}$$

and by substituting (8.43)–(8.45) into the BCs, we find the following expressions:

$$\Phi_{1q,m} = \frac{e n_0 a}{\varepsilon_0} K_m(qa) \left(q^2 + \frac{m^2}{a^2} \right) \Psi_{q,m} , \tag{8.47}$$

$$\Phi_{2q,m} = \frac{en_0 a}{\varepsilon_0} I_m(qa) \left(q^2 + \frac{m^2}{a^2}\right) \Psi_{q,m} .$$ (8.48)

Since we are concerned with real valued fields in the total Lagrangian, $\Psi_{q,m}$ must satisfy the Hermitian relation $\Psi_{q,m}^* = \Psi_{-q,-m}$.

Now, the kinetic energy portion of $\mathcal{L}_{sp}$ can be calculated by using (8.45) in the first term of (8.41) and integration over the tube surface, and we find

$$a\frac{m_e n_0}{2} \sum_{q,m} \sum_{q',m'} \int_0^{2\pi} d\phi \int_{-L/2}^{L/2} dz$$

$$\times \left\|\left(q\mathbf{e}_z + \frac{m}{a}\mathbf{e}_\phi\right) \cdot \left(q'\mathbf{e}_z + \frac{m'}{a}\mathbf{e}_\phi\right)\right\| \dot{\Psi}_{k,m} \dot{\Psi}_{q',m'} e^{i(q+q')z} e^{i(m+m')\phi}$$

$$= \frac{m_e n_0 A}{2} \sum_{q,m} \left(q^2 + \frac{m^2}{a^2}\right) \dot{\Psi}_{q,m} \dot{\Psi}_{-q,-m} ,$$ (8.49)

where $A = 2\pi L a$ is the surface of the system. The portion of hydrodynamic compressional energy of $\mathcal{L}_{sp}$ can be calculated in the same manner. By using (8.44) in the second term of (8.41) and integration over the tube surface, we obtain

$$a\frac{m_e \alpha^2}{2n_0} \int_0^{2\pi} d\phi \int_{-L/2}^{L/2} n^2 dz = \frac{m_e n_0 A \alpha^2}{2} \sum_{q,m} \left(q^2 + \frac{m^2}{a^2}\right)^2 \Psi_{q,m} \Psi_{-q,-m} .$$ (8.50)

Finally, the potential electrical energy portion of $\mathcal{L}_{sp}$ can be written as

$$a\frac{e}{2} \int_0^{2\pi} d\phi \int_{-L/2}^{L/2} n\Phi|_{\rho=a} dz$$

$$= -\frac{e^2 n_0^2 a A}{2\varepsilon_0} \sum_{q,m} \left(q^2 + \frac{m^2}{a^2}\right)^2 I_m(qa) K_m(qa) \Psi_{q,m} \Psi_{-q,-m} .$$ (8.51)

Therefore, the Lagrangian of the system is expressed in terms of $\Psi_{k,m}$ and $\dot{\Psi}_{k,m}$ as

$$\mathcal{L}_{sp} = \frac{m_e n_0 A}{2} \sum_{q,m} \left(q^2 + \frac{m^2}{a^2}\right) \dot{\Psi}_{q,m} \dot{\Psi}_{-q,-m}$$

$$- \frac{m_e n_0 A}{2} \sum_{q,m} \left(q^2 + \frac{m^2}{a^2}\right)^2 \left[\alpha^2 + \frac{e^2 n_0 a}{m_e \varepsilon_0} I_m(qa) K_m(qa)\right] \Psi_{q,m} \Psi_{-q,-m} .$$ (8.52)

The canonical momentum conjugate to the variables, $\Psi_{q,m}$, is given by

$$\pi_{q,m} = \frac{\partial \mathcal{L}_{sp}}{\partial \dot{\Psi}_{q,m}} = m_e n_0 A \left(q^2 + \frac{m^2}{a^2} \right) \dot{\Psi}_{-q,-m} , \qquad (8.53)$$

and we find the Hamiltonian $\mathcal{H}_{sp}$ for the SP oscillations as

$$\mathcal{H}_{sp} = \sum_{q,m} \pi_{q,m} \dot{\Psi}_{q,m} - \mathcal{L}_{sp} = \frac{1}{2 m_e n_0 A} \sum_{q,m} \left(q^2 + \frac{m^2}{a^2} \right)^{-1} \pi_{q,m} \pi_{-q,-m}$$

$$+ \frac{m_e n_0 A}{2} \sum_{q,m} \left(q^2 + \frac{m^2}{a^2} \right)^2 \left[\alpha^2 + \frac{e^2 n_0 a}{m_e \varepsilon_0} I_m(qa) K_m(qa) \right] \Psi_{q,m} \Psi_{-q,-m} .$$

$$(8.54)$$

The classical SP oscillations field is now quantized by the Bose–Einstein commutation relations

$$\left[\Psi_{q,m}, \pi_{q',m'} \right] = i \hbar \delta_{qq'} \delta_{mm'} . \qquad (8.55)$$

Based on the Hamiltonian relation in (8.54), we introduce creation and annihilation operators, for both the positive and negative wavenumbers, as

$$\hat{a}^{\dagger}_{q,m} = -i (2 m_e n_0 A)^{-1/2} \left(q^2 + \frac{m^2}{a^2} \right)^{-1/2} (\hbar \omega_{q,m})^{-1/2} \pi_{q,m}$$

$$+ \left(\frac{m_e n_0 A}{2 \hbar \omega_{q,m}} \right)^{1/2} \left(q^2 + \frac{m^2}{a^2} \right)^{1/2} \left[\alpha^2 + \frac{e^2 n_0 a}{m_e \varepsilon_0} I_m(qa) K_m(qa) \right]^{1/2} \Psi_{-q,-m} ,$$

$$(8.56)$$

$$\hat{a}_{q,m} = i (2 m_e n_0 A)^{-1/2} \left(q^2 + \frac{m^2}{a^2} \right)^{-1/2} (\hbar \omega_{q,m})^{-1/2} \pi_{-q,-m}$$

$$+ \left(\frac{m_e n_0 A}{2 \hbar \omega_{q,m}} \right)^{1/2} \left(q^2 + \frac{m^2}{a^2} \right)^{1/2} \left[\alpha^2 + \frac{e^2 n_0 a}{m_e \varepsilon_0} I_m(qa) K_m(qa) \right]^{1/2} \Psi_{q,m} ,$$

$$(8.57)$$

$$\hat{a}^{\dagger}_{-q,-m} = -i (2 m_e n_0 A)^{-1/2} \left(q^2 + \frac{m^2}{a^2} \right)^{-1/2} (\hbar \omega_{q,m})^{-1/2} \pi_{-q,-m}$$

$$+ \left(\frac{m_e n_0 A}{2 \hbar \omega_{q,m}} \right)^{1/2} \left(q^2 + \frac{m^2}{a^2} \right)^{1/2} \left[\alpha^2 + \frac{e^2 n_0 a}{m_e \varepsilon_0} I_m(qa) K_m(qa) \right]^{1/2} \Psi_{q,m} ,$$

$$(8.58)$$

$$\hat{a}_{-q,-m} = i(2m_e n_0 A)^{-1/2} \left(q^2 + \frac{m^2}{a^2}\right)^{-1/2} (\hbar\omega_{q,m})^{-1/2} \pi_{q,m}$$

$$+ \left(\frac{m_e n_0 A}{2\hbar\omega_{q,m}}\right)^{1/2} \left(q^2 + \frac{m^2}{a^2}\right)^{1/2} \left[\alpha^2 + \frac{e^2 n_0 a}{m_e \varepsilon_0} I_m(qa) K_m(qa)\right]^{1/2} \Psi_{-q,-m} .$$

$$(8.59)$$

By their definitions, they must have unit real commutators and the following commutation rules are understood:

$$\left[\hat{a}_{q,m}, \hat{a}^\dagger_{q',m'}\right] = \delta_{qq'}\delta_{mm'} , \tag{8.60}$$

$$\left[\hat{a}_{q,m}, \hat{a}_{q',m'}\right] = 0 = \left[\hat{a}^\dagger_{q,m}, \hat{a}^\dagger_{q',m'}\right] . \tag{8.61}$$

The fields associated with the creation and annihilation operators are found via solving their definitions. Using (8.56)–(8.59), we obtain

$$\pi_{q,m} = \frac{i}{2}(2m_e n_0 A)^{1/2} \left(q^2 + \frac{m^2}{a^2}\right)^{1/2} (\hbar\omega_{q,m})^{1/2} \left(\hat{a}^\dagger_{q,m} - \hat{a}_{-q,-m}\right) , \tag{8.62}$$

$$\pi_{-q,-m} = \frac{i}{2}(2m_e n_0 A)^{1/2} \left(q^2 + \frac{m^2}{a^2}\right)^{1/2} (\hbar\omega_{q,m})^{1/2} \left(\hat{a}^\dagger_{-q,-m} - \hat{a}_{q,m}\right) , \tag{8.63}$$

$$\Psi_{q,m} = \frac{1}{2}\left(\frac{m_e n_0 A}{2\hbar\omega_{q,m}}\right)^{-1/2} \left(q^2 + \frac{m^2}{a^2}\right)^{-1/2}$$

$$\times \left[\alpha^2 + \frac{e^2 n_0 a}{m_e \varepsilon_0} I_m(qa) K_m(qa)\right]^{-1/2} \left(\hat{a}^\dagger_{-q,-m} + \hat{a}_{q,m}\right) , \tag{8.64}$$

$$\Psi_{-q,-m} = \frac{1}{2}\left(\frac{m_e n_0 A}{2\hbar\omega_{q,m}}\right)^{-1/2} \left(q^2 + \frac{m^2}{a^2}\right)^{-1/2}$$

$$\times \left[\alpha^2 + \frac{e^2 n_0 a}{m_e \varepsilon_0} I_m(qa) K_m(qa)\right]^{-1/2} \left(\hat{a}^\dagger_{q,m} + \hat{a}_{-q,-m}\right) . \tag{8.65}$$

By substituting (8.62)–(8.65) into (8.54), we find

$$\mathscr{H}_{sp} = \sum_{q,m} \frac{\hbar\omega_{q,m}}{2} \left[\hat{a}^\dagger_{q,m}\hat{a}_{q,m} + \frac{1}{2} + \hat{a}^\dagger_{-q,-m}\hat{a}_{-q,-m} + \frac{1}{2}\right] , \tag{8.66}$$

provided that $\omega_{q,m}$ satisfies (8.12). The sum is over all wavenumbers, and the positive and negative terms give the same total, so

$$\mathcal{H}_{sp} = \sum_{q,m} \hbar\omega_{q,m}\left[\hat{a}_{q,m}^{\dagger}\hat{a}_{q,m} + \frac{1}{2}\right].$$ (8.67)

To conclude the quantization procedure, let us note that Φ and Ψ are now operators and may be expressed in terms of the $\hat{a}_{q,m}$ and $\hat{a}_{q,m}^{\dagger}$ as follows:

$$\hat{\Phi}_1(\rho,\phi,z,t) = \frac{ea}{\varepsilon_0}\left(\frac{\hbar n_0}{2m_e A}\right)^{1/2}$$

$$\times \sum_{q,m} \frac{1}{\omega_{q,m}^{1/2}}\left(q^2 + \frac{m^2}{a^2}\right)^{1/2} K_m(qa)I_m(q\rho)e^{i(qz+m\phi)}e^{-i\omega_{q,m}t}\left(\hat{a}_{-q,-m}^{\dagger} + \hat{a}_{q,m}\right),$$

(8.68)

$$\hat{\Phi}_2(\rho,\phi,z,t) = \frac{ea}{\varepsilon_0}\left(\frac{\hbar n_0}{2m_e A}\right)^{1/2}$$

$$\times \sum_{q,m} \frac{1}{\omega_{q,m}^{1/2}}\left(q^2 + \frac{m^2}{a^2}\right)^{1/2} I_m(qa)K_m(q\rho)e^{i(qz+m\phi)}e^{-i\omega_{q,m}t}\left(\hat{a}_{-q,-m}^{\dagger} + \hat{a}_{q,m}\right),$$

(8.69)

$$\hat{\Psi}(\phi,z,t) = \left(\frac{\hbar}{2m_e n_0 A}\right)^{1/2}$$

$$\times \sum_{q,m} \frac{1}{\omega_{q,m}^{1/2}}\left(q^2 + \frac{m^2}{a^2}\right)^{-1/2} e^{i(qz+m\phi)}e^{-i\omega_{q,m}t}\left(\hat{a}_{-q,-m}^{\dagger} + \hat{a}_{q,m}\right).$$ (8.70)

8.3   Surface Plasmon Modes of Two Coaxial Cylindrical Two-Dimensional Electron Gas Layers

We consider a system of two infinitely long and infinitesimally thin coaxial cylindrical 2DEG layers, with radii $a_1 < a_2$. Let us assume that the density of free-electron fluid over the each cylindrical surface (per unit area) is n_{0j} with $j = 1$ and 2. Also, for simplicity we assume that in regions $0 < \rho < a_1$, $a_1 < \rho < a_2$, and $\rho > a_2$, there is no material. The inclusion of an insulator core and/or embedding medium in the formalism is straightforward. Assuming that $n_j(a_j,\phi,z,t)$ is the perturbed density (per unit area) of the homogeneous electron fluid on the jth layer, due to propagation of an electrostatic surface wave with angular frequency ω parallel to both axial and azimuthal directions of the system. Here, the electric potential is also represented in the form of (8.5), whereas (8.6) must be written as

$$n_j(\phi, z) = \tilde{n}_j \exp(iqz) \exp(im\phi) , \tag{8.71}$$

where

$$\tilde{n}_j = -\frac{en_{0j}}{m_e} \frac{\left(q^2 + \dfrac{m^2}{a_j^2}\right)}{\omega^2 - \alpha_j^2 \left(q^2 + \dfrac{m^2}{a_j^2}\right)} \tilde{\Phi}\big|_{\rho=a_j} , \tag{8.72}$$

with $\alpha_j = v_{Fj}/2$ (v_{Fj} is the Fermi velocity of the jth 2DEG layer) and the solution of (8.7) has the form

$$\tilde{\Phi}(\rho) = \begin{cases} A_1 I_m(q\rho) , & \rho \le a_1 , \\ A_2 I_m(q\rho) + A_3 K_m(q\rho) , & a_1 \le \rho \le a_2 , \\ A_4 K_m(q\rho) , & \rho \ge a_2 . \end{cases} \tag{8.73}$$

For the present BVP, the BCs are

$$\Phi_1\big|_{\rho=a_1} = \Phi_2\big|_{\rho=a_1} , \tag{8.74a}$$

$$\Phi_2\big|_{\rho=a_2} = \Phi_3\big|_{\rho=a_2} , \tag{8.74b}$$

$$\frac{\partial \Phi_2}{\partial \rho}\bigg|_{\rho=a_1} - \frac{\partial \Phi_1}{\partial \rho}\bigg|_{\rho=a_1} = \frac{en_1}{\varepsilon_0} , \tag{8.74c}$$

$$\frac{\partial \Phi_3}{\partial \rho}\bigg|_{\rho=a_2} - \frac{\partial \Phi_2}{\partial \rho}\bigg|_{\rho=a_2} = \frac{en_2}{\varepsilon_0} , \tag{8.74d}$$

where subscripts 1 and 3 denote the regions inside and outside the cylinders having radii a_1 and a_2, respectively, while subscript 2 denotes the region between the two cylinders. Using (8.73) and (8.74), one arrives at the following homogeneous linear system of equations:

$$\begin{pmatrix} I_m(qa_1) & -I_m(qa_1) & -K_m(qa_1) & 0 \\ 0 & I_m(qa_2) & K_m(qa_2) & -K_m(qa_2) \\ M_{31} & qI'_m(qa_1) & qK'_m(qa_1) & 0 \\ 0 & -qI'_m(qa_2) & -qK'_m(qa_2) & M_{44} \end{pmatrix} \begin{pmatrix} A_1 \\ A_2 \\ A_3 \\ A_4 \end{pmatrix} = 0 , \tag{8.75}$$

where

$$
M_{31} = \frac{a_1 \omega_{p1}^2 \left(q^2 + \frac{m^2}{a_1^2}\right) I_m(qa_1)}{\left[\omega^2 - \alpha_1^2 \left(q^2 + \frac{m^2}{a_1^2}\right)\right]} - q I_m'(qa_1) \,,
$$

$$
M_{44} = \frac{a_2 \omega_{p2}^2 \left(q^2 + \frac{m^2}{a_2^2}\right) K_m(qa_2)}{\left[\omega^2 - \alpha_2^2 \left(q^2 + \frac{m^2}{a_2^2}\right)\right]} + q K_m'(qa_2) \,,
$$

with $\omega_{pj}^2 = n_{0j} e^2 / \varepsilon_0 m_e a_j$. The dispersion relation of SPP modes in a system of two coaxial 2DEG layers can be obtained by solving $\det \mathbf{M} = 0$, where $\mathbf{M}$ is the matrix figuring in (8.75). After some algebra, one arrives to the following two branches of the dispersion relations for each m, as:

$$
\omega_{\pm}^2 = \frac{\omega_1^2 + \omega_2^2}{2} \pm \sqrt{\left(\frac{\omega_1^2 - \omega_2^2}{2}\right)^2 + \Delta^2} \,, \tag{8.76}
$$

where

$$
\omega_j^2 = \alpha_j^2 \left(q^2 + \frac{m^2}{a_j^2}\right) + a_j^2 \omega_{pj}^2 \left(q^2 + \frac{m^2}{a_j^2}\right) I_m(qa_j) K_m(qa_j) \,, \tag{8.77}
$$

are the squares of the plasmon dispersion on the cylinders $j = 1$ and 2 and

$$
\Delta = a_1 a_2 \omega_{p1} \omega_{p2} \left(q^2 + \frac{m^2}{a_1^2}\right)^{1/2} \left(q^2 + \frac{m^2}{a_2^2}\right)^{1/2} I_m(qa_1) K_m(qa_2) \,, \tag{8.78}
$$

describes the electrostatic interaction between the two coaxial 2DEG tubes. This interaction or hybridization results in a splitting of the frequencies of the SP modes into a high-frequency ω_+ (symmetric) mode and a low-frequency ω_- (anti-symmetric) mode, as shown schematically in Fig. 8.6 for $m = 1$. It is interesting to note that this result is *vice versa* of the corresponding result for metallic nanotubes [15] and nanoshells [16] as discussed in Chap. 2. The allowed frequencies depend on wavenumber q, azimuthal quantum number m, and tube radii a_1 and a_2. Furthermore, we see that the coupling between both 2DEG tube play an important role on the frequency in the low-q range. In the special case, when $q = 0$ and $m \neq 0$ (8.76) can be reduced to the following equation:

Fig. 8.6 The energy diagram in the plasmon hybridization theory [16, 20, 21] of a system of two coaxial cylindrical 2DEG layers, when $m = 1$. The energy levels marked with frequencies ω_1 and ω_2 of two individual 2DEG tube with the radii a_1 and $a_2 > a_1$, respectively. The energy levels marked with ω_+ and ω_- represent the symmetric and anti-symmetric SP modes of the system, respectively, as derived from the surface charges distribution induced on the two surfaces of the system

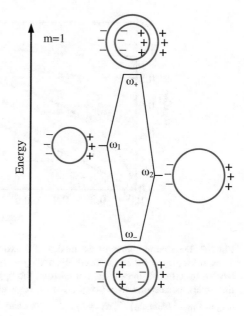

$$
\omega_{\pm} = \frac{1}{\sqrt{2}} \left\{ \left[m^2 \left(\frac{\alpha_1^2}{a_1^2} + \frac{\alpha_2^2}{a_2^2} \right) + \frac{m}{2} \left(\omega_{p1}^2 + \omega_{p2}^2 \right) \right] \right.
$$

$$
\left. \pm \sqrt{ \left[m^2 \left(\frac{\alpha_1^2}{a_1^2} - \frac{\alpha_2^2}{a_2^2} \right) + \frac{m}{2} \left(\omega_{p1}^2 - \omega_{p2}^2 \right) \right] + m^2 \omega_{p1}^2 \omega_{p2}^2 \left(\frac{a_1}{a_2} \right)^{2m} } \right\}^{1/2} .
$$

(8.79)

As an example, we choose a two-walled CNT with the inner and outer cylinders having radii $a_1 = 6.8$ Å and $a_2 = 10.2$ Å, respectively, and take the increments of the radii of the subsequent walls to be $a_2 - a_1 = a_0 = 3.4$ Å [17]. To see clearly behavior of the two groups of resonant SP dispersions from (8.76), we plot dimensionless frequency ω/ω_{p0}, versus dimensionless variable $q a_0$ in Fig. 8.7, for several modes $m = 0, 1, 2, 3$, and 4. We note that all curves of Fig. 8.7 tend to the same limit for the large value of q. The upper group of SP curves exhibits a characteristic dimensional crossover from a 1D to a 2D electron system [3], and lower group of SP curves exhibits weaker dispersions [6]. In particular, the $m = 0$ mode in the lower-energy plasmon group exhibits a quasi-acoustic (linear) dispersion at long wavelengths. This quasi-acoustic SP mode seems to be a common occurrence when a splitting of plasmon energies happens due to the electrostatic interaction, e.g. in the coupling between two parallel CNTs [18], or in the electron-hole plasma of a CNT [19].

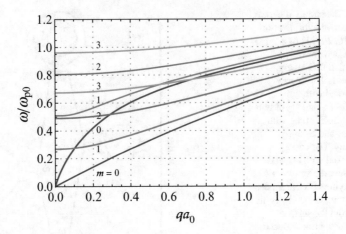

Fig. 8.7 Dispersion curves of SP modes of a two-walled CNT with the inner and outer radii $a_1 = 6.8$Å , $a_2 = 10.2$Å, respectively, and $a_2 - a_1 = a_0 = 3.4$ Å, for $m = 0, 1, 2, 3$, and 4, labeled by different line-styles. For each m, the upper and lower curves, plotted with the same line-styles, correspond to SP dispersion ω_+/ω_{p0} and ω_-/ω_{p0} from (8.76), respectively. Here $\omega_{p0} = \left(n_{0j}e^2/\varepsilon_0 m_e a_0\right)^{1/2}$, $\alpha_1 = v_F/\sqrt{2} = \alpha_2$, and $n_{01} = 4 \times 38$ nm$^{-2} = n_{02}$

8.3.1 Multilayer Systems

Assume a multilayer 2DEG consists of N coaxial 2DEG tubes with radii $a_1 < a_2 < \cdots < a_N$. To find the dispersion relation of SP modes of the system, relevant solution of (8.7) in the space above and below the jth 2DEG tube may be written as

$$\tilde{\Phi}_j(\rho) = A_j \begin{cases} K_m(qa_j)I_m(q\rho) \, , \, \rho \leq a_j \, , \\ I_m(qa_j)K_m(q\rho) \, , \, \rho \geq a_j \, , \end{cases} \quad (8.80)$$

where A_j with $1 \leq j \leq N$ are constants. Also, for the present BVP, the BCs are

$$\tilde{\Phi}_j\big|_{\rho>a_j} = \tilde{\Phi}_j\big|_{\rho<a_j} \, , \quad (8.81a)$$

$$\frac{\partial \tilde{\Phi}_j}{\partial \rho}\bigg|_{\rho>a_j} - \frac{\partial \tilde{\Phi}_j}{\partial \rho}\bigg|_{\rho<a_j} = -\frac{\omega_{pj}^2 a_j \left(q^2 + \dfrac{m^2}{a_j^2}\right)}{\omega^2 - \alpha_j^2\left(q^2 + \dfrac{m^2}{a_j^2}\right)} \sum_{i=1}^{N} \tilde{\Phi}_i \, , \quad (8.81b)$$

where the symbols have the same meaning as in previous section. Substitution of (8.80) into (8.81) leads to a set of linear homogeneous equations with unknown A_j. This set has non-trivial solutions that can be ascertained by solving the dispersion equation

$$\det \mathbf{M} = 0 \, , \tag{8.82}$$

where the element M_{ji} of the $N \times N$ matrix $\mathbf{M}$ is given by

$$M_{ji} = \begin{cases} I_m(qa_i)K_m(qa_j), & j > i, \\ I_m(qa_j)K_m(qa_i), & j < i, \\ \dfrac{\omega_j^2 - \omega^2}{\omega_{pj}^2 a_j^2 \left(q^2 + \dfrac{m^2}{a_j^2} \right)}, & j = i, \end{cases} \tag{8.83}$$

where ω_j is defined in (8.77). From (8.82) we obtain N positive roots for the frequencies of the SPs which are clearly separated into a high-frequency, ω_+, and a low-frequency, ω_-, group for each m. We note that (8.82) allows one to vary the number of layers at will. For example, for a system of two coaxial 2DEG tubes, i.e., $N = 2$, we have

$$\begin{vmatrix} \dfrac{\omega_1^2 - \omega^2}{\omega_{p1}^2 a_1^2 \left(q^2 + \dfrac{m^2}{a_1^2} \right)} & I_m(qa_1)K_m(qa_2) \\[4mm] I_m(qa_1)K_m(qa_2) & \dfrac{\omega_2^2 - \omega^2}{\omega_{p2}^2 a_2^2 \left(q^2 + \dfrac{m^2}{a_2^2} \right)} \end{vmatrix} = 0 \, , \tag{8.84}$$

that is equal with (8.76).

8.4   Surface Plasmon Modes of Two Parallel Cylindrical Two-Dimensional Electron Gas Layers

Let us consider a pair of parallel cylindrical 2DEG layers with radii a_1 and a_2 which density free-electron fluid over the each cylindrical surface (per unit area) is n_{01} and n_{02}, respectively. The distance between the two axes will be labeled d, where $d > a_1 + a_2$, and the used coordinates are illustrated by Fig. 8.8. The origin of the cylindrical coordinates $\mathbf{r}_1 = (\rho_1, \phi_1, z)$ be located at the point $z = 0$ on the axis of the 1st 2DEG tube. Assuming that the n_j be the perturbed density (per unit area) of the homogeneous electron fluid on the jth 2DEG tube (with $j = 1, 2$), due to propagation of a SP mode with angular frequency ω parallel to the surfaces of the system. Using the coordinates illustrated by Fig. 8.8, the eigensolutions of Laplace's equation, i.e., (8.3) in the three regions can be written as

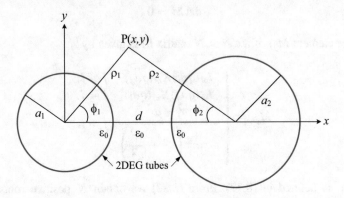

Fig. 8.8 Schematic representation of a pair of parallel 2DEG tubes with radii a_1 and a_2, the axis-to-axis separation begin d

$$\Phi(\mathbf{r}_1, \mathbf{r}_2) = e^{iqz} \sum_{m=-\infty}^{+\infty} \begin{cases} A_{1m} \dfrac{I_m(q\rho_1)}{I_m(qa_1)} e^{im\phi_1} , & \rho_1 \leq a_1, \\[2mm] A_{2m} \dfrac{K_m(q\rho_1)}{K_m(qa_1)} e^{im\phi_1} + A_{3m} \dfrac{K_m(q\rho_2)}{K_m(qa_2)} e^{im\phi_2}, & a_1 \leq \rho_1 \& a_2 \leq \rho_2, \\[2mm] A_{4m} \dfrac{I_m(q\rho_2)}{I_m(qa_2)} e^{im\phi_2}, & \rho_2 \leq a_2. \end{cases}$$

(8.85)

Also, for the present BVP, (8.6) must be written as

$$n_j = -\frac{en_{0j}}{m_e} e^{iqz} \sum_{m=-\infty}^{+\infty} e^{im\phi_j} \frac{\left(q^2 + \dfrac{m^2}{a_j^2}\right)}{\omega^2 - \alpha_j^2 \left(q^2 + \dfrac{m^2}{a_j^2}\right)} \Phi(\mathbf{r})\Big|_{\rho=a_j} .$$

(8.86)

The BCs at the surface of the first and second 2DEG tubes are

$$\Phi_1\big|_{\rho_1=a_1} = \Phi_3\big|_{\rho_1=a_1} ,$$

(8.87a)

$$\Phi_2\big|_{\rho_2=a_2} = \Phi_3\big|_{\rho_2=a_2} ,$$

(8.87b)

$$\frac{\partial \Phi_3}{\partial \rho_1}\bigg|_{\rho_1=a_1} - \frac{\partial \Phi_1}{\partial \rho_1}\bigg|_{\rho_1=a_1} = \frac{en_1}{\varepsilon_0} ,$$

(8.87c)

$$\frac{\partial \Phi_3}{\partial \rho_2}\bigg|_{\rho_2=a_2} - \frac{\partial \Phi_2}{\partial \rho_2}\bigg|_{\rho_2=a_2} = \frac{en_2}{\varepsilon_0} ,$$

(8.87d)

where subscripts 1 and 2 denote the regions inside the cylinders having radii a_1 and a_2, respectively, while subscript 3 denotes the region outside the two cylinders.

In order to apply the above BCs at the surface of the first tube, it is useful to express the term depending on ρ_2 and ϕ_2 in the outer potential in terms of ρ_1 and ϕ_1, using an addition theorem for modified Bessel functions, as [2, 22]

$$K_m(q\rho_2)e^{im\phi_2} = \sum_{n=-\infty}^{+\infty} K_{m+n}(qd)I_n(q\rho_1)e^{in\phi_1} . \tag{8.88}$$

Thus, for the external potential equation in the region $a_1 \leq \rho_1$ and $a_2 \leq \rho_2$, we have

$$\Phi_3\,(\mathbf{r}_1, \mathbf{r}_2) = e^{iqz} \sum_{m=-\infty}^{+\infty} \left[B_m \frac{K_m(q\rho_1)}{K_m(qa_1)} + \sum_{n=-\infty}^{+\infty} C_n \frac{K_{m+n}(qd)I_m(q\rho_1)}{K_n(qa_2)} \right] e^{in\phi_1} . \tag{8.89}$$

The continuity of the electrical potential at $\rho_1 = a_1$, i.e., (8.87a) gives

$$A_m = B_m + \sum_{n=-\infty}^{+\infty} C_n \frac{K_{m+n}(qd)I_m(qa_1)}{K_n(qa_2)} , \tag{8.90}$$

which we can write in matrix form,

$$\mathbf{a} = \mathbf{b} + \mathbf{Pc} , \tag{8.91}$$

where $\mathbf{a}$, $\mathbf{b}$, and $\mathbf{c}$ are column matrices whose components are A_m, B_m, and C_m, respectively, and $\mathbf{P}$ is a matrix whose elements are

$$P_{mn} = \frac{K_{m+n}(qd)I_n(qa_1)}{K_m(qa_2)} . \tag{8.92}$$

Similarly, the discontinuity of the normal component of the displacement vector at $\rho_1 = a_1$, i.e., (8.87c), yields

$$A_m \left\{ q \frac{I'_m(qa_1)}{I_m(qa_1)} - \frac{a_1\omega_{p1}^2\left(q^2 + \frac{m^2}{a_1^2}\right)}{\omega^2 - \alpha_1^2\left(q^2 + \frac{m^2}{a_1^2}\right)} \right\} = B_m q \frac{K'_m(qa_1)}{K_m(qa_1)}$$

$$+ \sum_{n=-\infty}^{+\infty} C_n \frac{K_{m+n}(qd)I'_m(qa_1)}{K_n(qa_2)} . \tag{8.93}$$

Eliminating A_m between (8.90) and (8.92) yields a set of relations between the coefficients B_m and C_m which can be cast into the matrix form

$$\mathbf{b} = \mathbf{Mc} \,, \tag{8.94}$$

where the matrix $\mathbf{M}$ has the elements

$$M_{mn} = \frac{a_1^2 \omega_{p1}^2}{\omega^2 - \omega_m^2} \left(q^2 + \frac{m^2}{a_1^2} \right) \frac{K_{m+n}(qd) K_m(qa_1) I_m^2(qa_1)}{K_n(qa_2)} \,, \tag{8.95}$$

where ω_m is the frequency of a SP mode of a single 2DEG tube with radius a_1 given in (8.77) by replacing m with j. In an analogous way, we use the BCs at the surface of the second tube, at $\rho_2 = a_2$, obtaining vector relations

$$\mathbf{d} = \mathbf{c} + \mathbf{Qb} \,, \tag{8.96}$$

and

$$\mathbf{c} = \mathbf{Nb} \,, \tag{8.97}$$

where the matrices $\mathbf{Q}$ and $\mathbf{N}$ are obtained from $\mathbf{P}$ and $\mathbf{M}$ through permutation of the indices 1 and 2, where ω_{pm} has been defined before in Sect. 8.3. From (8.94) and (8.97), one obtains

$$(\mathbf{MN} - I)\mathbf{b} = 0 \,, \tag{8.98}$$

where I is an identity matrix. The dispersion relations for the SP modes are obtained by equating the determinant of this set of equation to zero,

$$\det(\mathbf{MN} - I) = 0 \,. \tag{8.99}$$

If the two 2DEG tubes have equal radii, then $\mathbf{M} = \mathbf{N}$ and $\mathbf{P} = \mathbf{Q}$, and (8.99) subdivides into two sets corresponding to

$$\det(\mathbf{M} + I) = 0 \,, \tag{8.100}$$

$$\det(\mathbf{M} - I) = 0 \,. \tag{8.101}$$

For a pair of parallel 2DEG tubes with equal radii, the problem which arises now is to find the SP modes from (8.100) and (8.101), where the matrices are of infinite dimension. Solving (8.100) and (8.101) numerically require retaining for each interacting 2DEG tube only the modes for which $|m|$ is lower than a cutoff value m_{max}. This limits the matrices $\mathbf{M} + I$ and $\mathbf{M} - I$ to the dimension $2m_{max} + 1$, leading to a total of modes equal to $2(2m_{max} + 1)$. In each case convergence tests have been made with increasing values of m_{max} in order to obtain convergence for the SP modes. Limiting, for example, the dimensions of $\mathbf{M}$ to $m_{max} = 0$, one obtains two coupled monopole-monopole frequencies as

$$\omega_{\pm} = \left\{ \omega_0^2 \pm (qa_1)^2 \, \omega_{p1}^2 K_0(qd) I_0^2(qa_1) \right\}^{1/2} , \tag{8.102}$$

for $m, n = 0$. For coupling of dipole-dipole modes, $m, n = \pm 1$ and $(\mathbf{M} \pm \mathbf{I})$ is a 2×2 matrix [23], which leads to four solutions for ω, as

$$\omega_{\pm}^{\{+,-\}} = \left\{ \omega_1^2 \{-,+\} a_1^2 \omega_{p1}^2 \left[q^2 + \frac{1}{a_1^2} \right] I_1^2(qa_1) \left[K_2(qd) \{\mp, \pm\} K_0(qd) \right] \right\}^{1/2} . \tag{8.103}$$

Also, when, $m, n = \pm 2$, one obtains the coupled quadrupole-quadrupole modes, as

$$\omega_{\pm}^{\{+,-\}} = \left\{ \omega_2^2 \{-,+\} a_1^2 \omega_{p1}^2 \left[q^2 + \frac{4}{a_1^2} \right] I_2^2(qa_1) \left[K_4(qd) \{\mp, \pm\} K_0(qd) \right] \right\}^{1/2} . \tag{8.104}$$

Figure 8.9 shows the ten lowest coupled SP frequencies of a system of two parallel single-walled CNTs, as a function of the ratio d/a_1. In the limit of infinite separation the two lowest modes reduce to the $m = 0$ isolated single-walled CNT modes and the four upper modes to the $m = 1$ nanotube modes. Finally, we note here that the method employed here is not applicable at $a_1/d = 0.5$, i.e., for touching nanotubes [24, 25]. In fats, when $2a_1/d$ approaches unity, it is found that larger and larger values of m_{max} are needed to ensure the convergence of first few eigenvalues.

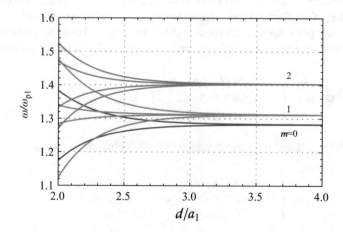

Fig. 8.9 The dimensionless frequencies ω/ω_{p1} of SP modes of a pair of parallel single-walled CNTs as a function of the ratio d/a_1, when $a_2 = a_1$, $qa_1 = 3$, and $\alpha_j = 0$. Only the frequencies reducing to the $m = 1$, 2, and 3 modes at infinite separation are shown

Fig. 8.10 A cylindrical 2DEG layer of radius a_2 and relative dielectric functions inside $\rho < a_1$ of ε_{wire} and outside $\rho > a_3$ of ε_{chan}

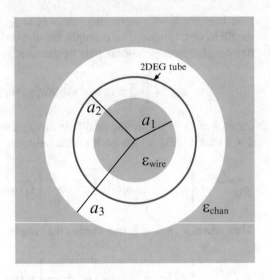

8.5   Surface Plasmon Modes of Cylindrical Two-Dimensional Electron Gas Layers in Dielectric Media

Let us finish this chapter by taking a brief look at SP dispersion relation of a cylindrical 2DEG layer that either is encapsulated in an EG channel or encapsulates an EG wire. Here, we may expect new excitations for the new configuration. As shown in Fig. 8.10, our system consists of a cylindrical 2DEG layer of radius a_2. The inner region has relative dielectric function ε_{wire} for $0 < \rho < a_1$, while the outer region has relative dielectric function ε_{chan} for $\rho > a_3$. Again, we consider the propagation of an electrostatic surface wave parallel to the surfaces of the system, so the electric potential is represented in the form of (8.5). Then the solution of (8.7) for each space region of a 2DEG tube encapsulating an EG wire is given by

$$
\tilde{\Phi}(\rho) = \begin{cases} A_1 I_m(q\rho) , & \rho \leq a_1 , \\ A_2 I_m(q\rho) + A_3 K_m(q\rho) , & a_1 \leq \rho \leq a_2 , \\ A_4 K_m(q\rho) , & \rho \geq a_2 . \end{cases} \tag{8.105}
$$

Similarly, for a 2DEG tube encapsulated by an EG channel, we have

$$
\tilde{\Phi}(\rho) = \begin{cases} B_1 I_m(q\rho) , & \rho \leq a_1 , \\ B_2 I_m(q\rho) + B_3 K_m(q\rho) , & a_1 \leq \rho \leq a_2 , \\ B_4 K_m(q\rho) , & \rho \geq a_2 . \end{cases} \tag{8.106}
$$

To solve Eq. (8.105) and (8.106) we have to provide appropriate BCs. With the induced density n, for a 2DEG tube encapsulating an EG wire, these BCs can be written as

$$\left.\Phi_1\right|_{\rho=a_1} = \left.\Phi_2\right|_{\rho=a_1} , \tag{8.107a}$$

$$\left.\Phi_2\right|_{\rho=a_2} = \left.\Phi_3\right|_{\rho=a_2} , \tag{8.107b}$$

$$\varepsilon_{wire} \left.\frac{\partial \Phi_1}{\partial \rho}\right|_{\rho=a_1} = \left.\frac{\partial \Phi_2}{\partial \rho}\right|_{\rho=a_1} , \tag{8.107c}$$

$$\left.\frac{\partial \Phi_3}{\partial \rho}\right|_{\rho=a_2} - \left.\frac{\partial \Phi_2}{\partial \rho}\right|_{\rho=a_2} = \frac{en}{\varepsilon_0} , \tag{8.107d}$$

where subscript 1 denotes the region inside the EG wire, while subscripts 2 and 3 denote the regions inside and outside the cylindrical 2DEG layer, respectively. Also in the case, where a 2DEG tube encapsulated by an EG channel, we have

$$\left.\Phi_1\right|_{\rho=a_1} = \left.\Phi_2\right|_{\rho=a_1} , \tag{8.108a}$$

$$\left.\Phi_2\right|_{\rho=a_2} = \left.\Phi_3\right|_{\rho=a_2} , \tag{8.108b}$$

$$\left.\frac{\partial \Phi_2}{\partial \rho}\right|_{\rho=a_1} - \left.\frac{\partial \Phi_1}{\partial \rho}\right|_{\rho=a_1} = \frac{en}{\varepsilon_0} , \tag{8.108c}$$

$$\left.\frac{\partial \Phi_2}{\partial \rho}\right|_{\rho=a_2} = \varepsilon_{chan} \left.\frac{\partial \Phi_3}{\partial \rho}\right|_{\rho=a_2} , \tag{8.108d}$$

where subscripts 1 and 2 denote the regions inside and outside the cylindrical 2DEG layer, respectively, while subscript 3 denotes the region inside the EG channel.

After doing some algebra and using a Drude EG wire and channel with $\varepsilon(\omega) = 1 - \Omega_p^2/\omega^2$,[5] from the above equations for a 2DEG encapsulating an EG wire or encapsulated by an EG channel, we obtain two branches for ω defining the dispersion relation of the SP modes which are clearly separated into a high-frequency, ω_+, and a low-frequency, ω_-, as

$$\omega_{\pm}^2 = \frac{\omega_{tube}^2 + \omega_{wire/chan}^2}{2} \pm \sqrt{\left(\frac{\omega_{tube}^2 - \omega_{wire/chan}^2}{2}\right)^2 + \Delta_{wire/chan}^2}, \tag{8.109}$$

where

$$\omega_{wire}^2 = \Omega_p^2 \, (qa_1) \, I_m' \, (qa_1) \, K_m \, (qa_1) , \tag{8.110}$$

---

[5] Ω_p is the electron plasma frequency of the EG wire/channel, as mentioned in Chap. 1.

$$\omega_{chan}^2 = -\Omega_p^2 (qa_3) I_m (qa_3) K_m' (qa_3) , \tag{8.111}$$

are the squares of the SP frequencies of a Drude EG wire and EG channel,[6] respectively, and

$$\omega_{tube}^2 = \alpha^2 \left(q^2 + \frac{m^2}{a_2^2} \right) + a_2^2 \omega_p^2 \left(q^2 + \frac{m^2}{a_2^2} \right) I_m (qa_2) K_m (qa_2) , \tag{8.112}$$

is the square of the SP frequency of a 2DEG tube and

$$\Delta_{wire} = a_2 \omega_p \omega_{wire} \left(q^2 + \frac{m^2}{a_2^2} \right)^{1/2} K_m (qa_2) \left[\frac{I_m (qa_1)}{K_m (qa_1)} \right]^{1/2} , \tag{8.113}$$

and

$$\Delta_{chan} = a_2 \omega_p \omega_{chan} \left(q^2 + \frac{m^2}{a_2^2} \right)^{1/2} I_m (qa_2) \left[\frac{K_m (qa_3)}{I_m (qa_3)} \right]^{1/2} , \tag{8.114}$$

give the interaction between the SPs of 2DEG tube and EG wire/channel. The dispersion characteristics of the SPs of 2DEG tube in the presence of dielectric media can now be understood as the interaction of the SPs of 2DEG tube and EG wire/channel.

References

1. Y.-N. Wang, Z.L. Mišković, Interactions of fast ions with carbon nanotubes: self-energy and stopping power. Phys. Rev. A **69**, 022901 (2004)
2. M. Abramowitz, I.A. Stegum (eds.), *Handbook of mathematical functions* (Dover, New York, 1965)
3. C. Yannouleas, E.N. Bogachek, U. Landman, Collective excitations of multishell carbon microstructures: multishell fullerenes and coaxial nanotubes. Phys. Rev. B **53**, 10225–10236 (1996)
4. S. Iijima, Helical microtubules of graphitic carbon. Nature (London) **354**, 56–58 (1991)
5. D.J. Mowbray, Z.L. Mišković, F.O. Goodman, Y.-N. Wang, Interactions of fast ions with carbon nanotubes: two-fluid model. Phys. Rev. A **70**, 195418 (2004)
6. P. Longe, and S. M. Bose, Collective excitations in metallic graphene tubules, Phys. Rev. B **48**, 18239–18243 (1993)
7. G. Barton, C. Eberlein, Plasma spectroscopy proposed for C_{60} and C_{70}. J. Chem. Phys. **95**, 1512–1517 (1991)
8. X. Jiang, Collective plasmon excitations in graphene tubules. Phys. Rev. B **54**, 13487–13490 (1996)

---

[6]See Eqs. (2.66) and (2.69).

9. L. Calliari, S. Fanchenko, M. Filippi, Plasmon features in electron energy loss spectra from carbon materials. Carbon **45**, 1410–1418 (2007)

10. Z.L. Mišković, S. Segui, J.L. Gervasoni, N.R. Arista, Energy losses and transition radiation produced by the interaction of charged particles with a graphene sheet. Phys. Rev. B **94**, 125414 (2016)

11. H. Khosravi, D.R. Tilley, R. Loudon, Surface polaritons in cylindrical optical fibers. J. Opt. Soc. Am. A **8**, 112–122 (1991)

12. A. Moradi, Theory of energy and power flow of plasmonic waves on single-walled carbon nanotubes. J. Appl. Phys. **122**, 133103 (2017)

13. C.F. Bohren, D.R. Huffman, *Absorption and Scattering of Light by Small Particles* (Wiley, New York, 1983)

14. J.C. Maxwell-Garnett, Colours in metal glasses and in metallic films. Philos. Trans. R. Soc. London, Ser. B **203**, 385–420 (1904)

15. A. Moradi, Plasmon hybridization in tubular metallic nanostructures. Physica B **405**, 2466–2469 (2010)

16. E. Prodan, C. Radloff, N.J. Halas, P. Nordlander, A hybridization model for the plasmon response of complex nanostructures. Science **302**, 419–422 (2003)

17. D.J. Mowbray, S. Chung, Z.L. Mišković, F.O. Goodman, Y.-N. Wang, Dynamic interactions of fast ions with carbon nanotubes. Nucl. Instrum. Methods B **230**, 142–147 (2005)

18. G. Gumbs, A. Balassis, Effects of coupling on plasmon modes and drift-induced instabilities in a pair of cylindrical nanotubes. Phys. Rev. B **68**, 075405 (2003)

19. A. Moradi, Electron-hole plasma excitations in single-walled carbon nanotubes. Phys. Lett. A **372**, 5614–5616 (2008)

20. A. Moradi, Plasmon hybridization in coated metallic nanowires. J. Opt. Soc. Am. B **29**, 625–629 (2012)

21. A. Moradi, Geometrical tunability of plasmon excitations of double concentric metallic nanotubes. Phys. Plasmas **19**, 062102 (2012)

22. A. Moradi, Plasma wave propagation in a pair of carbon nanotubes. JETP Lett. **88**, 795–798 (2008)

23. K. Andersen, K.L. Jensen, N.A. Mortensen, K.S. Thygesen, Visualizing hybridized quantum plasmons in coupled nanowires: from classical to tunneling regime. Phys. Rev. B **87**, 235433 (2013)

24. M. Schmeits, Surface-plasmon coupling in cylindrical pores. Phys. Rev. B **39**, 7567–7577 (1989)

25. A. Moradi, Plasmon hybridization in parallel nano-wire systems. Phys. Plasmas **18**, 064508 (2011)

Chapter 9
Electromagnetic Problems Involving Two-Dimensional Electron Gases in Cylindrical Geometry

Abstract In this chapter, some optical properties of two-dimensional electron gas layers in cylindrical geometry are studied within the framework of classical electrodynamics. We use the standard hydrodynamic model to describe the dielectric response of a cylindrical electron gas layer. Explicit results are given for a collection of electromagnetic boundary-value problems. For brevity, throughout the chapter the $\exp(-i\omega t)$ time factor is suppressed. Furthermore, all media under consideration are nonmagnetic and attention is only confined to the linear phenomena.

9.1 Plasmonic Properties of Cylindrical Two-Dimensional Electron Gas Layers

9.1.1 Dispersion Relation

We consider an infinitely long 2DEG tube with a radius a, and assume that 2DEG distributed uniformly over the cylindrical surface, with the equilibrium density (per unit area) n_0. We take cylindrical polar coordinates $\mathbf{r} = (\rho, \phi, z)$ for an arbitrary point in space.

The electromagnetic modes of the system are solutions of Maxwell's equations with standard BCs at $r = a$. Dependence on z, ϕ, and time t enters only by means of the second derivation $\partial^2/\partial z^2$, $\partial^2/\partial\phi^2$, and $\partial^2/\partial t^2$, so it is sufficient to seek solutions in which all field components contain a common factor $\exp(iqz + im\phi - i\omega t)$. Since the solutions must be single-valued functions of ϕ, m is restricted to integer values. To determine all the field components for $\rho > a$ and $\rho < a$, we need to only calculate E_z and H_z, and it is readily shown that these satisfy

$$\frac{\mathrm{d}^2 E_z}{\mathrm{d}\rho^2} + \frac{1}{\rho}\frac{\mathrm{d}E_z}{\mathrm{d}\rho} - \left(\kappa_j^2 + \frac{m^2}{\rho^2}\right)E_z = 0, \quad \rho \neq a, \tag{9.1}$$

where $\kappa_j^2 = q^2 - \varepsilon_j\omega^2/c^2$ with $j = 1$ for $\rho < a$ and $j = 2$ for $\rho > a$. The same equation holds for H_z. Note that (9.1) applies for both $+m$ and $-m$ modes. The

A. Moradi, *Canonical Problems in the Theory of Plasmonics*, Springer Series in Optical Sciences 230, https://doi.org/10.1007/978-3-030-43836-4_9

solutions for the fields involve the modified Bessel equation, since in this section we restrict attention to non-radiative modes. When all field components contain a common factor $\exp(iqz + im\phi - i\omega t)$, so that $\partial/\partial t \to -i\omega$, $\partial/\partial z \to iq$, $\partial/\partial\phi \to im$, and (1.33c) and (1.33d), take the forms

$$m E_z - q\rho E_\phi = \omega\mu_0\rho H_\rho \,, \tag{9.2a}$$

$$-\frac{dE_z}{d\rho} + iq E_\rho = i\omega\mu_0 H_\phi \,, \tag{9.2b}$$

$$E_\phi + \rho\frac{dE_\phi}{d\rho} - im E_\rho = i\omega\mu_0\rho H_z \,, \tag{9.2c}$$

and

$$m H_z - q\rho H_\phi = -\omega\varepsilon_0\varepsilon_j\rho E_\rho \,, \tag{9.3a}$$

$$-\frac{dH_z}{d\rho} + iq H_\rho = -i\omega\varepsilon_0\varepsilon_j E_\phi \,, \tag{9.3b}$$

$$H_\phi + \rho\frac{dH_\phi}{d\rho} - im H_\rho = -i\omega\varepsilon_0\varepsilon_j\rho E_z \,. \tag{9.3c}$$

The following expressions for E_ϕ and H_ρ in terms of E_z and H_z follow from (9.2a) and (9.3b)

$$E_\phi = \frac{qm}{\rho\kappa_j^2}E_z + \frac{i\omega\mu_0}{\kappa_j^2}\frac{dH_z}{d\rho} \,, \tag{9.4}$$

$$H_\rho = -\frac{m\omega\varepsilon_0\varepsilon_j}{\rho\kappa_j^2}E_z - \frac{iq}{\kappa_j^2}\frac{dH_z}{d\rho} \,. \tag{9.5}$$

Likewise, (9.2b) and (9.3a) give

$$H_\phi = \frac{qm}{\rho\kappa_j^2}H_z - \frac{i\omega\varepsilon_0\varepsilon_j}{\kappa_j^2}\frac{dE_z}{d\rho} \,, \tag{9.6}$$

$$E_\rho = \frac{m\omega\mu_0}{\rho\kappa_j^2}H_z - \frac{iq}{\kappa_j^2}\frac{dE_z}{d\rho}. \tag{9.7}$$

When (9.4)–(9.7) are substituted in (9.3c), the terms involving H_z cancel out, and the resulting equation for E_z is (9.1). Likewise, on substitution in (9.2c), the terms in E_z cancel out, leaving an equation for H_z which is identical in form to (9.1).

Now, we assume a SPP mode propagates parallel to both axial and azimuthal directions of a cylindrical 2DEG layer. As can be seen from (9.4)–(9.7), SPP modes are no longer TM^z or TE^z modes in general. Based on the linearized SHD theory, the electronic excitations of a cylindrical 2DEG layer may be described by the following linearized equations:

$$\left(\omega^2 - \alpha^2 q^2\right) J_z - \alpha^2 q \frac{m}{a} J_\phi = i\omega \frac{n_0 e^2}{m_e} E_z , \tag{9.8a}$$

$$-\alpha^2 q \frac{m}{a} J_z + \left(\omega^2 - \alpha^2 \frac{m^2}{a^2}\right) J_\phi = i\omega \frac{n_0 e^2}{m_e} E_\phi , \tag{9.8b}$$

where $J_z = -en_0 v_z$ and $J_\phi = -en_0 v_\phi$ are the polarization current density along the z- and ϕ-directions, respectively, due to the motion of 2DEG. Using (9.8a) and (9.8b), we can find the conductivity of the system that is a 2D tensor, consisting of four factors or components, as

$$\underline{\sigma} = \begin{pmatrix} \sigma_{zz} & \sigma_{z\phi} \\ \sigma_{\phi z} & \sigma_{\phi\phi} \end{pmatrix} , \tag{9.9}$$

with σ_{zz} ($\sigma_{\phi\phi}$) and $\sigma_{z\phi}$ ($\sigma_{\phi z}$) being the diagonal and off-diagonal components of conductivity tensor, respectively, given below [1]

$$\sigma_{zz} = \frac{in_0 e^2}{m_e} \frac{\omega^2 - \alpha^2 m^2 / a^2}{\omega \left[\omega^2 - \alpha^2 \left(q^2 + \frac{m^2}{a^2}\right)\right]} , \tag{9.10a}$$

$$\sigma_{z\phi} = \sigma_{\phi z} = \frac{in_0 e^2}{m_e} \frac{\alpha^2 q m / a}{\omega \left[\omega^2 - \alpha^2 \left(q^2 + \frac{m^2}{a^2}\right)\right]} , \tag{9.10b}$$

$$\sigma_{\phi\phi} = \frac{in_0 e^2}{m_e} \frac{\omega^2 - \alpha^2 q^2}{\omega \left[\omega^2 - \alpha^2 \left(q^2 + \frac{m^2}{a^2}\right)\right]} . \tag{9.10c}$$

For simplicity, in the following we consider a cylindrical 2DEG layer in vacuum, where $\kappa_1 = \kappa = \kappa_2$. Therefore, the relevant solution of (9.1) and the corresponding solution for H_z are

$$E_z = E_{0z} \begin{cases} \dfrac{I_m(\kappa\rho)}{I_m(\kappa a)} , & \rho \leq a , \\[2ex] \dfrac{K_m(\kappa\rho)}{K_m(\kappa a)} , & \rho \geq a , \end{cases} \tag{9.11}$$

and

$$H_z = H_{0z} \begin{cases} \dfrac{I_m(\kappa\rho)}{I'_m(\kappa a)} \,, & \rho \leq a \,, \\[3mm] \dfrac{K_m(\kappa\rho)}{K'_m(\kappa a)} \,, & \rho \geq a \,, \end{cases} \tag{9.12}$$

that involve amplitudes E_{0z} and H_{0z}, respectively. All the other components can be obtained from (9.4)–(9.7). The field components have to satisfy the usual BCs at $\rho = a$ that is, continuity of E_ϕ and E_z, as

$$E_{1\phi}|_{\rho=a} = E_{2\phi}|_{\rho=a} \,, \tag{9.13}$$

$$E_{1z}|_{\rho=a} = E_{2z}|_{\rho=a} \,, \tag{9.14}$$

and discontinuity of H_z and H_ϕ, due to the conductivity tensor $\underline{\sigma}$ associated with the cylindrical 2DEG layer, as

$$H_{2\phi}|_{\rho=a} - H_{1\phi}|_{\rho=a} = \sigma_{zz}E_z(a) + \sigma_{z\phi}E_\phi(a) \,, \tag{9.15}$$

$$H_{2z}|_{\rho=a} - H_{1z}|_{\rho=a} = -\sigma_{\phi z}E_z(a) - \sigma_{\phi\phi}E_\phi(a) \,, \tag{9.16}$$

where subscript 1 denotes the region inside the cylindrical 2DEG layer and subscript 2 denotes the region outside the system. Also $E_\phi(a) = E_{1\phi}|_{\rho=a} = E_{2\phi}|_{\rho=a}$ and $E_z(a) = E_{1z}|_{\rho=a} = E_{2z}|_{\rho=a}$. The first two are already satisfied in the forms of E_z and H_z chosen in (9.10) and (9.11). Using (9.15) and (9.16), we can write

$$\frac{H_{0z}}{E_{0z}} = -\frac{\sigma_{z\phi} + \sigma_{\phi\phi}\dfrac{qm}{a\kappa^2}}{\sigma_{\phi\phi}\dfrac{i\omega\mu_0}{\kappa} + \dfrac{1}{(\kappa a)I'_m(\kappa a)K'_m(\kappa a)}} \,, \tag{9.17}$$

$$\frac{E_{0z}}{H_{0z}} = \frac{\sigma_{z\phi}\dfrac{i\omega\mu_0}{\kappa} - \dfrac{qm}{a\kappa^2}\dfrac{1}{(\kappa a)I'_m(\kappa a)K'_m(\kappa a)}}{\dfrac{i\omega\varepsilon_0}{\kappa}\dfrac{1}{(\kappa a)I_m(\kappa a)K_m(\kappa a)} - \left(\sigma_{zz} + \sigma_{z\phi}\dfrac{qm}{a\kappa^2}\right)} \,. \tag{9.18}$$

Here, the Wronskian property $I'_m(x)K_m(x) - I_m(x)K'_m(x) = 1/x$ has been used. The multiplication of the right-hand side (9.17) and (9.18) should be equal. This leads to the dispersion equations of SPP modes of the system, as [1]

$$\frac{\sigma_{z\phi} + \sigma_{\phi\phi}\dfrac{qm}{a\kappa^2}}{\sigma_{\phi\phi}\dfrac{i\omega\mu_0}{\kappa} + \dfrac{1}{(\kappa a)I'_m(\kappa a)K'_m(\kappa a)}}$$

$$+\frac{\dfrac{i\omega\varepsilon_0}{\kappa}\dfrac{1}{(\kappa a)I_m(\kappa a)K_m(\kappa a)}-\left(\sigma_{zz}+\sigma_{z\phi}\dfrac{qm}{a\kappa^2}\right)}{\sigma_{z\phi}\dfrac{i\omega\mu_0}{\kappa}-\dfrac{qm}{a\kappa^2}\dfrac{1}{(\kappa a)I'_m(\kappa a)K'_m(\kappa a)}}=0\,,\qquad(9.19)$$

that is a transcendental equation and needs to be solved numerically in order to obtain the spectrum of SPP modes. This dispersion relation describes modes in which the field amplitude is maximum at $\rho=a$ and decreases with distance from the surface of the system. This behavior corresponds to the SPPs of a planar 2DEG layer (see Sect. 7.2), so it might be expected that the dispersion equation for this mode would appear as the $a\to\infty$ limit of (9.19). In this case we obtain

$$\kappa\sigma_{zz}(m=0)-2i\omega\varepsilon_0=0\,,\qquad i\omega\mu_0\sigma_{\phi\phi}(m=0)-2\kappa=0\,.\qquad(9.20)$$

It can be seen that the first of these corresponds to the standard form of the SPPs dispersion relation (see Sect. 7.2), while the second dispersion relation has no solution for real ω and q. We note the dispersion relation of non-radiative TM^z mode of a 2DEG tube is[1]

$$\frac{i\omega\varepsilon_0}{\kappa}\frac{1}{(\kappa a)I_0(\kappa a)K_0(\kappa a)}-\sigma_{zz}(m=0)=0\,.\qquad(9.21)$$

Figure 9.1, panels (a)–(d), shows the dependence of the dimensionless frequency ω/ω_{p0} on the dimensionless variable qa_0 for a single-walled CNT characterized by $n_0=4\times 38$ nm^{-2} in the $\sigma+\pi$ electron fluid model and different values of the nanotube radius a with $m=0,1,2$, and 3, where $\omega_{p0}=\sqrt{n_0e^2/\varepsilon_0 m_e a_0}$ and $a_0=1$nm. We see that in each of the four surface mode series, the $m=0$ mode is the lowest and the frequency increases with m. On the other hand, from panels (a)–(d), one can see that for long-wavelength region (i.e., $q\to 0$), as increasing the tube radius the frequency of SPP modes decreases, while the frequency of SPP modes approaches each other for short-wavelength region.

9.1.2 Excitation of Modes

In order to survey the excitation mechanism of SPP modes of a 2DEG tube, in panel (c) of Fig. 9.1 we have plotted the speed line of light in vacuum and speed line of an electron beam[2] by considering the expression $\omega=v_0 q$, where v_0 is the

---

[1] Set $m=0$ in (9.19).

[2] An electron beam is a stream of moving electrons in vacuum, emitted from a cathode, accelerated by the electric field between the cathode and the anode, confined by a longitudinal magnetic field, and finally collected by a collector [2].

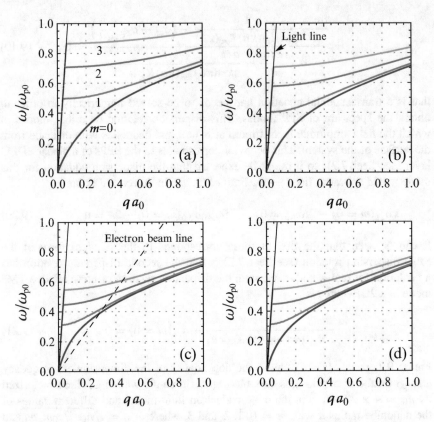

Fig. 9.1 Dispersion curves of SPP modes of a single-walled CNT as given by (9.19), when $m = 0, 1, 2,$ and 3. The different panels refer to (**a**) $a = 2$ nm, (**b**) $a = 3$ nm, (**c**) $a = 4$ nm, and (**d**) $a = 5$ nm. In panel (**c**) the dashed line is dispersion line of a fast electron beam with speed 1×10^7 m/s

electron beam speed. It is easy to find that we cannot excite a SPP mode with a suitable frequency and wavelength by exposing laser pulses from vacuum on the top of the system (see Sect. 3.3); however, when an electron beam velocity located in the appropriate range (i.e., the velocity of the electron beam can be equal to the phase velocity of the SPP modes), the electron beam is in synchronization with the surface waves, and they interact with each other and instability occurs between them [3, 4]. Therefore, we may conclude that SPP modes of a single-walled CNT characterized by $n_0 = 4 \times 38$ nm^{-2} in the $\sigma + \pi$ electron fluid model and $a = 4$nm (as an example) can be excited by applying some fast electron beams with the speed of about $v_0 = 10^7$m/s [5, 6].

9.1.3 Power Flow and Energy Distribution

We now consider a SPP mode that propagates along the z-axis of a 2DEG tube and is a periodic wave in the azimuthal direction. As mentioned in Sect. 8.1.2, for applications to power flow and energy density, it is convenient to work in terms of real functions $\cos m\phi$ and $\sin m\phi$ [7, 8]. We note that specific phase relations between the different field components exist. The relevant solution of E_z and H_z in the region $\rho \leq a$, are

$$E_{1z} = E_{0z} \frac{I_m(\kappa\rho)}{I_m(\kappa a)} \begin{Bmatrix} \cos m\phi \\ i \sin m\phi \end{Bmatrix} , \tag{9.22}$$

$$H_{1z} = H_{0z} \frac{I_m(\kappa\rho)}{I_m'(\kappa a)} \begin{Bmatrix} i \sin m\phi \\ \cos m\phi \end{Bmatrix} . \tag{9.23}$$

With the upper choice of field solution, we have

$$E_{1\rho} = -\frac{i}{\kappa^2} \left[q\kappa \frac{I_m'(\kappa\rho)}{I_m(\kappa a)} E_{0z} + im \frac{\omega\mu_0}{\rho} \frac{I_m(\kappa\rho)}{I_m'(\kappa a)} H_{0z} \right] \cos m\phi , \tag{9.24}$$

$$E_{1\phi} = \frac{i}{\kappa^2} \left[m\frac{q}{\rho} \frac{I_m(\kappa\rho)}{I_m(\kappa a)} E_{0z} + i\omega\mu_0\kappa \frac{I_m'(\kappa\rho)}{I_m'(\kappa a)} H_{0z} \right] \sin m\phi , \tag{9.25}$$

$$H_{1\rho} = -\frac{i}{\kappa^2} \left[m\frac{\omega\varepsilon_0}{\rho} \frac{I_m(\kappa\rho)}{I_m(\kappa a)} E_{0z} + iq\kappa \frac{I_m'(\kappa\rho)}{I_m'(\kappa a)} H_{0z} \right] \sin m\phi , \tag{9.26}$$

$$H_{1\phi} = -\frac{i}{\kappa^2} \left[\omega\varepsilon_0\kappa \frac{I_m'(\kappa\rho)}{I_m(\kappa a)} E_{0z} + im\frac{q}{\rho} \frac{I_m(\kappa\rho)}{I_m'(\kappa a)} H_{0z} \right] \cos m\phi . \tag{9.27}$$

By interchanging $\cos m\phi$ and $i \sin m\phi$ in (9.24)–(9.27) the field components for the lower choice of solutions are obtained. The energy density U_1 and power flow density $\mathbf{S}_1(\rho, \phi, z)$ associated with the SPP modes in the region $\rho \leq a$ are

$$U_1 = \frac{1}{2} \left(\varepsilon_0 |\mathbf{E}_1|^2 + \mu_0 |\mathbf{H}_1|^2 \right) , \tag{9.28}$$

$$\mathbf{S}_1 = \mathbf{E}_1 \times \mathbf{H}_1 . \tag{9.29}$$

The same equations hold for U_2 and $\mathbf{S}_2(\rho, \phi, z)$ for $\rho \geq a$. Let us note that Poynting vector $\mathbf{S}(\rho, \phi, z)$ have components in the all directions, but their ρ- and ϕ-components vanish on averaging over a cycle of oscillation of the fields. Thus, the time-averaged quantity $\mathbf{S}$ has only z-component, i.e., S_z. Furthermore, in the absence of the internal interaction or nonlocal effects in the 2DEG, i.e., $\alpha = 0$, the energy density U_{2D} and power flow density S_{2Dz} on the surface of the cylindrical

2DEG layer are, as $S_{2Dz} = 0$ and $U_{2D} = m_e n_0 \mathbf{v}^2/2$, where subscript 2D refers to the surface of the cylindrical 2DEG layer $\rho = a$. Note that, for the present system we have $\sigma_{z\phi} = 0 = \sigma_{\phi z}$, and

$$\sigma_{zz} = \sigma_{\phi\phi} = \sigma_0 = \frac{i n_0 e^2}{\omega m_e} . \tag{9.30}$$

By eliminating $\mathbf{v} = v_z \mathbf{e}_z + v_\phi \mathbf{e}_\phi$ with $v_z = -\sigma_0 E_z/e n_0$ and $v_\phi = -\sigma_0 E_\phi/e n_0$ in equation $U_{2D} = m_e n_0 \mathbf{v}^2/2$, the cycle-averaged of energy density on a cylindrical 2DEG layer may be expressed, in complex notation, by

$$U_{2D} = \frac{m_e}{4 n_0 e^2} \left\{ \left(|\sigma_0|^2 \cos^2 m\phi + \left| \sigma_0 \frac{qm}{a\kappa^2} \right|^2 \sin^2 m\phi \right) E_{0z}^2 \right.$$
$$\left. + \left(\frac{\omega\mu_0}{\kappa} \right)^2 |\sigma_0|^2 \sin^2 m\phi |H_{0z}|^2 - 2 \frac{i\omega\mu_0}{\kappa} \frac{qm}{a\kappa^2} \sigma_0^2 \sin^2 m\phi H_{0z} E_{0z} \right\} ,$$
$$\tag{9.31}$$

where E_{0z} has been defined to be real. We note that (9.31) applies for the upper choice of trigonometric functions in the field components (9.22) and (9.23); for the lower choice $\cos m\phi$ and $\sin m\phi$ are interchanged throughout. Note that the ratio H_{0z}/E_{0z}, i.e., (9.17) is such that the final term in (9.31) is real. Also, in the region $\rho \leq a$, we have

$$S_{1z} = \frac{\varepsilon_0}{2} \frac{q\omega}{\kappa^2} \left\{ \left[\frac{I'_m(\kappa\rho)}{I_m(\kappa a)} \right]^2 \cos^2 m\phi + \frac{m^2}{\rho^2\kappa^2} \left[\frac{I_m(\kappa\rho)}{I_m(\kappa a)} \right]^2 \sin^2 m\phi \right\} E_{0z}^2$$
$$+ \frac{\mu_0}{2} \frac{q\omega}{\kappa^2} \left\{ \left[\frac{I'_m(\kappa\rho)}{I'_m(\kappa a)} \right]^2 \sin^2 m\phi + \frac{m^2}{\rho^2\kappa^2} \left[\frac{I_m(\kappa\rho)}{I'_m(\kappa a)} \right]^2 \cos^2 m\phi \right\} |H_{0z}|^2$$
$$+ i \frac{m}{\rho\kappa^3} \left(q^2 + \frac{\omega^2}{c^2} \right) \frac{I_m(\kappa\rho)}{I_m(\kappa a)} \frac{I'_m(\kappa\rho)}{I'_m(\kappa a)} E_{0z} H_{0z} , \tag{9.32}$$

and

$$U_1 = \frac{1}{4} \left(\varepsilon_0 \mathbf{E} \cdot \mathbf{E}^* + \mu_0 \mathbf{H} \cdot \mathbf{H}^* \right) , \tag{9.33}$$

$$\mathbf{E} \cdot \mathbf{E}^* = \left(\left[\frac{I_m(\kappa\rho)}{I_m(\kappa a)} \right]^2 + \frac{q^2}{\kappa^2} \left[\frac{I'_m(\kappa\rho)}{I_m(\kappa a)} \right]^2 \right) E_{0z}^2 \cos^2 m\phi$$
$$+ \frac{q^2 m^2}{\rho^2\kappa^4} \left[\frac{I_m(\kappa\rho)}{I_m(\kappa a)} \right]^2 E_{0z}^2 \sin^2 m\phi$$

$$+ \frac{\omega^2 \mu_0^2}{\kappa^2} \left\{ \left[\frac{I'_m(\kappa\rho)}{I'_m(\kappa a)} \right]^2 \sin^2 m\phi + \frac{m^2}{\rho^2 \kappa^2} \left[\frac{I_m(\kappa\rho)}{I'_m(\kappa a)} \right]^2 \cos^2 m\phi \right\} |H_{0z}|^2$$

$$+ 2 \frac{qm}{\rho \kappa^2} \frac{i\omega\mu_0}{\kappa} \frac{I_m(\kappa\rho)}{I_m(\kappa a)} \frac{I'_m(\kappa\rho)}{I'_m(\kappa a)} E_{0z} H_{0z} , \qquad (9.34a)$$

$$\mathbf{H} \cdot \mathbf{H}^* = \frac{\omega^2 \varepsilon_0^2}{\kappa^2} \left\{ \left[\frac{I'_m(\kappa\rho)}{I_m(\kappa a)} \right]^2 \cos^2 m\phi + \frac{m^2}{\rho^2 \kappa^2} \left[\frac{I_m(\kappa\rho)}{I_m(\kappa a)} \right]^2 \sin^2 m\phi \right\} E_{0z}^2$$

$$+ \left\{ \frac{q^2 m^2}{\rho^2 \kappa^4} \left[\frac{I_m(\kappa\rho)}{I'_m(\kappa a)} \right]^2 \cos^2 m\phi + \left(\left[\frac{I_m(\kappa\rho)}{I'_m(\kappa a)} \right]^2 + \frac{q^2}{\kappa^2} \left[\frac{I'_m(\kappa\rho)}{I'_m(\kappa a)} \right]^2 \right) \sin^2 m\phi \right\} |H_{0z}|^2$$

$$+ 2 \frac{qm}{\rho \kappa^2} \frac{i\omega\varepsilon_0}{\kappa} \frac{I_m(\kappa\rho)}{I_m(\kappa a)} \frac{I'_m(\kappa\rho)}{I'_m(\kappa a)} E_{0z} H_{0z} . \qquad (9.34b)$$

In the region $\rho \geq a$, we use (9.32)–(9.34) with transformations I_m and I'_m to K_m and K'_m, respectively. The total power flow and energy density (per unit length of cylinder) associated with the SPP modes are determined by an integration over ϕ and ρ. We have

$$\langle S_z \rangle = \int_0^a \rho \, d\rho \int_0^{2\pi} S_{1z} \, d\phi + \int_a^\infty \rho \, d\rho \int_0^{2\pi} S_{2z} \, d\phi . \qquad (9.35)$$

$$\langle U \rangle = \int_0^a \rho \, d\rho \int_0^{2\pi} U_1 \, d\phi + a \int_0^{2\pi} U_{2D} \, d\phi + \int_a^\infty \rho \, d\rho \int_0^{2\pi} U_2 \, d\phi . \qquad (9.36)$$

Again, for general m the integrated power flow $\langle S_z \rangle$ and energy density $\langle U \rangle$ can only be evaluated numerically. For $m = 0$ (TM^z mode), however, analytic expressions can be found. After doing some algebra, the total power flow and energy density of a SPP with TM^z mode are given by

$$\langle S_z \rangle = \pi a^2 q \omega \frac{\varepsilon_0 E_{oz}^2}{2\kappa^2} \left\{ \left[\frac{K'_1(\kappa a)}{K_0(\kappa a)} \right]^2 - \left[\frac{I'_1(\kappa a)}{I_0(\kappa a)} \right]^2 \right.$$

$$\left. - \left(1 + \frac{1}{\kappa^2 a^2} \right) \left[\left[\frac{K_1(\kappa a)}{K_0(\kappa a)} \right]^2 - \left[\frac{I_1(\kappa a)}{I_0(\kappa a)} \right]^2 \right] \right\} , \qquad (9.37)$$

and

$$\langle U \rangle = \pi a^2 \frac{\varepsilon_0 E_{oz}^2}{4} \left\{ \left[\frac{K_1(\kappa a)}{K_0(\kappa a)} \right]^2 - \left[\frac{I_1(\kappa a)}{I_0(\kappa a)} \right]^2 + 2 \frac{\omega_p^2}{\omega^2} \right.$$

$$+ \left(q^2 + \frac{\omega^2}{c^2} \right) \left[\left[\frac{1}{\kappa} \frac{K_1'(\kappa a)}{K_0(\kappa a)} \right]^2 - \left[\frac{1}{\kappa} \frac{I_1'(\kappa a)}{I_0(\kappa a)} \right]^2 \right]$$

$$+ \left(1 + \frac{\omega^2}{c^2 q^2} \right) \left(1 + \frac{1}{\kappa^2 a^2} \right) \left[\left[\frac{I_1(\kappa a)}{I_0(\kappa a)} \right]^2 - \left[\frac{K_1(\kappa a)}{K_0(\kappa a)} \right]^2 \right] \right\} , \qquad (9.38)$$

where $\omega_p^2 = n_0 e^2 / \varepsilon_0 m_e a$, as introduced in the previous chapter. To proof the correctness of our findings, we note that the velocity of energy transport $v_e(q, \omega) = \langle S_z \rangle / \langle U \rangle$, and it can be shown numerically that it is equal to the group velocity $v_g(q, \omega) = \partial \omega / \partial q$.

9.1.4 Extinction Property

Let us consider an infinitely long cylindrical 2DEG layer with a radius a that is aligned along the z-axis in the cylindrical coordinates (ρ, ϕ, z). Here, the regions $\rho > a$ and $\rho < a$ are vacuum and we set $\alpha = 0$ (the nonlocal effects are neglected). The system is illuminated by a plane wave at oblique incidence, as shown in Fig. 9.2.

Again, there are two possible polarization states of the incident light: TE^z polarization and TM^z polarization. The vector cylindrical harmonic functions can be defined according to Sect. 3.7.2. For the present BVP, k and κ are given by $k_0 = \omega/c$ and $\kappa_0 = \sqrt{k_0^2 - q^2}$ inside and outside of the system, where $q = -k_0 \cos \zeta$ and ζ is the angle between the incident light and the axis of system. Also $Z_m(\kappa \rho)$ represents

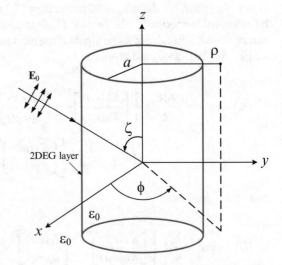

Fig. 9.2 An infinitely long cylindrical 2DEG layer and a cylindrical polar coordinate system. The z-axis lies along the axis of the cylindrical 2DEG layer. The system is illuminated by electromagnetic plane wave TM^z polarization at incidence angle ζ. The z-axis coincides with the axis of the tube and the xz plane shows the incidence plane

a cylindrical Bessel or Hankel function, and is chosen as follows. Inside the cylinder $J_m(\kappa_0\rho)$ is used and outside the cylinder $J_m(\kappa_0\rho)$ and $H_m(\kappa_0\rho)$ are used for the incident and scattered waves, respectively.

9.1.4.1 TE^z Polarization

For TE^z polarization, the incident electric field $\mathbf{E}_i$ can be shown by (3.109), when $\kappa_2 = \kappa_0$. The transmitted and scattered electric fields can be represented by (3.110) and (3.111), respectively, by replacing κ_2 and κ_1 by κ_0. The unknown expansion coefficients $a_{m\perp}$ and $b_{m\perp}$ are determined by the BCs at the surface of the system, i.e., (9.13)–(9.16). After doing some algebra and using the Wronskian property of the Bessel functions as $J_m(x)Y'_m(x) - J'_m(x)Y_m(x) = 2/\pi x$, one can find

$$a_{m\perp} = \frac{D_m E_m}{A_m D_m - B_m C_m}, \tag{9.39a}$$

$$b_{m\perp} = -\frac{C_m E_m}{A_m D_m - B_m C_m}, \tag{9.39b}$$

where

$$A_m = -i\omega H'_m(\kappa_0 a) \left[\sigma_0 + \frac{2(1/a\omega\mu_0)}{\pi J'_m(\kappa_0 a) H'_m(\kappa_0 a)} \right],$$

$$B_m = -c\sigma_0 \frac{qm}{a\kappa_0} H_m(\kappa_0 a),$$

$$C_m = -i \frac{qm}{a^2\mu_0\kappa_0^2} \frac{2}{\pi J'_m(\kappa_0 a)},$$

$$D_m = c\kappa_0 H_m(\kappa_0 a) \left[\sigma_0 + \frac{2(\omega\varepsilon_0/a\kappa_0^2)}{\pi J_m(\kappa_0 a) H_m(\kappa_0 a)} \right],$$

$$E_m = -i\omega\sigma_0 J'_m(\kappa_0 a),$$

and

$$\sigma_0 = \frac{in_0 e^2}{m_e} \frac{1}{\omega + i\gamma}. \tag{9.40}$$

Now, replacing k_2 by k_0, the extinction width can be expressed by (3.114).

9.1.4.2   TM^z Polarization

For TM^z polarization, the incident electric field $\mathbf{E}_i$ can be shown by (3.116), when $\kappa_2 = \kappa_0$. The transmitted and scattered electric fields can be represented by (3.117) and (3.118), respectively, by replacing κ_2 and κ_1 by κ_0. The coefficients of the scattered field can be written in the form

$$a_{m\parallel} = \frac{D_m V_m - B_n W_m}{A_m D_m - B_m C_m} ,$$
(9.41a)

$$b_{m\parallel} = \frac{A_m W_m - C_n V_m}{A_m D_m - B_m C_m} ,$$
(9.41b)

where A_m, B_m and so on were defined in the preceding section and

$$V_m = -c\sigma_0 \frac{qm}{a\kappa_0} J_m(\kappa_0 a) ,$$

$$W_m = c\kappa_0 \sigma_0 J_m(\kappa_0 a) .$$

For TM^z polarization, if we replace k_2 by k_0, the extinction width can be expressed by (3.121).

As mentioned before, the denominators of $a_{m\perp}$, $a_{m\parallel}$, $b_{m\perp}$, and $b_{m\parallel}$ vanish at the frequencies of the mixed TE^z and TM^z polaritons modes. The dispersion relation associated with SPPs propagating along z-axis having the factor $\exp(iqz + im\phi)$ is $A_m D_m - B_m C_m = 0$. For non-radiative SPPs the field amplitudes decay in an exponential fashion as one moves away from the interface into either medium. Therefore κ_0 has to be imaginary. A special case of $A_m D_m - B_m C_m = 0$ is $q = 0$, for which pure TE^z and TM^z modes can exist. In this case, we find

$$\sigma_0(q = 0) + \frac{2}{a\omega\mu_0} \frac{1}{\pi J_m'(k_0 a) H_m'(k_0 a)} = 0 ,$$
(9.42a)

$$\sigma_0(q = 0) + \frac{\omega\varepsilon_0}{ak_0^2} \frac{2}{\pi J_m(k_0 a) H_m(k_0 a)} = 0 ,$$
(9.42b)

for the TE^z and TM^z modes, respectively. Since q is the wavenumber in the direction of the 2DEG tube axis, polariton waves with $q = 0$ will be excited in an optical experiment in which the light wave is incident normally on the system. Another case for which pure TE^z and TM^z modes exist is $m = 0$ and q arbitrary. In this case $A_m D_m - B_m C_m = 0$ simplifies and we obtain

$$\sigma_0(m = 0) + \frac{2}{a\omega\mu_0} \frac{1}{\pi J_0'(\kappa_0 a) H_0'(\kappa_0 a)} = 0 ,$$
(9.43a)

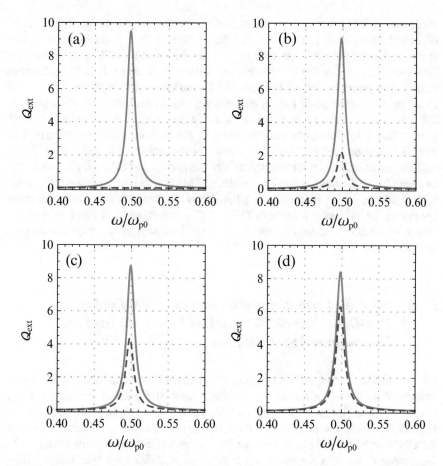

Fig. 9.3 Calculated extinction width (in units of the geometric width) of a single-walled CNT in vacuum for both TE^z and TM^z polarizations, and different values of the incidence angle, when $a = 2$ nm and $\gamma = 0.01\omega_p$. The different panels refer to (**a**) $\zeta = \pi/2$, (**b**) $\zeta = \pi/3$, (**c**) $\zeta = \pi/4$, and (**d**) $\zeta = \pi/6$. The red solid curves are for TE^z (or $\mathbf{E} \perp \mathbf{e}_z$) and the blue dashed ones for TM^z (or $\mathbf{E} \parallel \mathbf{e}_z$). Here, $\omega_{p0} = \sqrt{n_0 e^2/\varepsilon_0 m_e a_0}$, $a_0 = 1$ nm, and $n_0 = 4 \times 38$ nm^{-2}

$$\sigma_0(m = 0) + \frac{\omega \varepsilon_0}{a \kappa_0^2} \frac{2}{\pi J_0(\kappa_0 a) H_0(\kappa_0 a)} = 0 \,, \tag{9.43b}$$

for the pure TE^z and TM^z modes, respectively.

In Fig. 9.3, the extinction spectra of a single-walled CNT [9] in vacuum with various parameter ζ are plotted for both TE^z and TM^z polarizations, when $a = 2$ nm. We choose the friction coefficients to be small but finite parameters, $\gamma = 0.01\omega_{p0}$, with $\omega_{p0} = \sqrt{n_0 e^2/\varepsilon_0 m_e a_0}$, where $n_0 = 4 \times 38$ nm^{-2} and $a_0 = 1$ nm. The value employed here enables us to compare our results with the analogous dispersion relations spectra plotted in panels (a)–(d) of Fig. 9.1. Because of the smallness of the system radius, the extinction spectra may be interpreted in terms of the nonretarded limit and we see that the position of sharp extinction peak is due to absorption by the SP mode with $m = 1$, as shown in Fig. 9.1. This means that for a nano-2DEG tube the first three terms with $m = 0$, and ± 1, in the extinction efficiency expressions, i.e., (3.114) and (3.121) are often sufficient terms [10]. Also, by decreasing the angle of incidence wave with the system, the extinction peak of TE^z polarization shows a decrease in value. In particular, for the incidence angle $\zeta = \pi/2$, no SP resonance peak is found in the far-field extinction spectrum of TM^z polarization [see panel (a)]. However for the incidence angle $\zeta \neq \pi/2$, the incident polarization is not along the axis of the system and SP resonance peak can be found in the far-field extinction spectra of TM^z polarization, as shown in panels (b)–(d) of Fig. 9.1. Furthermore, Fig. 9.4 shows the dependence of the extinction spectra on the parameter a for both TE^z and TM^z polarizations, when $\zeta = \pi/4$. As shown in panels (a)–(d), when a increases, the extinction peak of the system shows a red-shift.

9.1.5   Effective Permittivity of a Composite of Cylindrical Two-Dimensional Electron Gas Layers: Beyond the Electrostatic Approximation

Let us consider a composite consisting of infinitely long cylindrical 2DEG layers (shown by number 1) of radius a and volume fraction f parallel to each other and randomly embedded in vacuum (shown by number 2) as sketched in panel (a) of Fig. 9.5. To study the effective permittivity ε_{eff} of the whole composite beyond the electrostatic limit [11], we invoke the standard effective medium theory [12–14], which deals with the scattering problem of a 2DEG tube with radius a that is enclosed by an insulator tube with external radius b within an effective medium with permittivity ε_{eff}, as seen in panel (b) of Fig. 9.5. The volume between a and b represents the average envelope per 2DEG tube and $f = a^2/b^2$. The effective dielectric permittivity of the system ε_{eff} may be determined by requiring that this 2DEG tube causes no perturbation of the electromagnetic waves in the surrounding medium.

We take an electromagnetic wave normally incident on the structure. For this propagation direction there are two different values of ε_{eff} corresponding to different polarizations. As mentioned in the previous section, in the case of TM^z polarization, no SP resonance peak is found in the far-field extinction spectrum of a 2DEG tube, because the polarization direction along the axis of 2DEG tubes cannot excite the collective motions of the conduction electrons. Therefore, we consider only the case of TE^z polarization.

For TE^z polarization, the magnetic field is parallel to the 2DEG tubes axis and the solutions in the effective medium can be written as

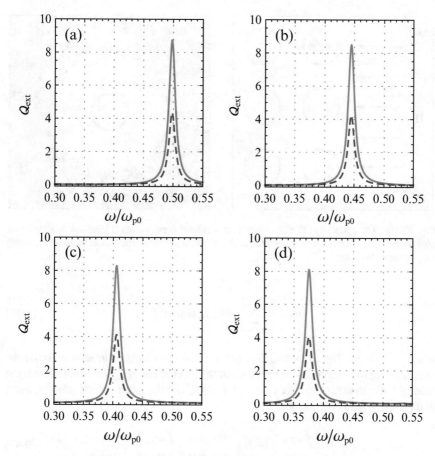

Fig. 9.4 Calculated extinction width (in units of the geometric width) of a single-walled CNT in vacuum for both TE^z and TM^z polarizations, and different values of the incidence angle, when $\zeta = \pi/4$ and $\gamma = 0.01\omega_p$. The different panels refer to (a) $a = 2$ nm, (b) $a = 2.5$ nm, (c) $a = 3$ nm and (d) $a = 3.5$ nm. The red solid curves are for TE^z (or $\mathbf{E} \perp \mathbf{e}_z$) and the blue dashed ones for TM^z (or $\mathbf{E} \parallel \mathbf{e}_z$). Here $\omega_{p0} = \sqrt{n_0 e^2 / \varepsilon_0 m_e a_0}$, $a_0 = 1$ nm and $n_0 = 4 \times 38$ nm^{-2}

$$H_{3z} = \sum_{m=-\infty}^{+\infty} \left[a_{3m} J_m(k_{eff}\rho) + b_{3m} H_m(k_{eff}\rho) \right] e^{im\phi} , \qquad (9.44)$$

where $k_{eff} = \omega \sqrt{\varepsilon_{eff}} \sqrt{\mu_0}$ and ω is angular frequency of the electromagnetic wave. Similarly, the magnetic fields in the background medium of the coated 2DEG tube and inside the 2DEG tube, i.e., media 2 and 1, respectively, can be shown by

$$H_{2z} = \sum_{m=-\infty}^{+\infty} \left[a_{2m} J_m(k_0\rho) + b_{2m} H_m(k_0\rho) \right] e^{im\phi} , \qquad (9.45)$$

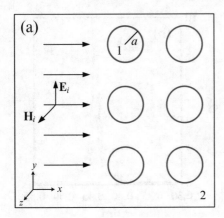
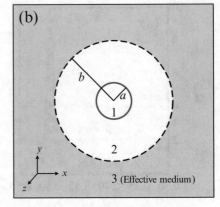

Fig. 9.5 (**a**) The cylindrical 2DEG layers in an insulator (here vacuum) host. (**b**) A 2DEG tube plus surrounding vacuum layer embedded in the effective medium. The 2DEG tubes are infinitely long in the z-direction

$$H_{1z} = \sum_{m=-\infty}^{+\infty} a_{1m} J_m(k_0\rho)e^{im\phi} \,, \tag{9.46}$$

where $k_0 = \omega/c$. The usual two BCs at the insulator–insulator interface require the continuity of the tangential components of the electric and magnetic fields across the interface. By matching the above-mentioned BCs on the interface of the background and effective medium, we obtain

$$\begin{pmatrix} a_{3m} \\ b_{3m} \end{pmatrix} = F \begin{pmatrix} A_{11} & A_{12} \\ A_{21} & A_{22} \end{pmatrix} \begin{pmatrix} a_{2m} \\ b_{2m} \end{pmatrix} \,, \tag{9.47}$$

where

$$F = \frac{1}{\varepsilon_0 k_{eff}} \frac{1}{J_m\left(k_{eff}b\right) H'_m\left(k_{eff}b\right) - J'_m\left(k_{eff}b\right) H_m\left(k_{eff}b\right)} \,,$$

$$A_{11} = \varepsilon_0 k_3 J_m\left(k_0 b\right) H'_m\left(k_{eff}b\right) - \varepsilon_{eff} k_0 J'_m\left(k_0 b\right) H_m\left(k_{eff}b\right) \,,$$

$$A_{12} = \varepsilon_0 k_3 H_m\left(k_0 b\right) H'_m\left(k_{eff}b\right) - \varepsilon_{eff} k_0 H'_m\left(k_0 b\right) H_m\left(k_{eff}b\right) \,,$$

$$A_{21} = \varepsilon_{eff} k_2 J_m\left(k_{eff}b\right) J'_m\left(k_0 b\right) - \varepsilon_2 k_{eff} J_m\left(k_0 b\right) J'_m\left(k_{eff}b\right) \,,$$

$$A_{22} = \varepsilon_{eff} k_2 J_m\left(k_{eff}b\right) H'_m\left(k_0 b\right) - \varepsilon_2 k_{eff} H_m\left(k_0 b\right) J'_m\left(k_{eff}b\right) \,.$$

Also, by matching the appropriate BCs on the interface of the 2DEG tube and using the Wronskian property of the Bessel functions, as $J_m(x)Y'_m(x) - J'_m(x)Y_m(x) = 2/\pi x$, we find

$$\begin{pmatrix} a_{2m} \\ b_{2m} \end{pmatrix} = a_{1m} \begin{pmatrix} 1 + \sigma_0 \dfrac{\pi \omega a \mu_0}{2} J'_m(k_0 a) H'_m(k_0 a) \\ -\sigma_0 \dfrac{\pi \omega a \mu_0}{2} \left[J'_m(k_0 a) \right]^2 \end{pmatrix}, \tag{9.48}$$

where σ_0 is the conductivity of the electron fluid along the ϕ-direction and may be written as (9.40). For $m = 1$, by using the condition $b_{31} = 0$ [12–14] from (9.47) and (9.48), we get

$$\frac{A_{21}}{A_{22}} = -\frac{b_{21}}{a_{21}} = a_{1\perp}(\omega, q = 0), \tag{9.49}$$

where $a_{1\perp}(\omega, q = 0)$ represents the Mie scattering coefficient of a cylindrical 2DEG layer, i.e., (9.39a) for $m = 1$, and has the form

$$a_{1\perp}(\omega, q = 0) = \frac{\sigma_0 \left[J'_1(k_0 a) \right]^2}{2/(\pi \omega a \mu_0) + \sigma_0 J'_1(k_0 a) H'_1(k_0 a)}. \tag{9.50}$$

By considering the limit of $k_3 b \ll 1$, the Bessel and Hankel functions of the first kind in A_{21} and A_{22} can be approximated as $J_1(x) \cong x/2$, $J'_1(x) \cong 1/2$, $H_1(x) \cong (x/2) - i(2/\pi x)$, and $H'_1(x) \cong (1/2) + i(2\pi x^2)$ with $x = k_3 b$. Therefore, using these approximations, (9.49) reduces to

$$\frac{\varepsilon_{eff} - \varepsilon_0 \dfrac{J_1(k_0 b)}{k_0 b J'_1(k_0 b)}}{\varepsilon_{eff} - \varepsilon_0 \dfrac{Y_1(k_0 b)}{k_0 b Y'_1(k_0 b)}} = -\frac{Y'_1(k_0 b)}{i J'_1(k_0 b)} \left(\frac{a_{1\perp}}{1 - a_{1\perp}} \right). \tag{9.51}$$

This equation is the (complex) effective dielectric function of a composite of aligned 2DEG tubes, beyond the electrostatic limit.

9.2   Surface Plasmon Polariton Modes of Coaxial Cylindrical Two-Dimensional Electron Gas Layers

Here, we consider a system of two coaxial cylindrical 2DEG layers with radii $a_1 < a_2$ and assume that each wall is occupied by a 2D electron fluid with the equilibrium density (per unit area) n_0. We use cylindrical coordinates $\mathbf{r} = (\rho, \phi, z)$ for an arbitrary point in space. For simplicity, we assume that in regions $0 < \rho < a_1$, $a_1 < \rho < a_2$, and $\rho > a_2$, there is no material and also we chose α to be zero.

Now, we consider the propagation of a SPP mode parallel to both axial and azimuthal directions of the system, so the wave equation is represented in the form of (9.1). Again, the field components for $0 < \rho < a_1$, $a_1 < \rho < a_2$, and $\rho > a_2$ can be expressed in terms of E_z and H_z. The relevant solution of (9.1) and the corresponding solution for H_z, are

$$E_z = \begin{cases} A_{1m} I_m(\kappa\rho) , & \rho \leq a_1 , \\ A_{2m} I_m(\kappa\rho) + A_{3m} K_m(\kappa\rho) & a_1 \leq \rho \leq a_2, \\ A_{4m} K_m(\kappa\rho) , & \rho \geq a_2 , \end{cases} \tag{9.52}$$

and

$$H_z = \begin{cases} B_{1m} I_m(\kappa\rho) , & \rho \leq a_1 , \\ B_{2m} I_m(\kappa\rho) + B_{3m} K_m(\kappa\rho) , & a_1 \leq \rho \leq a_2 , \\ B_{4m} K_m(\kappa\rho) , & \rho \geq a_2 , \end{cases} \tag{9.53}$$

where $\kappa^2 = q^2 - \omega^2/c^2$. The field components have to satisfy the usual BCs at $\rho = a_j$, that is, continuity of E_ϕ and E_z, as

$$E_{1z}|_{\rho=a_1} = E_{2z}|_{\rho=a_1} , \tag{9.54a}$$

$$E_{2z}|_{\rho=a_2} = E_{3z}|_{\rho=a_2} , \tag{9.54b}$$

$$E_{1\phi}|_{\rho=a_1} = E_{2\phi}|_{\rho=a_1} , \tag{9.54c}$$

$$E_{2\phi}|_{\rho=a_2} = E_{3\phi}|_{\rho=a_2} , \tag{9.54d}$$

and discontinuity of H_z and H_ϕ, due to the conductivity σ_0 [see (9.30)] associated with the cylindrical 2DEG layer, as

$$H_{2z}|_{\rho=a_1} - H_{1z}|_{\rho=a_1} = -\sigma_0 E_\phi(a_1) , \tag{9.55a}$$

$$H_{3z}|_{\rho=a_2} - H_{2z}|_{\rho=a_2} = -\sigma_0 E_\phi(a_2) , \tag{9.55b}$$

$$H_{2\phi}|_{\rho=a_1} - H_{1\phi}|_{\rho=a_1} = \sigma_0 E_z(a_1) , \tag{9.55c}$$

$$H_{3\phi}|_{\rho=a_2} - H_{2\phi}|_{\rho=a_2} = \sigma_0 E_z(a_2) , \tag{9.55d}$$

where subscripts 1 and 3 denote the regions inside and outside the cylinders having radii a_1 and a_2, respectively, while subscript 2 denotes the region between the two tubes. Also $E_\phi(a_1) = E_{1\phi}|_{\rho=a_1} = E_{2\phi}|_{\rho=a_1}$, $E_\phi(a_2) = E_{2\phi}|_{\rho=a_2} = E_{3\phi}|_{\rho=a_2}$, $E_z(a_1) = E_{1z}|_{\rho=a_1} = E_{2z}|_{\rho=a_1}$, and $E_z(a_2) = E_{2z}|_{\rho=a_2} = E_{3z}|_{\rho=a_3}$. These BCs on H_z, E_z, H_ϕ, and E_ϕ give eight linear equations and solvability condition takes the following dispersion relation for the coupled SPP mode frequencies, as

$$
\begin{vmatrix}
I_m(\kappa a_1)\Pi_1 & 0 & M_{13} & \dfrac{-1}{(\kappa a_1)K'_m(\kappa a_1)} \\[2ex]
\dfrac{I_m(\kappa a_1)K_m(\kappa a_2)\Pi_2}{K_m(\kappa a_1)} & K_m(\kappa a_2)\varXi\,\Pi_2 & M_{23} & M_{24} \\[2ex]
M_{31} & \dfrac{-i\varepsilon_0\omega}{\kappa^2 a_1 K_m(\kappa a_1)} & \dfrac{mq}{a_1^2\kappa^3 K'_m(\kappa a_1)} & \dfrac{-mq}{a_1^2\kappa^3 K'_m(\kappa a_1)} \\[2ex]
-\dfrac{I_m(\kappa a_1)K_m(\kappa a_2)\Pi_2}{K_m(\kappa a_2)} & M_{42} & 0 & \dfrac{mq}{a_2^2\kappa^3 K'_m(\kappa a_2)}
\end{vmatrix} = 0 ,
$$

(9.56)

where

$$
M_{13} = \frac{1}{(\kappa a_1)K'_m(\kappa a_1)} + \frac{i\omega\mu_0\sigma_0 I'_m(\kappa a_1)}{\kappa} ,
$$

$$
M_{23} = \frac{i\omega\mu_0\sigma_0}{\kappa}\,\frac{I'_m(\kappa a_1)K'_m(\kappa a_2)}{K'_m(\kappa a_1)} ,
$$

$$
M_{24} = \frac{1}{(\kappa a_2)K'_m(\kappa a_2)} + \frac{i\omega\mu_0\sigma_0 K'_m(\kappa a_2)\Gamma}{\kappa} ,
$$

$$
M_{31} = \frac{i\varepsilon_0\omega}{\kappa^2 a_1 K_m(\kappa a_1)} - I_m(\kappa a_1)\Pi_1 ,
$$

$$
M_{42} = \frac{i\varepsilon_0\omega}{\kappa^2 a_2 K_m(\kappa a_2)} - K_m(\kappa a_2)\varXi\,\Pi_2 ,
$$

$$
\varXi = \frac{I_m(\kappa a_2)}{K_m(\kappa a_2)} - \frac{I_m(\kappa a_1)}{K_m(\kappa a_1)} ,
$$

$$
\Gamma = \frac{I'_m(\kappa a_2)}{K'_m(\kappa a_2)} - \frac{I'_m(\kappa a_1)}{K'_m(\kappa a_1)} ,
$$

$$
\Pi_j = \sigma_0\frac{mq}{a_j\kappa^2} .
$$

From (9.56), we obtain two branches for ω defining the resonant frequencies of the SPP modes which are clearly separated into a high-frequency, $\omega_+(q, m)$, and a low-frequency, $\omega_-(q, m)$, for each m. To see clearly behavior of the two groups of SPP dispersions from (9.56), we plot dimensionless frequency $\omega_\pm/\omega_{p0}$ of SPP modes of a two-walled CNT, versus dimensionless variable qa_0 in Fig. 9.6, for different values of the azimuthal quantum number m, where $\omega_{p0} = (e^2 n_0/\varepsilon_0 m_e a_0)^{1/2}$, $a_0 = 1\text{nm}$, $a_1 = 2\text{nm}$, and $a_2 = 2.34\text{nm}$. In the above example, the parameter $\omega_{p0}^2 a_0^2/c^2$ is

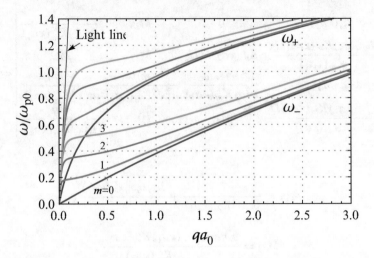

Fig. 9.6 The dimensionless frequencies ω_+/ω_{p0} and ω_-/ω_{p0} of SPP modes of a two-walled CNT, versus dimensionless variable qa_0 for $m = 0, 1, 2$, and 3, when $a_1 = 2$ nm and $a_2 = 2.34$ nm. For each m, the upper and lower curves, plotted with the same line-styles, correspond to ω_+/ω_{p0} and ω_-/ω_{p0}, respectively

about 0.005, so, for the sake of convenience we have neglected the term $\omega^2 a_0^2/c^2$ in $\kappa a_0 = \sqrt{q^2 a_0^2 - \omega^2 a_0^2/c^2}$ [15, 16].

References

1. A. Moradi, Surface plasmon-polariton modes of metallic single-walled carbon nanotubes. Photon Nanostruct. Fundam. Appl. **11**, 85–88 (2013)
2. K. Zhang, D. Li, *Electromagnetic Theory for Microwaves and Optoelectronics* (Publishing House of Electronics Industry, Beijing, 2001)
3. K.G. Batrakov, S.A. Maksimenko, P.P. Kuzhir, C. Thomsen, Carbon nanotube as a Cherenkov-type light emitter and free electron laser. Phys. Rev. B **79**, 125408 (2009)
4. Y. Zhang, M. Hu, Y. Yang, R. Zhong, S. Liu, Terahertz radiation of electron beam-cylindrical mimicking surface plasmon wave interaction. J. Phys. D: Appl. Phys. **42**, 045211 (2009)
5. C. Javaherian, B. Shokri, Guided dispersion characteristics of metallic single-wall carbon nanotubes. J. Phys. D: Appl. Phys. **42**, 055307 (2009)
6. A. Moradi, Fast electron beam-plasma interaction in single-walled carbon nanotubes. Appl. Phys. B **111**, 127–130 (2013)
7. H. Khosravi, D.R. Tilley, R. Loudon, Surface polaritons in cylindrical optical fibers. J. Opt. Soc. Am. A **8**, 112–122 (1991)
8. A. Moradi, Theory of energy and power flow of plasmonic waves on single-walled carbon nanotubes. J. Appl. Phys. **122**, 133103 (2017)
9. A. Moradi, Extinction properties of single-walled carbon nanotubes: two-fluid model. Phys. Plasmas **21**, 032106 (2014)

10. A. Moradi, Oblique incidence scattering from single-walled carbon nanotubes. Phys. Plasmas **17**, 033504 (2010)
11. A. Moradi, Maxwell-Garnett effective medium theory: quantum nonlocal effects. Phys. Plasmas **22**, 042105 (2015)
12. Y. Wu, J. Li, Z.-Q. Zhang, C.T. Chan, Effective medium theory for magnetodielectric composites: Beyond the long-wavelength limit. Phys. Rev. B **74**, 085111 (2006)
13. Y. Huang, L. Gao, Equivalent permittivity and permeability and multiple Fano resonances for nonlocal metallic nanowires. J. Phys. Chem. C **117**, 19203–19211 (2013)
14. A. Moradi, H.R. Zangeneh, F.K. Moghadam, Effective permittivity of single-walled carbon nanotube composites: two-fluid model. Phys. Plasmas **22**, 122104 (2015)
15. A. Moradi, Light conduction of metallic two-walled carbon nanotubes. Appl. Phys. A **113**, 97–100 (2013)
16. A. Moradi, Coupled surface plasmon-polariton modes of metallic single-walled carbon nanotubes. Plasmonics **8**, 1509–1513 (2013)

Chapter 10
Boundary-Value Problems Involving Two-Dimensional Electron Gases in Spherical Geometry

Abstract In this chapter, we study some electrostatic and electromagnetic boundary-value problems involving two-dimensional electron gas shells in spherical geometry. We use the standard hydrodynamic model to describe the dielectric response of a spherical electron gas shell. The main interest and the key first applications of presented boundary-value problems concern C_{60} molecule, while keeping in mind that the analysis can be applied to the other two-dimensional electron gas shells with spherical geometry. For brevity, in many sections of this chapter the $\exp(-i\omega t)$ time factor is suppressed. Furthermore, all media under consideration are nonmagnetic and attention is only confined to the linear phenomena.

10.1 Plasmonic Properties of Spherical Two-Dimensional Electron Gas Shells: Electrostatic Approximation

10.1.1 Dispersion Relation

Let us consider a 2DEG constrained to the surface of a sphere with a radius a, and assume that 2DEG is distributed uniformly over the spherical surface, with the equilibrium density (per unit area) n_0. We take spherical coordinates $\mathbf{r} = (r, \theta, \varphi)$ for an arbitrary point in space. We assume that $n(\theta, \varphi, t)$ is the first-order perturbed density (per unit area) of the homogeneous electron fluid on the spherical 2DEG shell, due to the oscillation of an electrostatic surface wave parallel to the spherical surface $r = a$.

Based on the SHD theory (see Sect. 1.2), in the linear approximation, the electronic excitations of this 2DEG may be described by

---

The original version of this chapter was revised. The correction to this chapter is available at https://doi.org/10.1007/978-3-030-43836-4_11

A. Moradi, *Canonical Problems in the Theory of Plasmonics*, Springer Series in Optical Sciences 230, https://doi.org/10.1007/978-3-030-43836-4_10

$$\frac{\partial \mathbf{v}}{\partial t} = \frac{e}{m_e} \nabla_\| \Phi \Big|_{r=a} - \frac{\alpha^2}{n_0} \nabla_\| n \, , \tag{10.1}$$

$$\frac{\partial n}{\partial t} + n_0 \nabla_\| \cdot \mathbf{v} = 0 \, , \tag{10.2}$$

$$\nabla^2 \Phi(\mathbf{r}) = \begin{cases} 0 \, , & r \neq a \, , \\ en/\varepsilon_0 \, , & r = a, \end{cases} \tag{10.3}$$

where the symbols have the same meaning as in Chap. 8 and $\mathbf{v}(\theta, \varphi, t) = v_\theta \mathbf{e}_\theta + v_\varphi \mathbf{e}_\varphi$ is the first-order perturbed values of electrons velocity parallel to the spherical 2DEG surface $r = a$. Also $\nabla_\| = a^{-1} \mathbf{e}_\theta (\partial/\partial\theta) + (a \sin\theta)^{-1} \mathbf{e}_\varphi (\partial/\partial\varphi)$ differentiates only tangentially to the 2DEG shell. Now, we obtain from (10.1) and (10.2) after the elimination of the velocity $\mathbf{v}(\theta, \varphi, t)$

$$\left(\omega^2 + \alpha^2 \nabla_\|^2 \right) n = \frac{en_0}{m_e} \nabla_\|^2 \Phi \Big|_{r=a} \, . \tag{10.4}$$

Due to the spherical symmetry of the present system, one can replace the quantities n and Φ in (10.3) and (10.4) by expressions of the form

$$\Phi(r, \theta, \varphi) = \tilde{\Phi}(r) Y_{\ell m}(\theta, \varphi) \, , \tag{10.5}$$

$$n(\theta, \varphi) = \tilde{n} \, Y_{\ell m}(\theta, \varphi) \, . \tag{10.6}$$

Then, we obtain

$$\frac{d^2 \tilde{\Phi}(r)}{dr^2} + \frac{2}{r} \frac{d \tilde{\Phi}(r)}{dr} - \frac{\ell(\ell+1)}{r^2} \tilde{\Phi}(r) = 0 \, , \quad r \neq a \tag{10.7}$$

and

$$\tilde{n} = -\frac{en_0}{m_e} \frac{\ell_a^2}{\omega^2 - \alpha^2 \ell_a^2} \tilde{\Phi}(r = a) \, , \tag{10.8}$$

where $\ell_a^2 = \ell(\ell+1)/a^2$. The solution of (10.7) has the form

$$\tilde{\Phi}(r) = \begin{cases} A_1 r^\ell \, , & r \leq a \, , \\ A_2 r^{-(\ell+1)} \, , & r \geq a \, , \end{cases} \tag{10.9}$$

where the relations between the coefficients A_1 and A_2 can be determined from the matching BCs at the spherical 2DEG surface. For the present case, the BCs are

$$\Phi_1 \big|_{r=a} = \Phi_2 \big|_{r=a} \, , \tag{10.10a}$$

$$\varepsilon_2 \frac{\partial \Phi_2}{\partial r} \Big|_{r=a} - \varepsilon_1 \frac{\partial \Phi_1}{\partial r} \Big|_{r=a} = \frac{en}{\varepsilon_0} \, , \tag{10.10b}$$

where subscript 1 denotes the region inside the 2DEG shell and subscript 2 denotes the region outside the system. In writing (10.10b), we assumed that the relative dielectric constant of the region inside the spherical 2DEG shell is ε_1 and the relative dielectric constant of the surrounding medium is ε_2. On applying the electrostatic BCs at $r = a$, the SP dispersion relation of the system is given by

$$\omega^2 = \alpha^2 \ell_a^2 + \omega_p^2 \frac{\ell(\ell+1)}{\ell\varepsilon_1 + (\ell+1)\varepsilon_2} , \qquad (10.11)$$

where $\omega_p^2 = e^2 n_0 / \varepsilon_0 m_e a$ as mentioned before. When the spherical 2DEG shell is bounded by vacuum, i.e., $\varepsilon_1 = 1 = \varepsilon_2$, (10.11) becomes [1, 2]

$$\omega^2 = \alpha^2 \ell_a^2 + \omega_p^2 \frac{\ell(\ell+1)}{2\ell+1} . \qquad (10.12)$$

It can be seen from (10.12) that the SP dispersion relation depends on the radius of 2DEG shell a and the surface electron density n_0.

As an example, we choose a spherical 2DEG shell with the radius $a = 6.683 a_B$ (a_B being the Bohr radius) and $n_0 = 240/4\pi a^2$ as a simple model of a C_{60} molecule[1] in the $\sigma + \pi$ electron fluid model.[2] Figure 10.1 shows the dimensionless

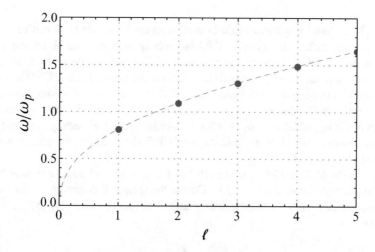

Fig. 10.1 Dispersion curve of SP modes of an isolated C_{60} molecule as given by (10.12)

---

[1]C_{60} molecule was discovered by Kroto et al. [3] in 1985. This molecule is a kind of highly stable and hollow quasi-spherical molecule with 240 valence electrons occupying σ and π orbitals that are typical of all sp^2 hybridized carbon nanostructures.

[2]The two electron fluid model divides both σ and π electron fluids [1, 4]. We note that the investigation of plasmonics properties of a C_{60} molecule by using the two fluid model makes an interesting BVP, but we leave it for the reader (see [1]).

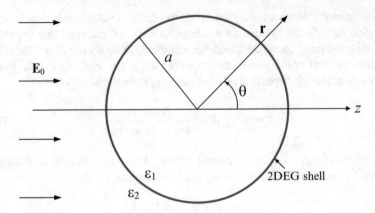

Fig. 10.2 Cross section of a spherical 2DEG shell placed into an electrostatic field

frequency ω/ω_p of SP oscillations of a C_{60} molecule versus the integer number ℓ. As increasing the integer number ℓ, the SP frequency of the system increases.

10.1.2 Surface Plasmon Resonance

Figure 10.2 shows a spherical 2DEG shell with radius a that is placed in a quasi-static electric field. This electric field is initially uniform and along the z-axis, $\mathbf{E} = \mathbf{e}_z E_0$, in the absence of the spherical 2DEG shell or far away from it. We use spherical polar coordinates with the origin at the center of the 2DEG shell. The relative dielectric constants of inside the 2DEG and the external medium are ε_1 and ε_2, respectively. Again, we assume that the equilibrium density (per unit area) of the system is n_0, while $n(\theta, \varphi, t)$ is the first-order perturbed density (per unit area) of the homogeneous electron fluid on the 2DEG shell, due to the presence of the external electric field.

We use the SHD model by using (10.1)–(10.3), where we add the damping term $-\gamma \mathbf{v}$ to the right-hand side of (10.1). Due to the spherical symmetry of the present system, one can replace the quantities n and Φ in (10.1)–(10.3), by expressions of the form

$$\Phi(r, \theta) = \tilde{\Phi}(r) \cos\theta \,, \tag{10.13}$$

$$n(\theta) = \tilde{n} \cos\theta \,. \tag{10.14}$$

Therefore, the present problem satisfies

$$\left[\frac{d^2}{dr^2} + \frac{1}{r} \frac{d}{dr} - \frac{2}{r^2} \right] \tilde{\Phi}(r) = 0 \,, \qquad r \neq a \,, \tag{10.15}$$

$$\tilde{n} = -2\frac{en_0}{m_e a^2} \frac{1}{\omega(\omega + i\gamma) - 2(\alpha^2/a^2)} \tilde{\Phi}\Big|_{r=a}.$$

(10.16)

For the present system, we look for the solution of (10.15) of the form

$$\tilde{\Phi}(r) = \begin{cases} A_1 r, & r \le a, \\ -E_0 r + A_2 r^{-2}, & r \ge a. \end{cases}$$

(10.17)

To determine the unknown coefficients A_1 and A_2, the BCs, i.e., (10.10a) and (10.10b), must be enforced. Then we obtain

$$A_1 = \frac{-3\varepsilon_2}{\varepsilon_1 + 2\varepsilon_2 - \dfrac{2\omega_p^2}{\omega(\omega + i\gamma) - 2(\alpha^2/a^2)}} E_0,$$

(10.18a)

$$A_2 = a^3 \frac{\varepsilon_1 - \varepsilon_2 - \dfrac{2\omega_p^2}{\omega(\omega + i\gamma) - 2(\alpha^2/a^2)}}{\varepsilon_1 + 2\varepsilon_2 - \dfrac{2\omega_p^2}{\omega(\omega + i\gamma) - 2(\alpha^2/a^2)}} E_0.$$

(10.18b)

Substituting the magnitude of coefficients A_1 and A_2 in (10.17) and using (10.13) give

$$\Phi(r,\theta) = E_0 r \cos\theta \begin{cases} -\dfrac{3\varepsilon_2}{\varepsilon_1 + 2\varepsilon_2 - \dfrac{2\omega_p^2}{\omega(\omega + i\gamma) - 2(\alpha^2/a^2)}}, & r \le a, \\[2em] -1 + \dfrac{a^3}{r^3}\dfrac{\varepsilon_1 - \varepsilon_2 - \dfrac{2\omega_p^2}{\omega(\omega + i\gamma) - 2(\alpha^2/a^2)}}{\varepsilon_1 + 2\varepsilon_2 - \dfrac{2\omega_p^2}{\omega(\omega + i\gamma) - 2(\alpha^2/a^2)}}, & r \ge a. \end{cases}$$

(10.19)

This result shows that the potential for $r > a$ consists of two terms. The first term is associated with the external field $\mathbf{E} = \mathbf{e}_z E_0$. The second term is the result of the introduction of the spherical 2DEG shell in the external field. This is a dipole field produced by a dipole moment located at the spherical shell center. We can rewrite Φ_2 by introducing the dipole moment $\mathbf{p}$ as

$$\Phi(r,\theta) = -E_0 r \cos\theta + \frac{\mathbf{p} \cdot \mathbf{r}}{4\pi\varepsilon_0\varepsilon_2 r^3},$$

(10.20a)

$$\mathbf{p} = 4\pi a^3 \varepsilon_0 \varepsilon_2 \frac{\varepsilon_1 - \varepsilon_2 - \dfrac{2\omega_p^2}{\omega(\omega + i\gamma) - 2(\alpha^2/a^2)}}{\varepsilon_1 + 2\varepsilon_2 - \dfrac{2\omega_p^2}{\omega(\omega + i\gamma) - 2(\alpha^2/a^2)}} \cdot \mathbf{E}_0 . \tag{10.20b}$$

This means that the applied field induces a dipole moment inside the spherical 2DEG shell of magnitude proportional to $|\mathbf{E}_0|$. Also, for the normalized polarizability of the system, we have

$$\alpha_{pol} = 3 \frac{\varepsilon_1 - \varepsilon_2 - \dfrac{2\omega_p^2}{\omega(\omega + i\gamma) - 2(\alpha^2/a^2)}}{\varepsilon_1 + 2\varepsilon_2 - \dfrac{2\omega_p^2}{\omega(\omega + i\gamma) - 2(\alpha^2/a^2)}} . \tag{10.21}$$

Finally, using the normalized polarizability formula of the system, we may write the scattering, absorption, and extinction cross sections (in units of the geometric cross section), as [5]

$$Q_{sca} = \frac{8}{3}\varepsilon_2^2 \left(\frac{a\omega}{c}\right)^4 \left| \frac{\varepsilon_1 - \varepsilon_2 - \dfrac{2\omega_p^2}{\omega(\omega + i\gamma) - 2(\alpha^2/a^2)}}{\varepsilon_1 + 2\varepsilon_2 - \dfrac{2\omega_p^2}{\omega(\omega + i\gamma) - 2(\alpha^2/a^2)}} \right|^2 , \tag{10.22a}$$

$$Q_{abs} = 4\varepsilon_2^{1/2} \left(\frac{a\omega}{c}\right) \mathrm{Im} \left[\frac{\varepsilon_1 - \varepsilon_2 - \dfrac{2\omega_p^2}{\omega(\omega + i\gamma) - 2(\alpha^2/a^2)}}{\varepsilon_1 + 2\varepsilon_2 - \dfrac{2\omega_p^2}{\omega(\omega + i\gamma) - 2(\alpha^2/a^2)}} \right] , \tag{10.22b}$$

$$Q_{ext} = Q_{sca} + Q_{abs} . \tag{10.22c}$$

Also, in the absence of the damping effects, the stored electrostatic energy[3] within this polarized spherical EG shell is

$$U = -\frac{1}{2V} \int \mathbf{p} \cdot \mathbf{E}_0 \, dv = -2\pi a^3 \varepsilon_0 \varepsilon_2 \frac{\varepsilon_1 - \varepsilon_2 - \dfrac{2\omega_p^2}{\omega^2 - 2(\alpha^2/a^2)}}{\varepsilon_1 + 2\varepsilon_2 - \dfrac{2\omega_p^2}{\omega^2 - 2(\alpha^2/a^2)}} E_0^2 . \tag{10.23}$$

---

[3] For the stored electromagnetic energy within a spherical EG shell see [6].

As mentioned before, this equation indicates that the resonance peaks may be appeared in the stored energy spectrum. Furthermore, the effective dielectric function ε_{eff} of a composite of spherical 2DEG shells embedded in a host matrix with relative dielectric constant ε_2 may be written as [see Sect. 2.7.4]

$$\frac{\varepsilon_{eff} - \varepsilon_2}{\varepsilon_{eff} + 2\varepsilon_2} = f \frac{\varepsilon_1 - \varepsilon_2 - \dfrac{2\omega_p^2}{\omega(\omega + i\gamma) - 2\left(\alpha^2/a^2\right)}}{\varepsilon_1 + 2\varepsilon_2 - \dfrac{2\omega_p^2}{\omega(\omega + i\gamma) - 2\left(\alpha^2/a^2\right)}} , \tag{10.24}$$

where f is the volume fraction of the embedded spherical 2DEG shells.

10.1.3 Multipolar Response

Here, we study the multiple response of a spherical 2DEG shell. As mentioned in Chap. 2, we may consider a general solution of (10.9) as

$$\tilde{\Phi}(r) = \begin{cases} A_{1+}r^\ell + A_{1-}r^{-(\ell+1)} , & r \leq a , \\ A_{2+}r^\ell + A_{2-}r^{-(\ell+1)} , & r \geq a . \end{cases} \tag{10.25}$$

Imposing the electrostatic BCs at $r = a$ leads to

$$\mathbf{M}_1 \begin{pmatrix} A_{1+} \\ A_{1-} \end{pmatrix} = \mathbf{M}_2 \begin{pmatrix} A_{2+} \\ A_{2-} \end{pmatrix} , \tag{10.26}$$

where

$$\mathbf{M}_1 = \begin{pmatrix} a^\ell & a^{-(\ell+1)} \\ \ell\varepsilon_1 a^{\ell-1} & -(\ell+1)\varepsilon_1 a^{-(\ell+2)} \end{pmatrix} , \tag{10.27}$$

$$\mathbf{M}_2 = \begin{pmatrix} a^\ell & a^{-(\ell+1)} \\ \dfrac{\ell}{a^{-\ell+1}}\left[\varepsilon_2 + \dfrac{(\ell+1)\omega_p^2}{\omega(\omega+i\gamma) - \alpha^2\ell_a^2}\right] & -\dfrac{\ell+1}{a^{\ell+2}}\left[\varepsilon_2 - \dfrac{\ell\omega_p^2}{\omega(\omega+i\gamma) - \alpha^2\ell_a^2}\right] \end{pmatrix} . \tag{10.28}$$

Then from (10.26), we find

$$\begin{pmatrix} A_{1+} \\ A_{1-} \end{pmatrix} = \begin{pmatrix} M_{11} & M_{12} \\ M_{21} & M_{22} \end{pmatrix} \begin{pmatrix} A_{2+} \\ A_{2-} \end{pmatrix} , \tag{10.29}$$

where

$$M_{11} = \frac{(\ell+1)\varepsilon_1 + \ell \left[\varepsilon_2 + \dfrac{(\ell+1)\omega_p^2}{\omega(\omega+i\gamma) - \alpha^2\ell_a^2} \right]}{(2\ell+1)\varepsilon_1} \, ,$$

$$M_{22} = \frac{\ell\varepsilon_1 + (\ell+1) \left[\varepsilon_2 - \dfrac{\ell\omega_p^2}{\omega(\omega+i\gamma) - \alpha^2\ell_a^2} \right]}{(2\ell+1)\varepsilon_1} \, ,$$

$$M_{12} = \frac{\ell+1}{2\ell+1} \frac{\varepsilon_1 - \varepsilon_2 + \dfrac{\ell\omega_p^2}{\omega(\omega+i\gamma) - \alpha^2\ell_a^2}}{\varepsilon_1} a^{-(2\ell+1)} \, ,$$

$$M_{21} = \frac{\ell}{2\ell+1} \frac{\varepsilon_1 - \varepsilon_2 - \dfrac{(\ell+1)\omega_p^2}{\omega(\omega+i\gamma) - \alpha^2\ell_a^2}}{\varepsilon_1} a^{2\ell+1} \, .$$

Now, using (2.144) we obtain

$$\alpha_{pol} = 3\ell a^{2(\ell-1)} \frac{\varepsilon_1 - \varepsilon_2 - \dfrac{(\ell+1)\omega_p^2}{\omega(\omega+i\gamma) - \alpha^2\ell_a^2}}{\ell\varepsilon_1 + (\ell+1) \left[\varepsilon_2 - \dfrac{\ell\omega_p^2}{\omega(\omega+i\gamma) - \alpha^2\ell_a^2} \right]} \, . \tag{10.30}$$

Equation (10.30) shows the (complex) multipole polarizability of spherical 2DEG shell. Also, $\ell\varepsilon_1 + (\ell+1)\varepsilon_2 - \dfrac{\ell(\ell+1)\omega_p^2}{\omega^2 - \alpha^2\ell_a^2} = 0$ shows the dispersion relation of undamped SP modes, when $\gamma = 0$.

10.2 Plasmonic Properties of Spherical Two-Dimensional Electron Gas Shells: Beyond the Electrostatic Approximation

10.2.1 Dispersion Relation

We now derive the equations for the frequencies of the various retarded surface modes. For TE^r polarization, the tangential components of fields inside and outside the spherical 2DEG shell are given by (3.141)–(3.144). To determine the unknown coefficients $A_{\ell m1}$ and $A_{\ell m2}$, the appropriate BCs must be enforced. In addition to the continuity of the tangential components of the electric field vector at the 2DEG shell, as

$$E_{1\theta}|_{r=a} = E_{2\theta}|_{r=a} \ , \tag{10.31}$$

$$E_{1\phi}|_{r=a} = E_{2\phi}|_{r=a} \ , \tag{10.32}$$

the second set of BCs must ensure that all tangential magnetic field components experience a jump due to the conductivity σ_0 [see (9.30) for the case, where the nonlocal effects are neglected, i.e., $\alpha = 0$] associated with the system as

$$H_{2\theta}|_{r=a} - H_{1\theta}|_{r=a} = \sigma_0 E_\phi(a) \ , \tag{10.33}$$

and

$$H_{2\phi}|_{r=a} - H_{1\phi}|_{r=a} = -\sigma_0 E_\theta(a) \ , \tag{10.34}$$

where subscript 1 denotes the region inside the 2DEG shell and subscript 2 denotes the region outside the system. Also $E_\phi(a) = E_{1\phi}|_{r=a} = E_{2\phi}|_{r=a}$ and $E_\theta(a) = E_{1\theta}|_{r=a} = E_{2\theta}|_{r=a}$. These BCs at $r = a$ yield the following homogeneous linear system of equations:

$$\begin{pmatrix} j_\ell(k_1 a) & -h_\ell(k_2 a) \\ [k_1 a j_\ell(k_1 a)]' - \sigma_0 j_\ell(k_1 a) & -[k_2 a h_\ell(k_2 a)]' \end{pmatrix} \begin{pmatrix} A_{\ell m 1} \\ A_{\ell m 2} \end{pmatrix} = 0 \ . \tag{10.35}$$

The equation for the frequencies of the TE^r modes is then obtained by solving the condition that the determinant of the matrix in (10.35) must be zero in order to exist a non-trivial solution, yielding

$$j_\ell(k_1 a) [k_2 a h_\ell(k_2 a)]' - h_\ell(k_2 a) \left\{ [k_1 a j_\ell(k_1 a)]' - \sigma_0 j_\ell(k_1 a) \right\} = 0 \ . \tag{10.36}$$

For TM^r modes, the tangential components of fields inside and outside the 2DEG shell are given by (3.148)–(3.151). Now, the BCs at $r = a$ yield the following homogeneous linear system of equations:

$$\begin{pmatrix} \dfrac{[k_1 a j_\ell(k_1 a)]'}{k_1 a} & -\dfrac{[k_2 a h_\ell(k_2 a)]'}{k_2 a} \\ k_1 a j_\ell(k_1 a) + \sigma_0 \dfrac{[k_1 a j_\ell(k_1 a)]'}{k_1 a} & -k_2 a h_\ell(k_2 a) \end{pmatrix} \begin{pmatrix} B_{\ell m 1} \\ B_{\ell m 2} \end{pmatrix} = 0 \ , \tag{10.37}$$

and equation for the frequencies of TM^r modes is

$$\varepsilon_2 h_\ell(k_2 a) [k_1 a j_\ell(k_1 a)]' - \left\{ \varepsilon_1 j_\ell(k_1 a) + \sigma_0 [k_1 a j_\ell(k_1 a)]' \right\} [k_2 a h_\ell(k_2 a)]' = 0 \ . \tag{10.38}$$

Fig. 10.3 A spherical polar coordinate system centered on a spherical 2DEG shell of radius a. Also the figure shows a uniform plane wave incident on the system

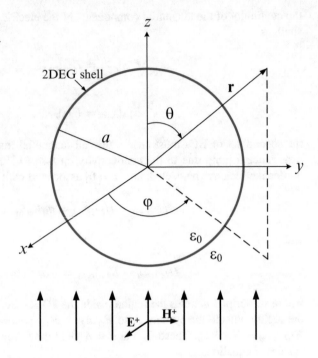

10.2.2   Extinction Property

Let us now consider a spherical 2DEG shell of radius a irradiated by a z-directed, x-polarized plane wave, as shown in Fig. 10.3. We assume that the regions $r > a$ and $r < a$ filled with vacuum. We consider the incident electric field as (3.155). Also, the transmitted and scattered fields can be represented as (3.157) and (3.159), respectively. In these equations, $\mathbf{M}_{e\ell 1}$ and $\mathbf{N}_{e\ell 1}$ are the vector spherical harmonics and can be represented by (3.139a) and (3.139b), where $z_\ell(kr)$ represents a spherical Bessel or spherical Hankel function of the first kind, and is chosen as follows. Inside the sphere $j_\ell(k_0 r)$ is used and outside the system $j_\ell(k_0 r)$ and $h_\ell(k_0 r)$ are used for the incident and scattered waves, respectively.

To determine the unknown coefficients a_ℓ, b_ℓ, c_ℓ, and d_ℓ, the appropriate BCs must be enforced, i.e., (10.31)–(10.34). Using the Wronskian property for spherical Bessel functions as $j_\ell(x) y_\ell'(x) - j_\ell'(x) y_\ell(x) = 1/x^2$ and $h_\ell = j_\ell + i y_\ell$, after doing some algebra, one can find [7]

$$a_\ell = \frac{\eta \sigma_0 [\psi_\ell'(x)]^2}{1 + \eta \sigma_0 \psi_\ell'(x) \xi_\ell'(x)} , \tag{10.39a}$$

$$b_\ell = \frac{\eta \sigma_0 [\psi_\ell(x)]^2}{1 + \eta \sigma_0 \psi_\ell(x) \xi_\ell(x)} , \tag{10.39b}$$

$$c_\ell = 1 - \frac{\xi_\ell(x)}{\psi_\ell(x)} b_\ell \,, \qquad (10.39c)$$

$$d_\ell = 1 - \frac{\xi'_\ell(x)}{\psi'_\ell(x)} a_\ell \,, \qquad (10.39d)$$

where $\eta = \sqrt{\mu_0/\varepsilon_0}$ is the wave or vacuum impedance, σ_0 is given by (9.40), and $\psi_\ell(x) = x j_\ell(x)$ and $\xi_\ell(x) = x h_\ell(x)$ with $x = k_0 a$. Also, as mentioned before, $\psi'_n(x)$ and $\xi'_n(x)$ denote the derivative of the Ricatti–Bessel functions with respect to argument $x = k_0 a$. The extinction cross section (in units of the geometric cross section) can be expressed as (3.167), by replacing k_2 with k_0.

Let us note that the denominators of a_ℓ and b_ℓ vanish at the frequencies of the TM^r and TE^r SPPs, respectively, so that the optical cross sections will have resonances at these frequencies. The equation for the frequencies of TE^r SPP modes is

$$1 + \eta\sigma_0\psi_\ell(x)\xi_\ell(x) = 0 \,, \qquad (10.40)$$

and the equation for TM^r SPP modes is

$$1 + \eta\sigma_0\psi'_\ell(x)\xi'_\ell(x) = 0 \,. \qquad (10.41)$$

One can see that a spherical 2DEG shell differs markedly from a cylindrical 2DEG layer. In the latter geometry there exist both radiative and non-radiative SPPs. However, all the SPP frequencies of a spherical 2DEG shell, i.e., the solutions of (10.40) and (10.41), are complex. This means that for real frequencies, as in an optical experiment, the denominators of a_ℓ and b_ℓ can never be exactly equal to zero. Every radiative SPP mode contributes a Lorentzian peak to the cross section. The peak is centered at the real part of the mode frequency, and has a half-width equal to the imaginary part of the mode frequency. It is easy to find that no surface modes of the TE^r mode exist and TM^r SPP modes do occur and for a small system (nano system), they coincide with the nonretarded SP modes discussed in Sect. 10.1.1.

10.2.3   Effective Permittivity of a Composite of Spherical Two-Dimensional Electron Gas Layers

As shown in panel (a) of Fig. 10.4, we suppose a system of spherical 2DEG shells (shown by number 1) of radius a and small volume fraction f embedded in a vacuum, which is illuminated by a z-directed, x-polarized plane wave. To study the effective permittivity ε_{eff} of the whole composite, we use the method discussed in Sect. 9.1.5. In this method we deal with the scattering problem of a spherical 2DEG shell that is enclosed within a vacuum shell of radius b in an effective medium with

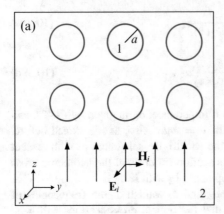 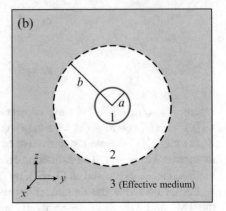

Fig. 10.4 (**a**) The spherical 2DEG shells in an insulator (here vacuum) host. (**b**) A spherical 2DEG shell plus surrounding vacuum shell embedded in the effective medium

relative permittivity ε_{eff}, as shown in panel (b) of Fig. 10.4. The volume between a and b represents the average envelope per 2DEG shell and small filling factor $f = a^3/b^3$. The effective permittivity of the system ε_{eff} can be determined by requiring that this 2DEG shell produces no perturbation of the electromagnetic wave in the surrounding medium.

We suppose that an electromagnetic wave is incident on a coated spherical shell with inner radius a and outer radius b, as shown in panel (b) of Fig. 10.4. For TM^r polarization, the expansion of the electric and magnetic fields of incident plane wave of angular frequency ω is given by

$$\mathbf{E}_i = -i \sum_{\ell=1}^{+\infty} E_\ell A_{\ell 1(eff)} \mathbf{N}_{e\ell 1}^{(1)} , \qquad (10.42)$$

where

$$E_\ell = i^\ell \frac{2\ell + 1}{\ell(\ell + 1)} E_0 .$$

The corresponding incident magnetic field is

$$\mathbf{H}_i = -\frac{k_{eff}}{\omega \mu_0} \sum_{\ell=1}^{+\infty} E_\ell A_{\ell 1(eff)} \mathbf{M}_{e\ell 1}^{(1)} , \qquad (10.43)$$

where $k_{eff} = \sqrt{\varepsilon_{eff}} \omega/c$. Here, we append the superscript (1) to vector spherical harmonics for which the radial dependence of the generating functions is specified by j_ℓ. The scattered field $(\mathbf{E}_s, \mathbf{H}_s)$ in the region $r \geq b$ can be represented as

$$\mathbf{E}_s = i \sum_{\ell=1}^{+\infty} E_\ell B_{\ell 1(eff)} \mathbf{N}_{e\ell 1}^{(2)} , \qquad (10.44)$$

$$\mathbf{H}_s = \frac{k_{eff}}{\omega \mu_0} \sum_{\ell=1}^{+\infty} E_\ell B_{\ell 1(eff)} \mathbf{M}_{e\ell 1}^{(2)} , \qquad (10.45)$$

where we append the superscript (2) to vector spherical harmonics for which the radial dependence of the generating functions is specified by h_ℓ. However, in the region $a \leq r \leq b$, both spherical Bessel and spherical Hankel functions, i.e., j_ℓ and h_ℓ are finite; as a consequence, the expansion of the fields $(\mathbf{E}_t, \mathbf{H}_t)$ in this region [i.e., region 2 in panel (b) of Fig. 10.4] can be written in the form

$$\mathbf{E}_{t2} = -i \sum_{\ell=1}^{+\infty} E_\ell \left(C_{\ell 1(0)} \mathbf{N}_{e\ell 1}^{(1)} + D_{\ell 1(0)} \mathbf{N}_{e\ell 1}^{(2)} \right) , \qquad (10.46)$$

$$\mathbf{H}_{t2} = -\frac{k_0}{\omega \mu_0} \sum_{\ell=1}^{+\infty} E_\ell \left(C_{\ell 1(0)} \mathbf{M}_{e\ell 1}^{(1)} + D_{\ell 1(0)} \mathbf{M}_{e\ell 1}^{(2)} \right) . \qquad (10.47)$$

At this stage, by using the usual BCs at the interface of the background and effective medium, i.e., the continuity of the tangential components of the electric and magnetic fields at $r = b$, we obtain

$$\begin{pmatrix} A_{\ell 1(eff)} \\ B_{\ell 1(eff)} \end{pmatrix} = \frac{i}{k_0} \begin{pmatrix} \Upsilon_{11} & \Upsilon_{12} \\ \Upsilon_{21} & \Upsilon_{22} \end{pmatrix} \begin{pmatrix} C_{\ell 1(0)} \\ D_{\ell 1(0)} \end{pmatrix} , \qquad (10.48)$$

where

$$\Upsilon_{11} = k_{eff} \xi_\ell \left(k_{eff} b \right) \psi_\ell' (k_0 b) - k_0 \xi_\ell' \left(k_{eff} b \right) \psi_\ell (k_0 b) ,$$

$$\Upsilon_{12} = k_{eff} \xi_\ell \left(k_{eff} b \right) \xi_\ell' (k_0 b) - k_0 \xi_\ell' \left(k_{eff} b \right) \xi_\ell (k_0 b) ,$$

$$\Upsilon_{21} = k_{eff} \psi_\ell \left(k_{eff} b \right) \psi_\ell' (k_0 b) - k_0 \psi_\ell' \left(k_{eff} b \right) \psi_\ell (k_0 b) ,$$

$$\Upsilon_{22} = k_{eff} \psi_\ell \left(k_{eff} b \right) \xi_\ell' (k_0 b) - k_0 \psi_\ell' \left(k_{eff} b \right) \xi_\ell (k_0 b) .$$

Now, by using the condition $B_{11}(eff) = 0$ [8, 9], from (10.48) we obtain

$$\frac{\Upsilon_{21}}{\Upsilon_{22}} = -\frac{D_{11(0)}}{C_{11(0)}} = A_{11} , \qquad (10.49)$$

where A_{11} represents the Mie scattering coefficient of a spherical 2DEG shell and has the form of (10.39a) but for $\ell = 1$.

By considering the limit of $k_{eff}b \ll 1$, the spherical Bessel and spherical Hankel functions of the first kind in Υ_{21} and Υ_{22} can be approximated as $j_1(x) \cong x/3$ and $h_1(x) \cong (x/3) - i/x^2$ with $x = k_{eff}b$. Therefore, using these approximations, (10.49) reduces to

$$\frac{\varepsilon_{eff} - 2\varepsilon_0 \dfrac{j_1(k_0b)}{j_1(k_0b) + k_0bj_1'(k_0b)}}{\varepsilon_{eff} - 2\varepsilon_0 \dfrac{y_1(k_0b)}{y_1(k_0b) + k_0by_1'(k_0b)}} = -\frac{\left[y_1(k_0b) + k_0by_1'(k_0b) \right]}{i\left[j_1(k_0b) + k_0bj_1'(k_0b) \right]} \left(\frac{A_{11}}{1 - A_{11}} \right).$$

(10.50)

The above equation is the (complex) effective dielectric function of a system of spherical 2DEG shells, beyond the electrostatic limit.

10.3   Surface Plasmon Modes of Concentric Spherical Two-Dimensional Electron Gas Shells

We consider a system of two concentric spherical 2DEG shells with radii $a_1 < a_2$ [10, 11], where the density of free-electron fluid over the each spherical shell (per unit area) is n_{0j} with $j = 1$ and 2. Here, we assume that in regions $0 < r < a_1$, $a_1 < r < a_2$, and $r > a_2$, there is no material. The inclusion of an insulator core and/or embedding medium in the formalism is straightforward. Assuming that $n_j(a_j, \theta, \varphi, t)$ is the perturbed density (per unit area) of the homogeneous electron fluid on the jth shell, due to oscillation of an electrostatic surface wave with angular frequency ω. Again, the electric potential is represented in the form of (10.5). In this case, (10.8) must be written as

$$\tilde{n}_j = -\frac{en_{0j}}{m_e} \frac{\ell_{a_j}^2}{\omega^2 - \alpha_j^2 \ell_{a_j}^2} \tilde{\Phi}(r = a_j),$$

(10.51)

where $\ell_{a_j}^2 = \ell(\ell+1)/a_j^2$ and the solution of (10.7) has the form

$$\tilde{\Phi}(r) = \begin{cases} A_1 r^\ell, & r \le a_1, \\ A_2 r^\ell + A_3 r^{-(\ell+1)}, & a_1 \le r \le a_2, \\ A_4 r^{-(\ell+1)}, & r \ge a_2. \end{cases}$$

(10.52)

For the present case, the BCs are

$$\Phi_j\big|_{r=a_j} = \Phi_{j+1}\big|_{r=a_j},$$

(10.53a)

$$\left.\frac{\partial \Phi_{j+1}}{\partial r}\right|_{r=a_j} - \left.\frac{\partial \Phi_j}{\partial r}\right|_{r=a_j} = \frac{en_j}{\varepsilon_0} \,, \tag{10.53b}$$

where subscripts j and $j+1$ for potentials denote the regions inside and outside the 2DEG shell with radius a_j, respectively. The application of these electrostatic BCs at $r = a_1$ and a_2 gives the eigenvalue equation for the resonant frequencies of the SP modes of two concentric spherical 2DEG shells, with the following two branches of the dispersion relations for each ℓ, as

$$\omega_{\pm}^2 = \frac{\omega_1^2 + \omega_2^2}{2} \pm \sqrt{\left(\frac{\omega_1^2 - \omega_2^2}{2}\right)^2 + \Delta^2} \,, \tag{10.54}$$

where

$$\omega_j^2 = \alpha_j^2 \ell_{a_j}^2 + \omega_{pj}^2 \frac{\ell(\ell+1)}{2\ell+1} \tag{10.55}$$

that are the squares of the SP dispersion on the spheres $j = 1, 2$ and

$$\Delta = \omega_{p1}\omega_{p2} \frac{\ell(\ell+1)}{(2\ell+1)} \left(\frac{a_1}{a_2}\right)^{\ell+1/2} \,, \tag{10.56}$$

describes the electrostatic interaction between the two concentric spherical 2DEG shells. One can see that in the limit $a_1 \ll a_2$, or $\ell \gg 1$, the two shells decouple and oscillate independently of each other with frequencies given by (10.55). However, to better understand the behavior of the SP oscillations in the coupled 2DEG shells, we plot dimensionless frequency ω/ω_s, versus variable ℓ in Fig. 10.5 for several values of ℓ. One can see that, as increasing the number ℓ, the SP frequencies of the system increase.

10.3.1 Multishell Systems

Assume a multishell structure consists of N concentric spherical 2DEG shells with radii $a_1 < a_2 < \cdots < a_N$. To find the dispersion relation of SP modes of the system, relevant solution of (10.7) in the space above and below the jth shell may be written as

$$\tilde{\Phi}_j(r) = A_j \begin{cases} a_j^{-(\ell+1)} r^\ell \,, & r \le a_j \,, \\ a_j^\ell r^{-(\ell+1)} \,, & r \ge a_j \,, \end{cases} \tag{10.57}$$

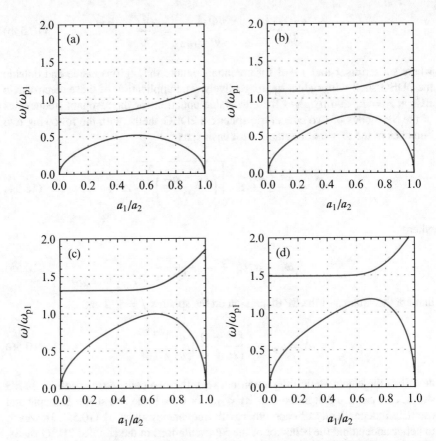

Fig. 10.5 Dispersion curves of SP modes of two concentric spherical 2DEG shells as given by (10.54), when $\alpha_j = 0$ (the nonlocal effects are neglected). The different panels refer to (**a**) $\ell = 1$, (**b**) $\ell = 2$, (**c**) $\ell = 3$, and (**d**) $\ell = 4$. For each ℓ, the upper and lower curves correspond to SP modes ω_+/ω_{p1} and ω_-/ω_{p1}, respectively

where A_j with $1 \leq j \leq N$ are constants. Also, for the present BVP, the BCs are

$$\tilde{\Phi}_j\big|_{r>a_j} = \tilde{\Phi}_j\big|_{r<a_j} , \tag{10.58a}$$

$$\frac{\partial \tilde{\Phi}_j}{\partial r}\bigg|_{r>a_j} - \frac{\partial \tilde{\Phi}_j}{\partial r}\bigg|_{r<a_j} = -\frac{\omega_{pj}^2 a_j \ell_{a_j}^2}{\omega^2 - \alpha_j^2 \ell_{a_j}^2} \sum_{i=1}^{N} \tilde{\Phi}_i , \tag{10.58b}$$

where the symbols have the same meaning as in previous section. Substitution of (10.57) into (10.58a) and (10.58b) leads to a set of linear homogeneous equations with unknown A_j. This set has non-trivial solutions that can be ascertained by solving the dispersion equation

$$\det \mathbf{M} = 0 , \tag{10.59}$$

where the element M_{ji} of the $N \times N$ matrix $\mathbf{M}$ is given by

$$M_{ji} = \begin{cases} a_j^{-(\ell+1)} a_i^\ell, & j > i, \\ a_i^{-(\ell+1)} a_j^\ell, & j < i, \\ \dfrac{2\ell + 1}{\ell(\ell+1)} \dfrac{\omega_j^2 - \omega^2}{a_j \omega_{pj}^2}, & j = i, \end{cases} \tag{10.60}$$

and ω_j is defined by (10.55). From (10.59) we obtain N positive roots for the frequencies of the SPs which are clearly separated into a high-frequency, ω_+, and a low-frequency, ω_-, group for each m. Note that (10.59) allows one to vary the number of layers at will. For example, for a system of two concentric 2DEG shells, i.e., $N = 2$, we have

$$\begin{vmatrix} \dfrac{2\ell + 1}{\ell(\ell+1)} \dfrac{\omega_1^2 - \omega^2}{a_1 \omega_{p1}^2} & a_1^\ell a_2^{-(\ell+1)} \\ a_1^\ell a_2^{-(\ell+1)} & \dfrac{2\ell + 1}{\ell(\ell+1)} \dfrac{\omega_2^2 - \omega^2}{a_2 \omega_{p2}^2} \end{vmatrix} = 0 \tag{10.61}$$

that is equal with (10.54).

10.4   Surface Plasmon Modes of a Pair of Spherical Two-Dimensional Electron Gas Shells

Finally, let us consider a pair of spherical 2DEG shells with radii a_1 and a_2 which density free-electron fluid over the each spherical shell (per unit area) is n_{01} and n_{02}, respectively. The distance between their centers will be labeled by d, where $d > a_1 + a_2$, as shown in Fig. 10.6. Using spherical coordinates centered on both shells with $\mathbf{r}_1 = (r_1, \theta_1, \varphi)$ and $\mathbf{r}_2 = (r_2, \theta_2, \varphi)$ for an arbitrary point in space (as defined by Fig. 10.6) the rotational invariance around the z-axis is maintained. As a consequence, the φ-dependent part of the potential will be $\exp(im\varphi)$. Assuming that n_j be the perturbed density (per unit area) of the homogeneous electron fluid on the jth EG spherical shell (with $j = 1, 2$), due to SP oscillations with angular frequency ω. Using the coordinates illustrated by Fig. 10.6, the eigensolutions of Laplace's equation, i.e., (10.3) in the three regions can be written as

Fig. 10.6 Definition of the polar coordinates used for solving Laplace's equation for a pair of spherical 2DEG shells with radii a_1 and a_2, the center-to-center separation being d

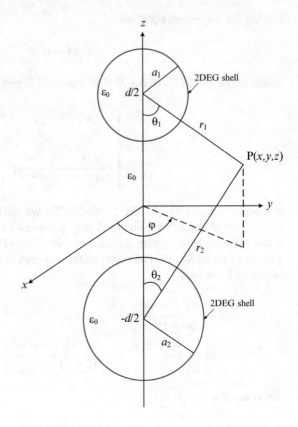

$$\Phi\left(\mathbf{r}_1, \mathbf{r}_2\right) = \sum_{\ell=0}^{+\infty} \sum_{m=-\ell}^{+\ell}$$

$$\times \begin{cases} A_{\ell m}\left(\dfrac{r_1}{a_1}\right)^{\ell} Y_{\ell m}\left(\theta_1, \varphi\right), & r_1 \leq a_1, \\[2ex] B_{\ell m}\left(\dfrac{a_1}{r_1}\right)^{\ell+1} Y_{\ell m}\left(\theta_1, \varphi\right) + C_{\ell m}\left(\dfrac{a_2}{r_2}\right)^{\ell+1} Y_{\ell m}\left(\theta_2, \varphi\right), & a_1 \leq r_1 \,\&\, a_2 \leq r_2, \\[2ex] D_{\ell m}\left(\dfrac{r_2}{a_2}\right)^{\ell} Y_{\ell m}\left(\theta_2, \varphi\right), & r_2 \leq a_2. \end{cases}$$

$$(10.62)$$

Also, in the present BVP, (10.8) must be written as

$$n_j = -\frac{e n_{0j}}{m_e} \frac{\ell_{a_j}^2}{\omega^2 - \alpha_j^2 \ell_{a_j}^2} Y_{\ell m}\left(\theta_j, \varphi\right) \Phi|_{r_j = a_j}. \qquad (10.63)$$

The BCs at the surface of the first and second shells are

$$\Phi_1\big|_{r_1=a_1} = \Phi_3\big|_{r_1=a_1} \,, \tag{10.64a}$$

$$\Phi_2\big|_{r_2=a_2} = \Phi_3\big|_{r_2=a_2} \,, \tag{10.64b}$$

$$\frac{\partial \Phi_3}{\partial r_1}\bigg|_{r_1=a_1} - \frac{\partial \Phi_1}{\partial r_1}\bigg|_{r_1=a_1} = \frac{en_1}{\varepsilon_0} \,, \tag{10.64c}$$

$$\frac{\partial \Phi_3}{\partial r_2}\bigg|_{r_2=a_2} - \frac{\partial \Phi_2}{\partial r_2}\bigg|_{r_2=a_2} = \frac{en_2}{\varepsilon_0} \,, \tag{10.64d}$$

where subscripts 1 and 2 denote the regions inside the shells having radii a_1 and a_2, respectively, while subscript 3 denotes the region outside the two shells.

In order to apply the mentioned BCs at the surface of the first shell, it is useful to express the term depending on r_2 and φ_2 in the outer potential in terms of r_1 and φ_1, using an addition theorem for Legendre functions, as [12, 13]

$$r_2^{-(\ell+1)} P_\ell^m(\cos\theta_2) = \sum_{k=|m|}^{+\infty} \frac{(\ell+k)!}{(\ell-m)!(k+m)!} \frac{r_1^k}{d^{\ell+k+1}} P_k^m(\cos\theta_1) \,. \tag{10.65}$$

Defining four column matrices **a**, **b**, **c**, and **d**, whose components are $A_{\ell m}$, $B_{\ell m}$, $C_{\ell m}$, and $D_{\ell m}$, respectively, the BCs for the surface $r_1 = a_1$ yield a set of equations that can be cast into the concentrated form

$$\mathbf{a} = \mathbf{b} + \mathbf{Pc} \,, \tag{10.66}$$

$$\mathbf{b} = \mathbf{Mc} \,, \tag{10.67}$$

where the matrices **P** and **M** are, respectively,

$$P_{k\ell}^m = \frac{\xi_{\ell m}}{\xi_{km}} \frac{(\ell+k)!}{(\ell-m)!(k+m)!} \frac{a_1^k a_2^{\ell+1}}{d^{\ell+k+1}} \,, \tag{10.68}$$

$$M_{k\ell}^m = \frac{\xi_{\ell m}}{\xi_{km}} \frac{(\ell+k)!}{(\ell-m)!(k+m)!} \frac{k(k+1)}{2k+1} \frac{\omega_{p1}^2}{\omega^2 - \omega_k^2} \frac{a_1^k a_2^{\ell+1}}{d^{\ell+k+1}} \,, \tag{10.69}$$

where ω_k is the frequency of a SP mode of a single spherical 2DEG shell with radius a_1 given by (10.55). Also ω_{pj} has been defined before in Sect. 10.3 and we have

$$\xi_{\ell m} = \left[\frac{2\ell+1}{4\pi} \frac{(\ell-m)!}{(\ell+m)!} \right]^{1/2} \,,$$

with $\ell, k \geq m$. The BCs written on the surface of the spherical shell of radius a_2 yield two equations equivalent to (10.66) and (10.67) where the index 1 and 2 have to be permuted. They may be written

$$\mathbf{d} = \mathbf{c} + \mathbf{Qb} , \tag{10.70}$$

and

$$\mathbf{c} = \mathbf{Nb} . \tag{10.71}$$

From (10.67) and (10.71), one obtains

$$(\mathbf{MN} - I)\mathbf{b} = 0 . \tag{10.72}$$

The dispersion relation for the SP modes is obtained by equating the determinant of this set of equation to zero,

$$\det(\mathbf{MN} - I) = 0 . \tag{10.73}$$

This yields the eigenfrequencies of the coupled SP modes. But in principle, $(\mathbf{MN} - I)$ is an infinite-dimensional matrix. Solving (10.73) numerically requires cutting the matrices $\mathbf{M}$ and $\mathbf{N}$ to a finite upper size which is imposed by the value ℓ_{max} at which the different sums in (10.62) are truncated. This is equivalent to considering in the interaction between the multipolar modes (ℓ, m) on each 2DEG shell, only those for which $\ell < \ell_{max}$. Let us note that for a spherical 2DEG shell, because of spherical symmetry for any given ℓ the $2\ell + 1$ modes, i.e., $m = 0, \pm 1, …, \pm \ell$, are degenerate.

In a system of a pair spherical 2DEG shells, when 2DEG shells are far apart and do not interact the SP frequencies are still given by (10.55), but now the degeneracy is doubled, so that the ℓth mode is $(4\ell + 2)$-fold degenerate [14, 15]. When the distance between the shells is reduced, this degeneracy is partially lifted. In general, there will exist two different frequencies for each (ℓ, m) pair. The (ℓ, m) modes will, however, be degenerate with the $(\ell, -m)$ modes. Thus, for any ℓ there will exist $2(\ell + 1)$ different frequencies and 2ℓ are doubly degenerate, so that the total number of modes is again $2(2\ell + 1)$. If the 2DEG shells have equal radii, then $\mathbf{M} = \mathbf{N}$ and $\mathbf{P} = \mathbf{Q}$, and (10.73) subdivides into two sets corresponding to

$$\det(\mathbf{M} + I) = 0 , \tag{10.74}$$

$$\det(\mathbf{M} - I) = 0 . \tag{10.75}$$

Limiting, for example, the dimensions of $\mathbf{M}$ to $\ell_{max} = 1$, one obtains for present system the coupled dipole frequencies as

$$\frac{\omega^2}{\omega_{p1}^2} = \frac{2}{3} \begin{cases} 1 \pm 2\left(\frac{a_1}{d}\right)^3 + 3\dfrac{\alpha^2}{a_1^2 \omega_{p1}^2}, & m = 0, \\[2ex] 1 \pm \left(\frac{a_1}{d}\right)^3 + 3\dfrac{\alpha^2}{a_1^2 \omega_{p1}^2}, & m = \pm 1. \end{cases} \tag{10.76}$$

Also, when $\ell_{max} = 2$ one obtains the coupled quadrupole-quadrupole frequencies, as

$$\frac{\omega^2}{\omega_{p1}^2} = \frac{6}{5} \begin{cases} 1 \pm 6\left(\frac{a_1}{d}\right)^5 + 5\dfrac{\alpha^2}{a_1^2 \omega_{p1}^2}, & m = 0, \\[2ex] 1 \pm 4\left(\frac{a_1}{d}\right)^5 + 5\dfrac{\alpha^2}{a_1^2 \omega_{p1}^2}, & m = \pm 1, \\[2ex] 1 \pm \left(\frac{a_1}{d}\right)^5 + 5\dfrac{\alpha^2}{a_1^2 \omega_{p1}^2}, & m = \pm 2. \end{cases} \tag{10.77}$$

Figure 10.7 shows the ten lowest coupled SP frequencies of a system of two spherical 2DEG shells, as a function of the ratio d/a_1, which we have labeled arbitrarily $\xi = 1, 2, \ldots, 10$. The modes 1–4 correspond to $m = 0$, 5–8 to $m = 1$, and 9 and 10 to $m = 2$. The odd indexes are symmetric modes and the even indexes are anti-symmetric modes. In the limit of infinite separation the four lowest modes reduce to the $\ell = 1$ mode of an isolated spherical 2DEG shell and the six upper modes to the $\ell = 2$. Finally, we note here that the method employed here is not

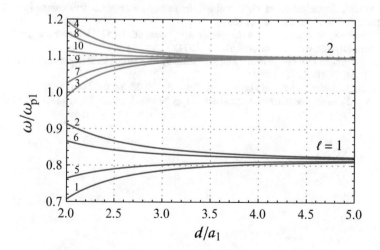

Fig. 10.7 The dimensionless frequencies ω/ω_{p1} of SP modes of a pair spherical 2DEG shells as a function of the ratio d/a_1. Only the frequencies reducing to the $\ell = 1$ and $\ell = 2$ modes at infinite separation are shown. Modes labeled $\xi = 1, 2, 3, 4$ are of $m = 0$ symmetry, $\xi = 5, 6, 7, 8$ are of $m = 1$ symmetry, and $\xi = 9, 10$ are of $m = 2$ symmetry

applicable at $d/a_1 = 2$, i.e., for touching shells. In fact, when $d/2a_1$ approaches unity, it is found that larger and larger values of ℓ_{max} are needed to ensure the convergence of first few eigenvalues.

References

1. G. Barton, C. Eberlein, Plasma spectroscopy proposed for C_{60} and C_{70}. J. Chem. Phys. **95**, 1512–1517 (1991)
2. A. Moradi, Quantum ion-acoustic wave oscillations in C_{60} molecule. Physica E **41**, 1338–1339 (2009)
3. H.W. Kroto, J.R. Heath, S.C. O'Brein, R.F. Curl, R.E. Smalley, C_{60}: buckminsterfullerene. Nature **318**, 162–163 (1985)
4. C.Z. Li, Z.L. Mišković, F.O. Goodman, Y.-N. Wang, Plasmon excitations in C_{60} by fast charged particle beams. J. Appl. Phys. **113**, 184301 (2013)
5. C.F. Bohren, D.R. Huffman, *Absorption and Scattering of Light by Small Particles* (Wiley, New York, 1983)
6. A. Moradi, Electromagnetic energy within an isolated C_{60} molecule, Optik **143**, 1–5 (2017)
7. A. Moradi, Extinction properties of an isolated C_{60} molecule. Solid State Commun. **192**, 24–26 (2014)
8. Y. Wu, J. Li, Z.-Q. Zhang, C.T. Chan, Effective medium theory for magnetodielectric composites: beyond the long-wavelength limit. Phys. Rev. B **74**, 085111 (2006)
9. A. Moradi, Effective medium theory for a system of C_{60} molecules. Phys. Plasmas **23**, 062120 (2016)
10. C. Yannouleas, E.N. Bogachek, U. Landman, Collective excitations of multishell carbon microstructures: multishell fullerenes and coaxial nanotubes. Phys. Rev. B **53**, 10225–10236 (1996)
11. A. Moradi, Investigation of high- and low-frequency electrostatic oscillations in multishell fullerenes. Phys. Scr. **81**, 055701 (2010)
12. D. Langbein, in *Theory of van der Waals Attraction*, ed. by G. Hohler. Springer Tracts in Modern Physics, vol. 72 (Springer, Berlin, 1974)
13. M. Schmeits, L. Dambly, Fast-electron scattering by bispherical surface-plasmon modes. Phys. Rev. B **44**, 12706–12712 (1991)
14. R. Ruppin, Surface modes of two spheres. Phys. Rev. B **26**, 3440–3444 (1982)
15. A. Moradi, Multipole plasmon excitations of C_{60} dimers. J. Chem. Phys. **141**, 024111 (2014)

Correction to: Canonical Problems in the Theory of Plasmonics

Correction to:
A. Moradi, *Canonical Problems in the Theory of Plasmonics*,
Springer Series in Optical Sciences 230,
https://doi.org/10.1007/978-3-030-43836-4

This book was inadvertently published without updating the following corrections and all the corrections has been updated in the book.

1) In page 5, after Eq. (1.2), the term "inverse Fermi" should be replaced by "Fermi".

2) To avoid misleading, after Eqs. (2.5), (2.44), (2.62), (2.101), (2.108), (3.18), (3.77), (3.87), (3.103), (4.12), (4.39), (4.56), (4.87), (5.13), (5.77), (6.9), (6.89), (6.103), (7.77), (10.9), the word "coefficients" should be replaced by "relations between the coefficients".

3) In footnote of page 96, the word "plane" should be replaced by "parallel".

4) In all of page 144, coefficients A_ℓ, B_ℓ, C_ℓ, and D_ℓ, should be replaced by a_ℓ, b_ℓ, c_ℓ, and d_ℓ, respectively.

The updated online versions of these chapters can be found at
https://doi.org/10.1007/978-3-030-43836-4_1
https://doi.org/10.1007/978-3-030-43836-4_2
https://doi.org/10.1007/978-3-030-43836-4_3
https://doi.org/10.1007/978-3-030-43836-4_4
https://doi.org/10.1007/978-3-030-43836-4_5
https://doi.org/10.1007/978-3-030-43836-4_6
https://doi.org/10.1007/978-3-030-43836-4_7
https://doi.org/10.1007/978-3-030-43836-4_8
https://doi.org/10.1007/978-3-030-43836-4_10

© The Editor(s) (if applicable) and The Author(s), under exclusive license to
Springer Nature Switzerland AG 2020
A. Moradi, *Canonical Problems in the Theory of Plasmonics*, Springer Series in
Optical Sciences 230, https://doi.org/10.1007/978-3-030-43836-4_11

5) In page 155, after Eq. (4.20), the term "$k_x < 0.01k_s$" should be replaced by "$k_x < 0.001k_s$" and "$\alpha/c \sim 0.01$" should be replaced by "$\alpha/c \sim 0.001$".

6) To avoid misleading, in pages 273, 292 and 340, after Eqs. (8.9), (8.80) and (10.57), respectively, the term "constants to be determined." should be replaced by "constants."

7) In all of pages 334 and 335, coefficients $A_{\ell 1}$, $B_{\ell 1}$, $C_{\ell 1}$, and $D_{\ell 1}$, should be replaced by a_ℓ, b_ℓ, c_ℓ, and d_ℓ, respectively.

8) In page 335, coefficients $A_{\ell m}$, $B_{\ell m}$ should be replaced by a_ℓ, b_ℓ, respectively.

Index

A
Absorbance coefficient, 244
Absorption, 59, 332
Additional boundary condition, 4, 153, 185
Anisotropic, 10
Annihilation operator, 288
Anti-symmetric mode, 44
Attenuated total reflection, 115

B
Band structure, 125, 277
Bessel equation, 127
Bipolar coordinates, 67
Bispherical coordinates, 86
Bloch theorem, 125, 238, 270
Bohm potential, 8, 160, 205, 278
Bohr velocity, 7, 155
Bose-Einstein commutation relation, 288
Bound electron, 200, 278

C
Carbon nanotube, 276, 309
Clausius–Mossotti, 77
Coherent state, 229
Collision frequency, 6, 218
C_{60} molecule, 329
Conductivity, 10, 16, 243, 307
Constitutive parameter, 9
Constitutive relation, 10
Continuity equation, 5
Creation operator, 288
Critical angle, 116, 245, 257

C (continued)
Cutoff, 19, 154, 219
Cyclotron frequency, 38
Cylindrical Bessel function, 50

D
Damping rate, 218
Debye screening length, 4
Degenerate, 3
Dielectric, 15
Dielectric function, 14
Dipole mode, 55
Dipole moment, 75
Dispersion relation, 18, 33, 68
Displacement potential, 286
Drift motion, 5
Drude-Lorentz dielectric function, 277
Drude model, 3, 18, 31, 88, 95

E
Electric energy, 20
Electric field lines, 102
Electron beam, 178, 309, 310
Elliptic polarization, 106
Energy density, 23, 35, 278, 311
Expectation value, 36
Extinction, 59, 132, 284, 315, 332

F
Fano wave, 101
Faraday configuration, 37
Fermi energy, 5